MW00513644

Coatings on Glass

THIN FILMS SCIENCE AND TECHNOLOGY

Advisory Editor: G. Siddall

THIN FILMS SCIENCE AND TECHNOLOGY, 6

Coatings on Glass

H.K. PULKER

Basic Research Laboratory, Balzers AG, Liechtenstein

ELSEVIER, Amsterdam — Oxford — New York — Tokyo 1984

ELSEVIER SCIENCE PUBLISHERS B.V.
Sara Burgerhartstraat 25,
P.O. Box 211, 1000 AE Amsterdam, The Netherlands

Distributors for the United States and Canada:

ELSEVIER SCIENCE PUBLISHING COMPANY INC.
52, Vanderbilt Avenue
New York, NY 10017

First edition 1984
Second impression 1985

ISBN 0-444-42360-5 (Vol. 6)
ISBN 0-444-41903-9 (Series)

Printed in The Netherlands

DEDICATION

This monograph is dedicated to three distinguished personalities who have played a decisive role in the development and spread of the industrial manufacture of thin films produced under vacuum by physical methods:

Prof. Dr. Dr.h.c. Max Auwärter
Founder and first President
of the BALZERS AG

Dr. Albert Ross
President of the BALZERS AG, ret.

Dr. Otto Winkler
Director of the BALZERS AG, ret.

A short time ago Max Auwärter celebrated his 75th, and Albert Ross his 70th birthday. I would like to take this opportunity to offer my very best wishes to them both on the occasion of their birthdays. My heartfelt thanks also go to Otto Winkler for his many valuable suggestions in the writing of this book.

Hans. K. Pulker

FOREWORD

The subject matter of this book *COATINGS ON GLASS* includes a wealth of information for physicists, chemists and engineers who need to know more about thin films for research purposes, or who want to use this special form of solid material to achieve a variety of application-oriented goals.

This particular publication is exceptional because the author makes available his extensive theoretical and practical experience which has been acquired over more than 20 years of intensive work on thin films. He has been concerned with all details that have an influence on the final product and can thus describe with great thoroughness the properties of all glass-type substrates, dealing also with very difficult questions concerning surface physics.

Glass can be produced by a variety of methods. The manufacturing process and the chemical composition determine how resistant a particular glass is to its environment. There are also different processes for finishing the surface of glass and this, together with the two aforementioned factors, determines the surface characteristics. Apart from inorganic glass also organic glass and plastic materials are considered.

Today there are two preferred groups of methods for the production of thin films: Chemical vapour deposition and physical vapour deposition under vacuum; the three major technologies of the latter being sputtering, evaporation, and ion plating. These are discussed in detail. The author's wide experience allows him to give many valuable tips in the discussion of how to produce a vacuum with a desired residual gas atmosphere using appropriate vacuum techniques. He has also studied mechanical and optical film properties as well as film thickness measurement methods, and this too is included in the book. Information on calculation methods which allow complex film systems to be developed is also given. Precise calculations and extremely accurate measurements are the basis for the production of thin films in computer controlled coating systems.

Applications of thin films are also given an important place in the book. The company in which the author works is world famous for its thin film products.

In summary, this work could be called a sort of formulary on the subject of glass and thin films written by a scientist for scientists and technical people. It goes beyond the subject matter indicated by the title, filling in the gap which has existed until now in the available technical literature.

M. Auwärter

PREFACE

Hans K. Pulker is an old friend for whom I have always had the highest regard. He is also a thin-film worker with a well-deserved international reputation for careful and original scientific work. I was delighted, therefore, when I heard that he was in the course of writing a book on coatings on glass. It is a source of great satisfaction and pleasure to me to write the preface so that I can be associated even in this minor way with what promises to be one of the standard works of reference in this important and ever-growing field.

All of us who work in the field of thin films have had more than our share of frustrating experiences with apparently inexplicable failures of a coating process. Frequently these involve well-developed techniques which have been running quietly and efficiently for some time. Suddenly coatings no longer adhere to the substrates although we are apparently following our standard procedure. Or stains appear in coatings on carefully prepared substrates. Sometimes absorption is high when it should be low, or cloudy scattering is apparent. Occasionally the problems are even more insidious, appearing only at a much later stage when components have already been supplied to a customer and are already in an optical system. I dwell on optical coatings because that is my own particular field but similar problems plague other areas of thin-film work. Apart from the weather, bad luck has often seemed the only reason for such troubles and even lucky charms have been seen in coating shops.

The past decade has seen considerable changes. We still suffer from unforeseen problems but gradually we are coming to understand the complex physics and chemistry governing the phenomena. We are used to dealing with thin films on a macroscopic scale and indeed it is their macroscopic properties in the main that we seek to produce with our coatings. These macroscopic properties, however, are entirely determined by film microstructure and it is only when we begin to understand the microstructure on an atomic scale and the physics and chemistry associated with it that we can appreciate the source of our problems and begin constructively to overcome them and to advance.

Films and substrates are held together by very short range forces, the bonds between one atom and the next. The mechanical properties of the films, including their intrinsic stress, are ultimately determined by these same bonds and their interaction with the microstructure. Their behaviour at surfaces can be greatly modified by minute quantities of impurity and they can be blocked completely by one molecular layer of contaminant. Adsorption is a particularly important process in thin films which must also be understood on an atomic scale.

Our understanding of film systems must begin with the nature of the surface on which they are deposited. Glass surfaces, the subject of this book, are particularly complex because of the nature of glass itself. Because of the gross effect of even minute contamination, cleaning is extremely important. Significant sections of the book are devoted to these topics followed by surveys of the bewilderingly extensive array of processes for thin-film deposition, techniques for measurement and characterization, fundamental film properties and optical coatings.

Hans K. Pulker himself has made notable contributions to the improvement of our understanding of the physics and chemistry of thin films and to deposition processes. The subject is vast and the literature extensive but scattered through journals in many different disciplines. It is difficult to know where to begin a study of the subject. I should say that it *was* difficult because the situation has changed with the publication of this book.

H. A. Macleod

AUTHOR'S PREFACE

When the Elsevier Scientific Publishing Company first invited me to write a monograph on *COATINGS ON GLASS*, I was hesitant because of the enormous amount of work involved. However, after critically examining my own collection of literature on thin films and glass, and considering the amount of pertinent information available in the scientific libraries of both the Balzers AG and the Swiss Federal Institute of Technology in Zürich, I changed my mind.

Dr. A. Ross, President of the Balzers AG until the end of 1982, encouraged me with the project, and his successor Dr. G. Zinsmeister, continued this support. I was thus able to notify the publisher that I could accept their invitation.

Much of the material in the present work was originally prepared for lectures presented at the Institute of Physical Chemistry at the University of Innsbruck in Austria. However, to limit the length of the work some elementary material has been eliminated, leaving room for new information and details on technical processes so that the topic could be covered comprehensively. It was not my intention to publish in-depth studies of any particular area but rather to provide a well founded background from which individual interests in various directions could be developed by the specialist or the newcomer to the field.

Following a history of glass and films presented as an introduction in Chapter 1, the Chapters 2 to 4 of this monograph attempt to outline the present-day knowledge of glass, with particular attention given to such factors as structure and composition and how they influence the properties of the material. The condition of the substrate surface is of primary importance in the coating process and therefore the generation, cleaning and properties of the glass surface is treated extensively. In addition to inorganic glass, organic glass and plastics are also discussed.

There has been a close interrelation between glass and thin films for a long time. Thin films on glass are used both for scientific and industrial purposes. One of the most important requirements placed on industrial coatings is that they adhere well to the substrate surface, particularly if they are to be subjected to extreme environmental conditions. This topic is treated in Chapter 5.

There are many chemical and physical methods for producing thin films with reproducible characteristics on a variety of substrates. A distinction is also made between wet and dry film formation methods. In some of these methods the depositions are carried out in air at normal atmospheric pressure or in the presence of protective gases and still others are carried out under vacuum. However, not all these methods are suitable for coating glass substrates because of their insulating nature and the relatively low thermal stability of many inorganic and practically all organic glasses. In Chapter 6 suitable coating methods and the plants for them are described.

The various methods for the determination of film thickness and deposition rate are treated in Chapter 7.

Thin films generally have large surface to volume ratios, and consequently the extended surface usually has a large influence on the film properties. Structure, microstructure, chemical composition, mechanical and optical properties, etc. of films deposited by different methods are described and discussed in Chapter 8.

Finally, in Chapter 9, relevant industrial and scientific applications for thin films on glass are given.

The bibliography for this monograph does not pretend to be more than a selected one. The references cited are mainly original investigations, but review articles and books containing supplementary references are also listed.

Acknowledgements are made to Prof. Dr. M. Auwärter and Dr. A. Ross (Balzers AG) as well as to Dr. H. Dislich (Schott Glaswerke) for valuable discussions and advice.
Further thanks are given to the scientists and technologists of various other industrial companies and scientific institutes who gave their support to the author, making available many figures and valuable information. Here particularly must be mentioned: Dr. Ian Seddon (Optical Coating Labs. Inc., Santa Rosa, CA, USA), Dr. G. Kienel and Ing. G. Deppisch (Leybold Heraeus GmbH, Köln, FRG), Dr. K. Deutscher (Leitz, Wetzlar, FRG), Dr. H. Bach (Schott, Mainz, FRG), Dr. F. Geotti-Bianchini (Stazione Sperimentale del Vetro, Murano-Venezia, Italy), Dr. C. Misiano (Selenia SpA, Roma, Italy), Prof. Dr. R. W. Hoffman (Case Western Reserve University, Cleveland, OH, USA), Prof. Dr. R. Th. Kersten (Technische Universität, Berlin), Prof. Dr. H. A. Macleod (Optical Sciences Center, University of Arizona, Tucson, AR, USA) and Dr. E. Pelletier (Ecole Nationale Supérieure de Physique, Marseille, France).

The author is also very much obliged to friends and colleagues for their engaged assistance during the preparation of this book. Special thanks are due to Dr. O. Winkler, Dr. E. Ritter, Dr. G. Trabesinger and Prof. H. A. Macleod for critical reading of the manuscript and for their stimulating comments, and to Mr. W. Frischknecht (Fri-Grafik, Vaduz, Liechtenstein) for drawing most of the figures and to Mr. and Mrs. L. Hilty (printing office, Schaan, Liechtenstein) for preparing the camera-ready manuscript.

Last not least it is a pleasure to acknowledge the assistance of the staff of Elsevier Science Publishers in Amsterdam and to thank Mrs. A. Ruhe-Lodge, Miss K. O'Day and Dr. A. J. Perry for language corrections.

Hans. K. Pulker

TABLE OF CONTENTS

xiv

CHAPTER 1

1. INTRODUCTION AND HISTORY

Glass is the general term for a number of special materials of varying composition and properties. In our time, glass is of considerable importance for the production of scientific, technical, architectural and decorative objects and many articles used in daily life. The material, however, was also very important in former times. Glass has been known since the earliest times in history. Molten minerals containing SiO_2, Al_2O_3, Fe_2O_3, FeO, MgO, CaO, Na_2O, K_2O, TiO_2 and H_2O extruded from inside the earth and quickly cooled down formed a natural glass, obsidian. The appearance of this mineral glass is translucent, usually coloured and sometimes also transparent. Worked objects such as spear heads and knives from the Stone Age and obsidian mirrors and objects for ceremonial purposes and jewelry from later periods have been found almost everywhere that the mineral occurs.

When and where glass was first made artificially is an open question. It would appear, however, that it took place in Asia Minor and Egypt about 4500 years ago. Including glazed pottery, it might be more than 7000 years ago.

Glass was used first as a gem and then later for hollow vessels, especially unguent jars and small vases, which were not blown but moulded. The Egyptians used mostly coloured glass with little brilliance because of its low refractive index. Transparent glass was rare and it always contained small bubbles as a consequence of insufficient melting temperatures. For a long period the art of making glass was limited by the primitive firing technique. The development of better glass melting furnaces brought about an improvement in glass quality.

With the invention of glass blowing, which is assumed to have occurred at the beginning of the Christian era, glass objects became more utilitarian than purely decorative and luxurious. About 100 years after the birth of Christ, in Alexandria colourless glass was obtained by the addition of manganese oxide to the glass melt. In the Roman Empire, glass had become a common household material. Besides tableware, toilet bottles and unguent vases, it was also used for storage of wine and other liquids and for seals and signets. The first glass windows are mentioned at the end of the third century. In this period, glass manufacture spread throughout the Empire. After the fall of Rome, glass manufacture was carried on in the Eastern Roman Empire.

In the Islamic world, besides tableware and decorative glass elements, the use of glass for weights was developed to high accuracy by about the year 780 A.D. At the time of the Crusades and after the fall of the Eastern Empire at the beginning of the eleventh century, Venice became famous for glass manufacture. Excellent glasses, the crackled ware and coloured types based on oxide coloration, were made in Murano. The highest point in Venetian glass manufacture at this time was the creation of an extraordinarily pure crystal glass. Although not invented there, the art of making mirrors with mercury layers was known by about 1369. Glass manufacture spread from Venice to many other European countries in the following centuries and the development of glass was now rapid.

In 1600, the art of cut glass was developed and many beautiful objects were made using this technique. Flint glass, having a higher refractive index and greater dispersive power than the crown glass, as a consequence of lead oxide content, was invented in England in 1675. The invention of gold-ruby glass and the development of a commercial process for the production of such glasses occurred in about 1680.

Progress in glass manufacture during the nineteenth century was made by the use of selected and purified raw materials and the construction of new and better melting furnaces.

New types of glass, known as optical glasses, were created in the glass factories of Germany, England and France. Compared with ordinary glasses, optical glasses are free from physical imperfections and have a wide range of refractive indices and dispersion. Very important studies of the properties of glass as a function of its composition were performed mainly in Germany in about 1890. Art and experience in glass-making were enlarged by scientific knowledge.

The production of hollow and flat glass, however, was still very difficult. The output depended greatly on the skill of the individual workman. More details concerning the history of early glass production can be found in refs. [1-8]. The revolution in glass manufacturing in the twentieth century was and is the trend towards mechanization of the glass industry. The Fourcault process (Belgium, 1905) was the basis for the first machine-drawn plate glass production in 1914 [8]. Some years later, Colburn in the USA invented another drawing method which became known in 1917 as the Libbey-Owens process [8]. The advantages of both older processes were combined in a very rapid new drawing technology used since 1928 and known as the Pittsburgh process (Pittsburgh Plate Glass Company, USA).

Most plate glass was drawn in continuous sheets directly from a melting tank using these processes.

At the beginning of 1960, a completely new and very powerful technology, the float glass process, became famous, replacing extensively the older drawing techniques. It was developed in England by Pilkington Brothers Ltd. [8]. Casting and pressing of glass is also entirely mechanized. The hand process was replaced step by step by machines, leading to today's computer-controlled automatic production plants.

Plastic materials must be included in the most important inventions of our century. Although plastics differ in chemical composition, method of manufacture and many properties from traditional inorganic glasses, some of them are transparent and in other optical properties similar to glass [9]. Thin foils can be made which are interesting for special applications. Organic glasses offer unique possibilities in design of optical elements [10]. They therefore become more and more a valuable complement to inorganic optical glass.

Finally, it must be stated that there are very few applications of glass in research and industry where the surface properties of the glass can be neglected. Thus the development of processes for surface treatments such as grinding, polishing, cleaning and coating has been of immense importance.

There are only few monographs available concerning the properties of glass surfaces [11, 12]. Particularly in the case of organic glasses, published data are rare. Marked differences between inorganic and organic materials, however, are one reason for different surface properties. Other factors influencing the nature of a surface, even with the same material, are the various methods used to produce a surface. The success of further surface treatment operations often depends on previous grinding, polishing and cleaning procedures. Proper coating includes all the procedures before film deposition.

Coatings on various glasses are made to produce controlled special variations of their optical, electrical, chemical and mechanical properties. The glass to be coated may have two different functions: it can be an active element (e. g. a lens) or a passive element (e. g. a substrate for an interference filter). The deposited films may be either functional or decorative in their technical applications.

It is interesting to note that the first antireflection layer on glass was made incidentally 1817 by Fraunhofer [13] in Germany by treating polished glass with concentrated sulphuric or nitric acid. But no technical application was tried at that time.

In 1935 Strong [14] in the USA and Smakula [15] in Germany developed the single layer antireflection coating produced by evaporation and condensation of calcium fluoride under vacuum onto glass surfaces. In 1938 Cartwright and Turner [16] suggested the use of a number of further suitable compounds. The list contained magnesium fluoride and cryolite among other fluorides. Although optically excellent, the mechanical and environmental stability of these coatings was poor. An exception was magnesium fluoride which could be rendered water resistant by baking at high temperature. In 1942, however, in the USA the first stable and abrasion resistant single layer antireflection coating of magnesium fluoride was made by deposition onto previously heated glass by Lyon [17]. After World War II, this material and deposition technique became standard procedure.

Attempts to develop multilayer a. r. coatings had already been made in 1938 in the USA [18] and a little later also in Europe [19]. But a technically satisfactory solution with a double layer system was first achieved in 1949 by Auwärter [20]. The superior quality of this antireflection double layer coating named «Transmax®» (Balzers) caused it to soon be in high demand.

Broadband, triple layer antireflection systems became available in 1965. Film systems could be designed easier and faster by that time with the aid of modern electronic computers.

The industrial development of highly reflective coatings started earlier than that of a. r. coatings. Great enhancement of glass reflectivity can be obtained using high reflecting metal films. Silver mirror films deposited by wet chemical processes were produced before 1835 [21]. The deposition of surface mirror films by the evaporation and condensation of various metals under vacuum was first attempted in 1912 by Pohl and Pringsheim [22]. They used ceramic crucibles as their evaporation sources. Evaporated silver mirrors were also made by Ritschl [23] in 1928. A technically simple but very effective solution for aluminium evaporation was found by Strong [24] in 1933 who used aluminium loaded helical tungsten filament evaporators, a technique which is still often used today. Mirrors made from aluminium by an evaporation technique are superior to silver in several respects. The reflectivity of aluminium is nearly the same as for silver in the visible, but much higher for the ultra violet [25]. Al mirrors adhere to glass more tenaciously than silver, and when exposed to atmosphere show no tarnish. Other features, such as its uniformity of reflectivity and transmission in the visible make Al films important for many technical and scientific applications. The first aluminized lamp, for instance, was made by Wright [26] at General Electric, USA, in 1937. During that period the first evaporated corrosion resistant and hard rhodium mirrors were also made by Auwärter [27] at W. C. Heraeus GmbH. Such mirrors became important for medical applications. Several years later in 1941 Walkenhorst [28] noticed increase in reflectivity of Al films for the visible with increasing deposition rate. Fast deposition was later also found to be necessary to achieve a high ultra violet reflectance.

At the same time Hass [29] invented the silicon oxide protected aluminium surface mirror. Protection against mechanical damage became important in mirror fabrication with the exception of Al mirrors for astronomical applications. And finally, for use in the near ultra violet range, pure aluminium films have to be protected primarily against degrading oxidation on exposure to air by covering them with layers of MgF_2 or LiF, as was found out by Hass et al. [30] in the time between 1955 and 1961.

At Schott und Genossen in Germany, Geffcken [31] deposited the first thin film metal dielectric narrow band interference filter of the Fabry-Perot type in 1939; Geffcken also made valuable contributions to the theory of thin film interference systems which were important for the later development of thin film interference system design. The practical realization of environmentally stable complex thin film interference systems, such as e. g. multilayer antireflection coatings, various interference mirror stacks, beam splitters and some high and low pass edge filters require hard and resistant coating materials. Such materials became available in the form of a variety of oxide films which can be easily deposited by the reactive gas deposition process invented by Auwärter [32] in 1952.

From that time there has been very intensive development. This has been concerned primarily with the conception and detailed development in the fields both of computer aided coating design as well as in coating production methods.

In general, coatings are desirable or may even be necessary for a variety of reasons. These include unique optical and electrical properties, materials conservation and economics or the engineering and design flexibility. Weight reduction is important in the automobile industry. Therefore heavy metallic parts such as grills are being replaced with light-weight plastic parts which have been coated with chromium, aluminium and other metals or alloys. Thin films are also used extensively in aluminium-coated plastic foils for heat insulation, decorative and packaging purposes. Another rather new application is to coat glass panes or plastic foils with indium tin oxide films or special cermet films which are used as architectural coatings to improve the energy efficiency and performance of buildings.

The hermetically sealing of sensitive surfaces from water vapour is of great practical importance and can be performed to a high degree of effectiveness with thin films. Using polymeric organic coatings of polytetrafluoroethylene and vinylidene chloride the best barriers are achieved with a water vapour transmission rate of only 4.8 g mil/m² 24hr. Whereas polymethyl methacrylate has a hundred times greater and cellulose acetate a more than even two hundred times greater value, for comparison.

These few examples may illustrate the broad spectrum of possibilities offered by coating glass and plastic with thin films.

It is the aim of this book to review the production and the nature of glass surfaces together with the various film deposition and measuring techniques and to describe the properties of the films. In a last passage a survey is presented about most of the typical technical thin film products.

REFERENCES

[1] G.W. Morey, The Properties of Glass, Reinhold, New York, 1945.
[2] I. I. Kitaigorodski, Technologie des Glases, R. Oldenbourg Verlag, München, 1957.
[3] W. A. Weyl, Coloured Glasses, Dawson of Pall Mall, London, 1959.
[4] Comptes Rendus VIIᵉ Congrès International du Verre, Art et Histoire, Bruxelles, 1961.
[5] H. Jebsen-Marwedel, Glas in Kultur und Technik, Verlag Aumann KG, Selb, 1976.
[6] S. Lohmeyer   (Ed.), Werkstoff Glas, Württ. Lexika Verlag, Grafenau, 1979.
[7] W. Vogel, Glas-Chemie, VEB Deutscher Verlag Grundstoffindustrie, Leipzig, 1979.
[8] H. G. Pfaender and H. Schröder, Schott Glaslexikon, Moderne Verlag GmbH, München, 1980.
[9] W. Holzmüller und K. Altenberg, Physik der Kunststoffe, Akademie Verlag, Berlin, 1961.
[10] Handbook of Plastic Optics, US Precision Lens Inc., Cincinnati, Ohio, 1973.
[11] L. Holland, The Properties of Glass Surfaces, Chapman & Hall, London, 1964.
[12] Comptes Rendus Symposium sur la Surface du Verre et ses Traitements Modernes, Luxembourg, 6-9 June 1967.
[13] Joseph von Fraunhofer, Versuche über die Ursachen des Anlaufens und Mattwerdens des Glases und die Mittel denselben zuvorzukommen (1817, Appendix 1819). J. v. Fraunhofer's gesammelte Schriften, München, 1888.
[14] J. Strong, J. Opt. Soc. Am. 26 (1936) 73.
[15] A. Smakula, German Patent Nr. 685767 (1935).
[16] C. H. Cartwright and A. F. Turner, Bull. Am. Phys. Soc. 13 (Dec. 12, 1938) 10.
[17] D. A. Lyon, US Patent No. 2.398.382 (Nov. 1942).
[18] Research Corp. N. Y. US Patent Application 27. 12. 1938 and Swiss Patent No. 221.992 (30. 6. 1942).
[19] A. Ross, private communication 1982:
     Zeiss (Smakula), Schott (Geffcken) and Steinheil (Schröder) 1940 – 1941
     See also: L. Holland, Vacuum Deposition of Thin Films, Chapman and Hall, London, 1956; and H. A. MacLeod, Thin Film Optical Filters, Adam Hilger, London, 1969.
[20] M. Auwärter (A. Vogt, Lawyer) Brit. Patent No. 697.403 (April 1950).
[21] J. Liebig, Ann. Pharmaz. 14 (1835) 134.
[22] R. Pohl and P. Pringsheim, Verh. dtsch. physik. Ges. 14 (1912) 506.
[23] R. Ritschl, Z. Physik 69 (1931) 578.
[24] J. Strong, J. Phys. Rev., 43 (1933) 498; and Astrophys. J., 83 (1936) 401.
[25] W. W. Coblentz and R. Stair, Bur. Standards J. Research, 4 (1930) 189.

6

[26] F. Adams, Proc. 23rd Ann. Techn. Conf. of the Society of Vacuum Coaters, Chicago, IL, USA, 1980, p. 15.
[27] M. Auwärter, Die Umschau, Heft 43 (1937); and Z. Techn. Physik 18 (1937) 457; and J. Appl. Phys. 10 (1939) 705.
[28] W. Walkenhorst, Z. Techn. Physik, 22 (1941) 14.
[29] G. Hass, German Patent No. 949 315 (1940).
[30] D. W. Angel, W. R. Hunter, R. Tousey and G. Hass, J. Opt. Soc. Am., 51 (1961) 913
[31] W. Geffcken, German Patent No. 716 153 (1939).
[32] M. Auwärter, Austrian Patent No. 192 650 (1952), U.S. Patent No. 2.920.002 (1960), and patents in many other countries.

CHAPTER 2

2. COMPOSITION, STRUCTURE AND PROPERTIES OF INORGANIC AND ORGANIC GLASSES

The scientific investigation of composition and properties of inorganic glass started in the last century. By comparison, the development of organic glass is just in its early stages. The question of the structure of glass led to a discussion of whether it exists in a microcrystalline or in an amorphous state. From the thermodynamic point of view, all condensed substances at zero temperature in equilibrium conditions should be crystalline. There are, however, also non-crystalline solids in a metastable state. Relaxation times for crystallization of these solids are extremely long and they therefore remain amorphous in practice. Crystals are rare and their structures are limited in number and can be reduced to only 14 Bravais lattices. The number of possible non-crystalline stuctural arrangements, however, is infinite. This diversity of the positions of atoms and molecules does not affect their thermodynamic and transport properties. On the whole, disordered structures are macroscopically presented as homogeneous and isotropic media. It is generally assumed today that glass belongs to the predominant non-crystalline solids.

2.1 GLASS-FORMING INORGANIC MATERIALS

The glassy state is known in some elements, notably selenium and tellurium. Selenium also forms glassy mixtures with phosphorus. There are also some semiconductive glasses of compounds like As_2Se_3. A number of salts may exist as glass. The best known is BeF_2. Complex types of glass have been prepared containing BeF_2 together with NaF, KF, LiF, CaF_2, MgF_2 and AlF_3. Some nitrates (Na, K), sulfates and chlorides have been obtained as small glassy droplets by spraying the molten material onto cold plates. More details on such glasses can be found in ref. [1]. Most of these types of glass, however, have little technical significance. This is not true of the glassy metals. Amorphous alloys or metallic glass containing two or three components such as PdSi, FeB, TiNi, NiPB etc., are new materials that have interesting mechanical and magnetic properties that are highly desirable in modern technical applications. They are produced in the form of strips by quenching the melt extremely rapidly [2]. The most important material termed as glass is formed, however, by oxides [1,3,4].

The central difference between metallic, semiconductive and oxide glasses lies in the relative strengths of their chemical bonds as measured by the energy gap between occupied and unoccupied electronic states. In oxide glasses this gap is more than 5 electron volts, that is, it lies in the vacuum ultraviolet, so that such glasses are transparent and colourless, apart from impurities. The semiconductive glasses have energy gaps near 1,5 eV and are coloured yellow or red, while in metallic glasses the energy gap is zero.

Typical and possible glass-forming oxides are listed in Table 1.

TABLE 1

GLASS-FORMING OXIDES

Typical glass formers:	B_2O_3	SiO_2	P_2O_5
	As_2O_3	GeO_2	As_2O_5
	Sb_2O_3		
Possible glass formers:	Bi_2O_3	ZrO_2	V_2O_5

Silica is the constituent material in technical glass. Commercial glass is almost exclusively silicate glass. Oxides that apparently do not form glass but may be included in glass to obtain special properties such as chemical durability, low electrical conductivity, high refractive index and dispersion, increase in hardness and melting point, etc. are listed in Table 2. To obtain glass that transmits infrared, some special components such as As_2S_3 and TeO_2 are used [5].

TABLE 2

GLASS-PROPERTY-MODIFYING OXIDES

Na_2O	ZnO	Al_2O_3	SnO_2
K_2O	BeO	La_2O_3	TiO_2
Pb_2O	PbO	Y_2O_3	ThO_2
Cs_2O	MgO	In_2O_3	
	CaO		
	CrO		
	SrO		
	BaO		
	CdO		

In term of bond strength betweeen the cation and the oxygen all glass formers have values greater than 5 eV and the typical modifiers have lower values in the range of about 2.8 eV [6].

2.1.1 CRYSTALLITE THEORY

When cooling down from the melting point, many materials pass through a temperature range in which the liquid becomes unstable with respect to one or more crystalline compounds. An increase in viscosity, however, may partially or completely prevent the discontinuous change into the crystalline phase. Studies of refractive index changes by heat treatment and investigations of other physical properties of glass led Lebedev [1,4,7] to the conclusion that glass contains ordered zones of small crystallites. In the crystallite hypothesis, it is assumed that glass may contain both amorphous and crystalline zones which are linked by an intermediate formation. These remarkably small crystallites of about 10 Å in size consisting of 3-6 atoms are assumed to be of irregular form with distortions in their lattice. Unfortunately, no stringent experimental evidence can be found to support this hypothesis because even X-ray and electron-diffraction structural analysis is unable to detect the possible existence of crystals in the range of about 10 Å.

2.1.2 RANDOM NETWORK THEORY

Extensive X-ray structural analyses of glass as well as studies of the melting process allowed Zachariasen [8] to explain glass as an extended molecular network without symmetry and periodicity. The glass-forming cations such as Si^{4+} and B^{3+} are surrounded by oxygen ions arranged in the shape of tetrahedra or triangles. Regarding the oxygen ions, a distinction must be made between bridging and non-bridging ions. In the first case, two polyhedra are linked together over an oxygen ion, and in the second case the oxygen ion belongs only to one polyhedron and has one remaining negative charge. In this way, a polymer structure consisting of long chains crosslinked at intervals is produced. The unbalanced negative charge is compensated by low charge and large size cations, e. g. Na^+, K^+, Ca^{++}, Ba^{++} located in the holes between the oxygen polyhedra. Substitution of silicon ions in the network by other large charge and small size cations is possible. The network theory was supported by further X-ray investigations by Warren in 1933 and 1937 [9a], and in investigations by other scientists. In Fig. 1, a two-dimensional drawing shows the crystalline state of SiO_2 (a), the glass network of SiO_2 (b) and the glass network of a sodium silicate glass (c).

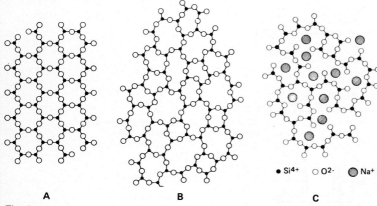

| A | B | C |

$\bullet$ Si^{4+} $\circ$ O^{2-} $\oslash$ Na^+

Fig. 1
Structure of crystallized silica (A), of fused silica (B) and of sodium silicate glass (C).

More recent investigations of chalcogenide glasses such as As_2Se_3 [9b] and also of oxide glass [9b] in transmission electron microscopy suggest that there exist structural domains, large macromolecules or clusters which are, in the case of silica and other oxide glasses, between 60 and 100 Å in diameter [9c]. The domain structure in oxide glass is difficult to observe because of the possible polymerization of the domain interfaces by ambient moisture.

In this sense, glass can be viewed as an assembly of subunits [10] which is not in opposition with the random network theory.

Zachariasen [8] has carefully distinguished between random orientation on a local level and cluster formation on a larger scale. The idea of continuous large scale random networks is thus merely a considerable oversimplification of his ideas.

Today the network theory is generally accepted for ordinary glass. It appears, however, that some complex multicomponent types of glass may also consist to some extent of very small ordered zones in an amorphous network matrix. This is especially true after heat treatment, which can induce phase separation and crystallization.

2.1.3 PHASE SEPARATION, DEVITRIFICATION

Glassy materials can be considered as frozen-in liquids, which consist, in the case of oxidic materials, of polymer chains with branches and cross linkages. With the exception of quartz glass, all types of industrial glass are multicomponent systems. The fact that glass is a multicomponent material leads, however, to the formation of very complicated structures. These are characterized by the pre-

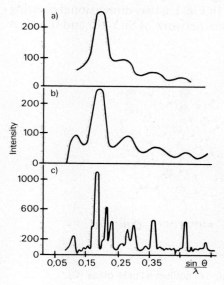

Fig. 2
Intensity curves of X-ray scattering of sodium silicate glass in various states according to Valenkov and Porai-Koshits [11].
a) Original glass,
b) glass annealed for 2 hours at 430°C,
c) glass after devitrification.

sence of glass-former skeletons of various shapes and also by a varied form of microheterogeneity. A variable short-range order in the distribution of ions and atoms exists, however, inside the microregions of the chemical and structural heterogeneities. The microheterogeneous structure of glass was discovered and studied first for two-component glass by Valenkov and Porai-Koshits [11]. They found that the X-ray diffraction pattern of sodium silicate glass depended on the thermal treatment of the sample, as seen in Fig. 2.

The interpretation of the diffraction patterns showed a clear deviation from the Zachariasen-Warren concept, according to which the Na^+ ions have a random distribution in the holes between the oxygen ions of the disorderly continuous silica network. The pattern indicated a micro-heterogeneous structure [12], that consisted of microregions with a sodium metasilicate composition embedded in the glassy silica structure. Similar results were also obtained with binary borosilicate glass, three-component sodium borosilicate glass [13] and other types of glass.

Phase separation in certain optically clear types of glass was also indicated in electron-optical investigations. The phase separation that occurs in some glass, however, does not provide evidence for the crystallite theory. In many types of glass, crystalline nuclei and crystallites can be found which appear as the result of imperfections in the production technology or of the subsequent devitrification process. If, during the working or annealing processes, the glass is held too long in the temperature region in which crystallization takes place most readily, it will devitrify and be destroyed. Devitrification is the main factor which limits the composition range of practical types of glass. It is an ever-present danger in all glass manufacture and working. The devitrification in ordinary glass takes place chiefly on the glass surface [14-17] and manifests itself in different ways: from almost indiscernible microcrystals to a fully developed crystallization. It appears, however, that devitrification does not always start on a surface; it seems to be much more dependent on surface pretreatment.

2.1.4 GLASS-FORMING ORGANIC MATERIALS

Organic glass or transparent plastics are synthetic solid materials consisting of polymer compounds that are formed mainly by the elements C, H, O and N. The polymer macromolecules are obtained by polymerization, polycondensation or polyaddition reactions between monomers [18]. We distinguish here between thermosets, which undergo a destructive chemical change upon application of heat, and thermoplastics, which can be resoftened repeatedly without any change in chemical composition. Thermoplastics are generally preferred for optical applications [19,20]. Like inorganic glass, they have no fixed melting point but rather a softening region. The plastics can be made fluid and shaped by the application of heat and pressure.

TABLE 3

GLASS-FORMING ORGANIC MATERIALS

Type	Composition	Refractive index
Acrylic	Polymethyl methacrylate	1.491
Styrene	Polystyrene	1.590
NAS	Methylmethacrylate (70%) Styrene (30%) Copolymer	1.562*
Polycarbonate	---	1.586*
CR 39	Allyldiglycolcarbonate	1.490
ABS	Acrylonitril-butadiene-styrene Copolymer	* = depending on composition.

Plastic is increasingly used as a substitute for inorganic glass and other materials. It is, however, often necessary to retain the appearance of the substituted material by special surface treatments. High-impact strength organic polymers have a rapidly expanding market in applications as diverse as ophthalmic lenses, architectural glass, electronic equipment packaging, various shaped form parts in automotive industry and in the form of foils as substrates for various types of thin films.

2.1.5 CRYSTALLINE AND AMORPHOUS BEHAVIOUR OF POLYMERS

Generally the plastic materials can exist in an amorphous, in a crystalline or in a mixed state. The crystalline state is rare because the mobility of the large molecules that is required for the formation of a periodic arrangement is very low. The complex chemical bonds and the predominant homeopolar character lead, in contrast to many inorganic materials, mostly to the formation of very complicated structures, e.g. monoclinic, rhombic or triclinic crystals.

Many plastics exist in an amorphous state. The growth of macromolecules produces chains and three-dimensional networks. The disordered polymer structure of organic glass is very similar to the network structure of silicate glass. Typical organic glass shows, however, practically no tendency to crystallization because of steric hindrance by sidegroups or other large substituents. Some plastics exist in a mixed structural type. The crystalline areas in the amorphous matrix are often so extended that they can easily be detected by light scattering. In this case, the material cannot be used for optical applications.

2.2 THERMAL BEHAVIOUR OF INORGANIC AND ORGANIC GLASSES

Vitrification and softening are second-order phase transitions for both inorganic and organic glasses. To obtain a homogeneous melt with inorganic glass, a temperature is required where the viscosity η of the melt is about 10^2 Poise. The softening point is at $\eta = 10^{7.6}$ Poise and the working point has a viscosity of $\eta \sim 10^4$

Poise. The temperature T_g in the transformation interval for solidification corresponds to a viscosity range for glass of between $\eta = 10^{13}$ Poise, the annealing point, and $\eta = 10^{14.5}$ Poise, the strain point.

It was first clearly shown by Bartenev [21] that when the cooling rate is decreased, the solidification temperature of silicate glass decreases proportionally. This behaviour was confirmed for many types of silicate glass and organic polymers in subsequent papers by various authors [22] and [23].

On the other hand, the softening temperature T_w is a function of the rate of heating. This observation is of great practical importance in glass technology because during annealing and tempering temperature changes may occur at very different speeds. Thus for different processes, differences in the vitrification temperature of up to 50 to 100°C may result. It follows from experimental and mathematical treatments [7] that, at a standard glass transition temperature T_g^{ST} or at a standard softening temperature T_w^{ST}, vitrification or softening occurs if the rate of cooling or heating is equal to 0.2 deg·s^{-1} for inorganic glass and to 0.1 deg s^{-1} for organic polymers.

Some experimental data obtained from [7] and [18] are listed in Table 4. The values for the linear expansion coefficient α were taken near but clearly below the softening temperature.

TABLE 4

SOFTENING TEMPERATURE AND LINEAR EXPANSION COEFFICIENT OF INORGANIC AND ORGANIC GLASSES

Material	T_w^{ST} (°C)	$\alpha\,10^{-5}$ (°C^{-1})
Fused silica	1580	0.056 – 0.08
Alkali silicate glass	536 – 696	1.15 – 0.96
Lead silicate glass	440 – 480	0.70 – 1.10
Aluminium silicate glass	582 – 842	0.46 – 0.65
Soda lime borosilicate glass	708 – 815	0.32 – 0.52
Acrylic	76	7–9
Polystyrene	72	6–8

It can be concluded from these investigations that the structure of glass depends on its thermal history. Annealing always increases the density of the glass.

Different pieces of glass each with a different structure will have different softening temperatures T_w when heated at the same rate. For technical applications, it is therefore useful to choose maximum annealing temperatures below a temperature $(T_g - 200)$°C to prevent unwanted deformations.

The expansion coefficient is a property of glass that is greatly affected by changes in composition. The linear expansion coefficient α in the glassy state does not, however, depend on the heating rate in the region below the softening point. It is assumed to be constant within this wide temperature range. As can be seen from Table 4, the thermal expansion of the plastics is much higher than that of inorganic glass. Generally, when glass is bonded with other materials having different rates of expansion, temperature changes create mainly undesirable forces in the two materials. This affects many properties such as adhesion in the case of deposited thin films.

In many problems concerning heat transfer, the thermal conductivity λ of the materials is an important factor. The rate at which heat is transmitted through glass by conduction depends on size and shape, on the difference in temperature between the two faces and on the composition of the material. Thermal conductivity is commonly expressed in calories per centimetre per second per degree. Some data are listed in Table 5. Compared with metals, the values for glass and plastic are low. Radiation is another heat-transfer process. It may be of greater importance than thermal conduction when the temperature is increased to higher ranges. Table 5 also contains some data on the specific heats of different types of glass and plastic [1, 18]. The specific heat c_p of glass, which is important in determining its heat capacity, is a nearly additive quantity and can be calculated from the composition by using the factors for the various oxides. The factors and some experimental data are reviewed in refs. [1] and [4].

TABLE 5

THERMAL CONDUCTIVITY AND SPECIFIC HEAT OF INORGANIC AND ORGANIC GLASSES (Typical mean values)

Material	$\lambda 10^4$			c_p		
	-100°	0°C	100°C	-100°	0°C	100°
Fused silica	28	31.5	35.4	0.11	0.16	0.20
Soda-lime silicate glass	19	23	27	–	0.20	–
Soda-lime borosilicate glass	21	26	30	–	0.27	–
Lead silicate glass	11	14	17	–	0.21	–
Aluminium silicate glass	–	22	24	–	0.19	–
Acrylic	–	4.7	–	–	0.35	–
Polystyrene	–	3.6	–	–	0.32	–

Thermal conductivity: λ (cal cm^{-1} s^{-1} deg^{-1}) 10^{-4}.
Specific heat: c_p (cal g^{-1} deg^{-1}).

The ability to withstand thermal shock resulting from sudden changes in temperature is important for technical applications of glass. The thermal endurance of inorganic glass, studied mainly by Schott and Winkelmann [24] (see also refs. [1] and [25]), is a very complex property. The investigations have led to the definition of a coefficient of thermal endurance F:

$$F = \frac{P}{\alpha E} \sqrt{\frac{\lambda}{\rho \cdot c_p}} \tag{1}$$

in which P is the tensile strength, E is Young's modulus and ρ is the density.

It is interesting to know that most types of glass can withstand much greater temperature changes when suddenly heated than when rapidly cooled.

2.3 MECHANICAL PROPERTIES OF INORGANIC AND ORGANIC GLASSES

At normal temperatures, glass usually behaves as a solid material. The most important properties of solids are elasticity, rheology and strength. These properties depend not only on the molecular mechanisms of the deformation process, but on the viscous flow and on the structural peculiarities of industrial glass samples. For the application of polymeric materials, it is important to know that they possess extraordinarily complex mechanical properties. The high deformability, the marked incompressibility and the high sensitivity to changes in temperature are typical.

In Table 6, some mechanical data on various types of glass are listed [7,18]. For grinding glass, the so-called grinding hardness is important. The value of this for glass is dependent on chemical composition. The silica base used in most inorganic glass compositions is essentially very hard, but additions of other materials to modify, for example, the optical properties in optical glass will reduce the hardness to a greater or lesser extent. The grinding hardness G. H. is defined as a quotient: G. H. = rate of removal of standard glass / rate of removal of sample glass. Values are published in the glass catalogues. Unfortunately, there is no exact relationship between hardness and grinding hardness.

2.4. CHEMICAL PROPERTIES OF INORGANIC AND ORGANIC GLASSES

Although many types of technical silicate glasses are highly resistant to chemical attack, inorganic glass cannot be treated as an inert material. Chemical reactions take place with water, acids, alkali, salt solutions and various vapours, e.g. SO_2. Even apparently chemically resistant glasses may be attacked locally, producing remarkable changes in the composition of their surfaces compared with the bulk. The effect is much stronger with some optical glasses. Generally, however, plastic materials are more resistant than inorganic glass.

The water attack starts with the diffusion of H^+ into the glass. This is a rapid process at higher temperatures. The water uptake increases with increasing pressure and the glass begins to swell. The quantity of incorporated water usually amounts to a very small percentage of the weight of the sample. Its presence promotes the tendency to crystallization. As can be seen in Table 7, silicate glasses are more strongly attacked in alkaline solutions than in neutral or acidic solutions because the alkali supplies hydroxyl ions, which react with the silica network. No protective layer forms during the corrosion.

The attack of acids differs from that of water because the dissolved alkali and basic oxide components are neutralized by the acid. In this way, a silica-rich surface layer is formed which reduces the rate of attack with time. When major

TABLE 6

MECHANICAL PROPERTIES OF INORGANIC AND ORGANIC GLASSES

Material	Density	Young's modulus (E, kp mm⁻²)	Hardness (HV, kp mm⁻²) (~100g)	Total extension at fracture %	Rupture strength (kp mm⁻²)	Bending strength (kp mm⁻²)	Impact strength (kg mm⁻²)
Fused silica	2.20	7220-8000	710		5.4-8.7	5-6	
Soda-lime silicate glass	2.44-2.47	6900-7400	650		6.5		
Soda-lime borosilicate glass	2.02-2.27	6100-7310	780		7	1.5	2
Lead silicate glass	2.4 -3.0		490-550		6		
Aluminosilicate glass	2.40-2.76	5700-6900					
Acrylic	1.19-1.22	300	17	2-8	6-8	11	
Polystyrene	1.06	300-400	~15	1-4	2-6	7-10	

TABLE 7

CHEMICAL PROPERTIES

Material	Chemical attack by:			Water vapour absorption in humid atmosphere* or immersed ** %
	H₂O 25°C, 4h	NaOH (5%) 100°C, 6h	HCl (5%) 100°C, 24h	
		weight loss in mg cm⁻²		
Fused silica	8×10⁻⁵	0.9	4×10⁻⁴	
Soda-lime silicate glass	1.2×10⁻³	1.1	1×10⁻²	0.65 – 1.7, 25°C, 4 h *
				2.6, 550°C
Soda-lime borosilicate glass	8×10⁻⁴	1.4	5×10⁻³	
Lead glass	2×10⁻⁴	3.6	2×10⁻²	
Aluminosilicate glass	2.4×10⁻⁴	0.4	3.5×10⁻¹	
Acrylic	Practically no weight loss			0.3, 25°C, 24 h**
Polystyrene	Practically no weight loss			0.2, 25°C, 24 h**

amounts of soluble oxides are present, which may occur with some types of highly refractive optical glass, the glass will disintegrate. The corrosion resistance is influenced by the glass composition. It increases with higher amounts of SiO_2 or of MeO, e.g. CaO, MgO, ZnO and PbO. Addition of even small amounts of Me_2O_3 impurities such as B_2O_3 and Al_2O_3 increase the resistance. Proper tempering of leached silica-rich surface films on technical glass decreases the porosity and increases the stability with regard to chemical attacks [1, 3, 18, 25].

Special durability tests have been made on optical glass. Qualitative gradings of durability were developed and listed in the glass catalogues. Quantifying such chemical attack is extremely difficult and has been the subject of international research.

2.5 ELECTRICAL PROPERTIES

The data used for the electrical characterization of glasses [1, 3, 18, 25] include the volume and surface resistivity, the dielectric constant, the dielectric loss and the dielectric strength. As may be seen in Table 8, fused silica has a high volume resistivity of 10^{19} Ω cm, but the addition of other oxides decreases the resistivity. Organic glass has values similar to inorganic multicomponent glass. Generally, the electrical resistivity, or its reciprocal value the conductivity, of multicomponent glasses depends on the chemical composition, on the temperature, and to some extent also on the atmospheric conditions. Inorganic glass is an ionic conductor. The transport of alkali ions is easier than that of alkaline earth ions. The mobility decreases as the ionic radius increases. The glasses are insulators at low temperatures and increase their conductivity as the temperature is raised because of the greater ease with which the ions can move when thermal energy has weakened the binding forces to the silica network. In the temperature range of $25 - 1200°$ C, the resistivity may vary between 10^{19} Ω cm and 1 Ω cm. At normal temperatures and in a humid atmosphere, the volume conductivity is surpassed by the much higher surface conductivity. The adsorbed H_2O, which forms an electrolyte with the dissolved alkali, is responsible for this phenomenon. The reciprocal of surface conductivity, surface resistivity, is usually defined in Ω cm^2.

Glass has a special position among solid dielectrics because of its extended range of dielectric constants. Organic glass has very low values, and high lead silicate glass shows the highest values. The dielectric constant of most types of glass decreases as the frequency of the applied field increases. For each frequency, the dielectric constant increases with increasing temperature, but the increase is less at high frequencies. The energy lost in a dielectric due to electrical conduction losses (the dielectric loss) is measured by the angle δ between the current and the charging potential. The value tan δ depends on frequency and temperature.

The dielectric strength of a glass is the voltage required to puncture it, usually expressed in V per cm. Unfortunately, it is not a true physical constant of a material because it is dependent on the thickness of the glass. Generally, the dielectric strength decreases with increasing sample thickness. The values listed

TABLE 8

ELECTRICAL PROPERTIES

Material	Electrical resistivity (Ω cm at °C)		Dielectric constant (E at 10^6 Hz and 29°C)	Dielectric loss (tg δ at 1 MHz, 25°C)	Dielectric strength (V cm^{-1})
Fused silica	10^{19}	25°	3.85	2×10^{-5}	5.4×10^6
	10^{13}	250°			
	10^{10}	400°			
Soda lime silicate glass	10^{10}-10^{13}	25°	6.4	1×10^{-2}	3.6×10^6
	10^6	300°			
Soda lime borosilicate glass	10^{15}	25°	5.1	2.6×10^{-3}	4.8×10^6
	10^9	250°			
	10^7	350°			
Lead silicate glass	10^{12}-10^{14}	25°	7-18	?	0.8-3.1×10^6
	10^8	300°			
Aluminosilicate glass	10^{15}	200°	5.84	1×10^{-3}	?
	10^{13}	300°			
	10^9	500°			
Acrylic	10^{12}-10^{15}	25°	2.6	8×10^{-2}	$<1\times10^6$
Polystyrene	10^{14}-10^{17}	25°	2.5	1×10^{-3}	$<1\times10^6$

in Table 8 are measured at glass thicknesses between 0.1 and 0.25 mm. At lower temperatures, the magnitude of the dielectric strength is relatively insensitive to temperature. However, after a higher temperature limit has been exceeded, the glass becomes progressively weaker. The decrease in strength when the frequency of the applied voltage is higher may be due to a heating effect.

The prevention of static charges on glasses is of great practical importance in avoiding attraction of dust particles as well as for other reasons. Two different dielectric materials brought into close contact and separated afterwards receive opposite charges whereby the material with the higher dielectric constant exhibits the positive charge. This phenomenon shows a dependence on surface conditions and is caused by the formation of an electrical double layer resulting from electron or ion migration during the contact. This produces high voltages after separation. The generation of a static charge on plastics is caused mainly by electron migration, but on inorganic glass it is caused by ion migration. According to Weyl [26], removal of only 1% of Na^+ ions from a monomolecular surface layer produces a negative charge of about -3×10^5 V. The size of the electrical charge formed on a glass is strongly dependent on its alkali content. A positively charged piece of glass can be transformed into a negatively charged one by heating. For many types of glass, this reversal point is at about 260°C [27]. The positive charge decreases in a continuous way during heating beyond the reserval point and a negative charge is formed which remains stable for many hours. Rapid cooling in liquid air also leads to the generation of a temporary negative charge. The intensity of the charge depends on the glass composition. Crown glass yields a higher charge than flint glass, and borosilicate glass is in between. The charge on a smooth surface will be higher than on a rough surface. Charged glass or plastic can be neutralized by blowing with ionized air.

2.6 OPTICAL PROPERTIES

The transmission of light is one of the first properties that comes to mind when glass is mentioned, and it is important for many of its applications. Sometimes it is sufficient that glass transmits most of the incident light, but in other cases the requirements are more stringent. There are many types of optical glass, besides ordinary window glass. For special needs, glass that is transparent in the visible but opaque in the infrared, or transparent for ultraviolet light and X-rays, is available. Special types of coloured glass and light-sensitive glass are also available.

The region of high transmission of all these types of glass is bounded at short wavelengths by the fundamental absorption edge and at the long wavelengths by the ir-absorption bands. With the usual colourless or white glass, the limits of transmission in the ultraviolet are determined mainly by the Fe_2O_3 content, which appears to have an absorption in the near ultraviolet. The infrared transmission limit is determined by the FeO content, which shows strong absorption at about 1 μm.

Today, the use of pure raw materials or of decolourizers enables the production of common window glass that has almost the same transparency as optical glass. The transmission of the untreated glass plates, however, is always limited by the reflection loss. The transmission characteristics for optical glass are dependent on the glass type, the purity of the raw material, and the melting schedules used in the production of the glass. Internal transmittance data for the various types of glass are tabulated for different wavelengths in the catalogues of the glass companies. Each glass type is identified by its mean refractive index n_d, and its constringence Abbé value v_d.

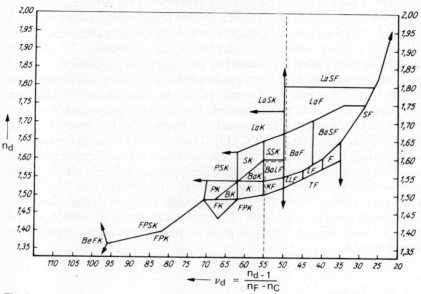

Fig. 3

n/v Diagram of various types of optical glass. Last symbol K means crown,
F means flint glass, e.g. BaF = Baryt flint, LaK = Lanthanum crown.

As may be seen in Fig. 3, a graph of n_d versus v_d shows the grouping of different types of glass according to their optical properties. An arbitrary distinction between crown and flint glasses is represented by the line $v_d = 55$ for glass with n_d less than 1.6, and $v_d = 50$ for glass with n_d greater than 1.6. Further subdivisions into different types of crown and flint glass are shown clearly on this diagram.

Abrupt discontinuities in glass cause striations. To obtain homogeneity of the glass, gradual changes in chemical composition and irregularities in heat treatment must be avoided. Optical glass should not contain undissolved solid or gaseous inclusions.

In Table 9, refractive index, temperature coefficient and Abbé value are listed for some typical types of inorganic and organic glasses [1, 3, 18, 19, 25, 28]. It is interesting to note that acrylic glass has a higher transparency than inorganic optical glass. Polystyrene, like acrylic glass, is very clear. Early styrene had impurities that decreased the transmittance at short wavelengths, causing the material to have a yellowish tint. Nowadays, this condition is only evident in extremely thick pieces.

Compared with acrylic glass, polystyrene has a lower light stability and resistance to ultraviolet radiation. The slight degradation when exposed to ultraviolet radiation is a disadvantage that occurs with all optical plastics. The higher refractive index of polystyrene permits its use in conjunction with acrylic lenses for the design of colour-corrected optical systems. Most types of organic optical glass transmit very well in the near infrared region. The energy absorption at wavelengths shorter than 2.1 μm is low.

TABLE 9

OPTICAL PROPERTIES

Material	Refractive index	Temperature coefficient of refraction dn/dT $(10^{-5}\,^{\circ}C^{-1})$	Abbé value
Fused silica	1.46	0.8	67.9
Soda lime silicate glass	1.52	0.2	58.9
Soda lime borosilicate glass	1.50	0.15	65.1
Lead glass	1.80	0.5 → 1.4 (for heavy flint)	25.4
Aluminosilicate glass	1.53		51.1
Acrylic	1.49	8.5	57.2
Polystyrene	1.59	12.0	30.9

Internal mechanical stresses may develop from the heat-treatment of inorganic and organic glass during its manufacture. These stresses cause the glass to have two indices of refraction. This phenomenon is called birefringence. Carefully designed temperature schedules can reduce these stresses to acceptable levels. For normal optical purposes, the limit of acceptable birefringence, expressed as a path difference, is 10 nm per cm. The stress optical coefficient, otherwise known as the Brewster constant, relates the mechanical stress to the optical retardation (nm cm^{-1}/kg cm^{-2}). This constant depends on glass composition. For various types of optical glass, the values can be found in the catalogues.

2.7 MATERIALS TRANSPARENT IN ULTRAVIOLET AND INFRARED

Materials that are transparent in the ultraviolet or in the infrared ranges are required for many scientific and technical instruments and apparatus. Such requirements exist, for example, in refracting optics and for special optical windows. For these applications, besides special types of glass and some plastics, many crystalline materials are in use [28,29]. The usefulness of a material depends on how far into the ultraviolet or infrared it transmits. The transmission edge in the ultraviolet is determined by the energy necessary for electron transfer (fundamental absorption) while the long wave absorption bands are caused by inner molecule vibrations or by lattice vibrations. Some fluorides and oxides are transparent relatively far into the ultraviolet range because the F^- ions and the O^{2-} ions possess the highest electron affinity among the anions.

A high infrared transmittance is observed with materials consisting of molecules composed of heavy atoms, because their valence vibrations have the lowest frequencies. Up to a wavelength of about 2.5 μm, practically all types of optical inorganic glass with a mean thickness of 10 mm have reasonable transmittance. The first limitation is caused by incorporated water. The O-H valence vibrations at 2.7 – 3 μm and also at 3.6 and 4.2 μm are responsible for absorption bands, while with far infrared transmitting glass such as As_2S_3-glass, the O-H flexural mode of vibration at 6 μm may also be responsible for absorption. Careful glass manufacture enables a reduction of the water content. A further limitation of infrared transmittance is the wide-band absorption caused by the natural oscillations of the glass network formers. Borate and phosphate glasses of 5mm thickness therefore become practically opaque at $\lambda = 3.5$ μm. The same happens with silicate glass between 4.4 and 4.8 μm and with Ca-aluminate and germanate glasses at wavelengths between 5.8 and 6 μm. There are types of Te-, Sb- and As-oxide glasses that are transparent up to about 6 μm [30]. Replacement of O by S yields glass such as As_2S_3 which is transparent even to wavelengths of about 13 μm. Newer 3 and 4-component glass systems enlarge the spectral range of transparency to about 20 μm. Of the crystalline materials, the alkali chlorides and bromides have been used for a long time in infrared optics, although their solubility in water is relatively high. With the exception of LiF, the fluorides CaF_2, MgF_2 and LaF_3 [31-37] are better suited because of their much higher stability towards the atmosphere. With new two- and multicomponent glasses of Ba-, Al-, La-, Zr-, Hf-, and Th- fluorides [36,37] an even further improved stability has been obtained. Materials like spinel ($Al_2O_3 \cdot MgO$), TiO_2, $SrTiO_3$, $BaTiO_3$ are actually insoluble in water. Single crystals are expensive to produce, sometimes optically anisotropic and sensitive to mechanical treatment because of cleavage and breakage. For this reason, pressure-sintered polycrystalline materials are often preferred. Although these materials show light scattering in the visible and near infrared, their optical properties are excellent in the far infrared. Last but not least, there are some element semiconductors and some compound semiconductors which, although opaque in the visible and near infrared, are nevertheless used in the spectral range in which they are transparent (usually between 1 μm and 22 μm).

Spectral characteristics of noncrystalline and crystalline materials are shown in Figs. 4 and 5 and data are listed in Table 10.

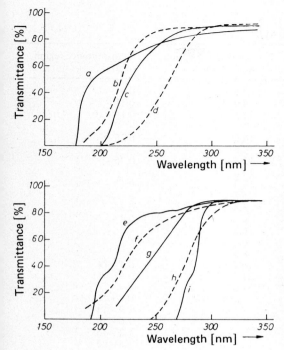

Fig. 4
Spectral transmittance of various
ultraviolet transmitting inorganic and
organic glasses of 1mm thickness
[29].
a) Boratglass (23),
b) Vycor (Corning),
c) WG 230 (Schott),
d) WG 280 (Schott),
e) Methylpolysiloxane,
f) Glass 8337,
g) Glass 8405,
h) Glass 9700,
i) Acrylglass.

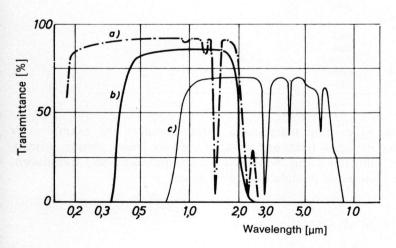

Fig. 5
Spectral transmittance of glasses for optical wave guides. Sample thickness: 200 mm [29]. a) Fused silica (suprasil), b) Flint glass, c) Arsenic
sulfide glass.

TABLE 10

INFRARED TRANSMITTING MATERIALS

Material	Type	Infrared absorption at: λ in μm	Transmittance region (μm)
LiF	single crystal	7.5	0.11 – 7
MgF_2	single crystal	8.0	0.11 – 7.5
CaF_2	single crystal	10.5	0.15 – 10
CdF_2	single crystal	11.5	–
LaF_3	single crystal	12.0	0.25 –
SrF_2	single crystal	12.5	–
PbF_2	single crystal	13.5	0.25 –
BaF_2	single crystal	14.0	–
SiO_2-glass	glass	4.5	0.15 –
Al_2O_3/MgO	single crystal	5.5	–
Al_2O_3	single crystal	5.8	0.2 –
TiO_2	single crystal	6.2	0.4 –
$SrTiO_3$	single crystal	6.8	
$BaTiO_3$	single crystal	6.9	
MgO	single crystal	8.5	
ZnS	polycrystalline	14.7	0.4 – 14
ZnSe	pressure sintered	21.8	0.55 – 15
CdTe		31.0	–
CdS	single crystal		0.8 – 16
InP	single crystal		1.0 – 14
GaAs	single crystal		1.0 – 16
Si	single crystal		1.2 – 15
Ge	single crystal		1.8 – 22
InAs	single crystal		3.5 – 9
$CaLa_2S_4$	single crystal	18.0	
Germanate (Corning 9754)	glass	5.0	
Ca-aluminate (Barr & Stroud BS 39 B)	glass	5.5	
Tellurite	glass	6.0	
ZrF_4-BaF_2	glass	8.0	
HfF_4-BaF_2	glass	8.0	
HfF_4-BaF_2-LaF_3-AlF_3	glass	8.0	
ZrF_4-BaF_2-ThF_4	glass	8.0	

The number of existing materials is large, therefore the examples are restricted to those materials (glass and crystals) used for lenses, prisms, mirrors, filters or other substrates and windows. Polarizing elements, transducers, optically-active devices and modulators are excluded.

2.8 PHOTOCHROMIC GLASSES

The reversible generation of colours or the reversible darkening with specially doped, colourless, transparent, inorganic glasses during irradiation with light is called photochromism.

Four different groups of light-sensitive glasses are known today:
— photochromic silicate glasses doped with rare earth ions
— photochromic borosilicate glasses doped with silver halides
— photochromic borosilicate glasses doped with silver molybdate or silver tungstate
— photochromic borosilicate glasses doped with copper or cadmium halides.

In 1962 Cohen and Smith [38] reported the observation of colour centers in irradiated silicate glasses doped with Ce^{3+} or Eu^{2+}-oxide. Soon after this investigation, photochromic light-sensitive glasses were developed in 1964 by Corning glass works [39,40]. The photochromic behaviour of these glasses is due to the formation of light-sensitive silver halide microcrystals of about 40 Å in size which are created by precipitation and subsequent heat-treatment at a temperature between the annealing strain point and the softening point of the borosilicate glass [41].

Photochromism consists of the three processes: optical darkening, optical bleaching and thermal bleaching. Darkening is induced by wavelengths ranging from the near ultraviolet and through the whole visible spectrum, depending on the chemical composition of the glass. Glasses containing silver chloride are sensitive in the region between 300 and 400 nm, and those containing mixtures of silver chloride and silver iodide are sensitive in the 300 – 650 nm wavelength-range. The transmission of a transparent photochromic glass before darkening may be about the same as window glass. Irradiation with light of a given intensity at constant temperature produces in the visible range an exponential increase in optical density to an equilibrium value which may be as low as 1% transmission. Basically the photochromic behaviour of the glasses can be characterized by the phototropic equilibrium constant k', defined by:

$$k' = \frac{1}{t} \log \frac{T_o}{T} \qquad (2)$$

t = glass thickness, T = percentage transmission of the darkened glass at the maximum value of the absorption band, T_o = percentage transmission of the bleached glass. The saturation value of k' obtained with equal irradiation is termed k'_o

The basic reaction in irradiation of silver halide–containing glasses is characterized by the following equation:

$$Ag^+ Cl^- \xrightarrow{h\nu} Ag^o + 1/2\, Cl_2^o \qquad (3)$$

It is necessary that the elemental halogen formed remains in the matrix. This would not be the case in a gelatine layer but the requirement is fulfilled with inorganic glass.

In the case of sensibilization with copper an additional reaction may occur:

$$Ag^+Cu^+ \underset{}{\overset{h\nu}{\rightleftarrows}} Ag^o + Cu^{2+} \tag{4}$$

It is important to mention here that the darkening is not directly proportional to light intensity. The effect occurs more strongly at higher intensities than at low ones.

Silver halide glasses are believed to be almost unique among photosensitive materials in being truly reversible and immune to fatigue [42-44]. The rate of fading in the dark increases with temperature and since the darkening rate is relatively insensitive to temperature, the steady state optical density of the photochromic glass decreases with increasing temperature.

Since light of different wavelengths may cause reaction in different ways, a change in transmittance can be achieved by darkening the glass with short-wavelength light or bleaching it after it has been darkened with light of longer wavelengths.

For sun-protection glasses a typical half-time of darkening may be about 20 to 30 seconds and the half-time of bleaching-out may range between 350 and 450 seconds.

The typical behaviour is shown in Figures 6 and 7.

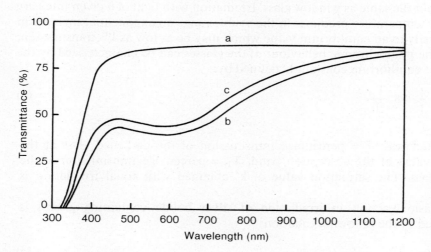

Fig. 6

Spectral transmittance of photochromic glass (thickness about 2 mm).
a) bleached glass in dark
b) irradiated with a high pressure mercury lamp
c) after 1 hour in sunlight at 25°C

More recent developments concern the photochromic borosilicate glasses containing silver-molybdate and tungstate [45] or copper- and cadmium-halides

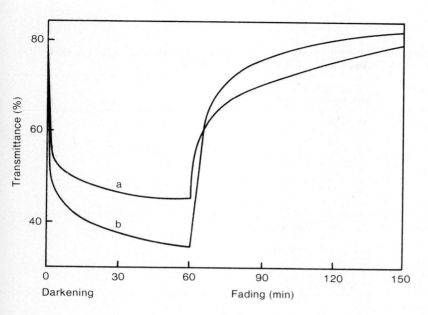

Fig. 7
Darkening and fading of photochromic glass (Corning, Photogray, thickness
2 mm) as function of time.
a) in sunlight at 25°C
b) in sunlight at 0°C

[46, 47] instead of silver halides. The glasses have a similar spectral sensitivity but show a stronger proportionality between light intensity and darkening.

Thermal treatment of photochromic glasses is generally possible, but the tempering process must be very closely controlled. The thermal stability depends on the type of the glass. If photochromic glasses are thermally overtreated they may become distorted and have a degraded photochromic performance. It is therefore recommended to keep the temperature below about 200° C and to use only short durations of tempering.

2.9 GLASS CERAMICS

As already mentioned, upon cooling down of a melt below the melting point of crystals having the same chemical composition as the glass, then all of the glass is in the state of undercooled liquid. The main reason for crystallization (devitrification) not taking place then, is mainly due to the fact that the crystal growth, controlled by the diffusion of the components, is far too slow. This is due either to the increasing viscosity of the melt with decreasing temperature, or to the fact that the number of nuclei from which crystallites (the smallest particles in which crystal structures can be detected) can form is too low. However, in the case

TABLE 11

PHOTOCHROMATIC SILVER HALIDE CONTAINING BOROSILICATE GLASSES [39]
AND THEIR SPECTRAL SENSITIVITY.

Glass components	Concentration in mass-%
SiO_2	40 ... 76
B_2O_3	4 ... 26
Al_2O_3	4 ... 26
Li_2O	2 ... 8
and/or Na_2O	4 ... 5
and/or K_2O	6 ... 20
and/or Rb_2O	8 ... 25
and/or Cs_2O	10 ... 30
Doping with silver halides AgX	0.2 ... 0.7 (transparent)
	0,8 ... 1,5 (opaque)
Halide sensitized with:	0,2 ... 0,4
As, Sn, Cu	0,005 ... 0,5
Spectral sensitivity AgCl-Doping	300 ... 400 nm
AgBr-Doping	300 ... 500 nm
AgI-Doping	300 ... 650 nm

of glass ceramics, the crystallite formation is forced in suitable glass systems, in order to obtain materials with special properties e.g. [4].

To start with, a glass melt is required from which the required object is shaped by pressing, blowing, rolling or casting. During subsequent thermal treatment, according to exactly pre-determined temperature-time-graphs shown schematically in Figure 8, sub-microscopic fine crystallites form; a pre-requisite for this is the addition of materials having a high melting point (usually TiO_2 and ZrO_2) to the melt, which start crystallization when they are exuded as nucleus-forming components. As is shown in Figure 9 it is important that the temperature range of the maximum nucleation frequency $T_{nuc.}$ is below the temperature range of the maximum crystal growth rate T_{cg}; then the glass cannot crystallize upon cooling of the melt as long as the nuclei are still absent. Only when these have formed in sufficient quantities at T_{nuc}, can the required tiny crystallites arise in large numbers (up to 10^{17} cm^{-3}) upon re-heating to the temperature T_{cg}. The percentage of crystals in the volume can finally amount to 50-90% depending on the properties strived for. The breakthrough in the development of technical glass ceramics was achieved by Corning glass factories in 1957 with the fundamental investigations of Stookey [48-50], which led to the standard types Pyroceram 9606 and 9608 based on glass compositions of Li_2O-Al_2O_3-SiO_2-TiO_2. Other companies followed with similar products. In Europe, Schott glass factories have manufactured glass ceramics since 1965. One of the newer highly transparent types is known as Zerodur [51]. Some of their properties are listed in Table 12.

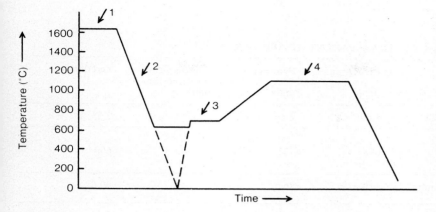

Fig. 8

Thermal treatment during glass ceramic production.
1 = melt
2 = shaping
3 = nucleation
4 = crystal growth

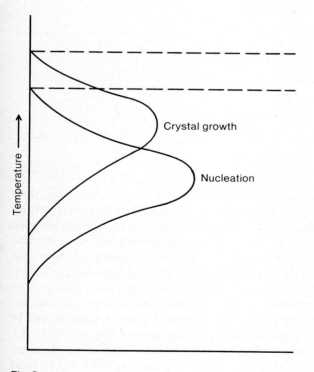

Fig. 9

Temperature dependence of nucleation rate and of crystal growth rate.

TABLE 12

PROPERTIES OF TRANSPARENT GLASS CERAMICS

Properties	Pyroceram 9606	Pyroceram 9608	Starting glass for Pyroceram 9608	Zerodur
Density (25°C)	2.60	2.50	2.403	2.53
Hardness Knoop kp mm^{-2}	698	705	–	600
Young's modulus 10^3 N mm^{-2}	119	86	–	90.6
Poisson's ratio	0.024	0.25	–	0.24
Thermal conductivity W m^{-1} K^{-1}	0.87	4.70	–	1.64
Linear expansion coefficient α 10^{-7} K^{-1} (0° ... 300°C)	56	4-11	33.6	1.2
Dielectric constant 1 MHz, 25°C	5.58	6.78	–	7.4
Dielectric losses tan δ 10^{-4} 1 MHz, 25°C	15	30	–	15.5

The technical importance of glass ceramics lies in the fact that their properties are determined not solely by the glass percentage, but also decisively by the types of crystals formed. Systems in which crystal phases of very low or even negative thermal expansion are formed (e.g. lithium-alumo-silicate) have become of great significance. Materials are obtained here with almost zero expansion over a wide temperature range, which are nondeformable up to approximately 800° C and completely insensitive to sudden changes in temperature, and thus can be used, for example, for hot plates and various domestic glass items and also for telescope mirror supports, laser mirror substrates, unit of length standards, etc. In the case of a different group of glass ceramics containing the components Si, Al, Mg, K, F and oxygen, during the ceramic process crystals similar to mica are formed [52-54]. These materials can be machined on a turning lathe due to their low brittleness, (e.g. Macor from Corning). Macor has the following composition (in wt%)

SiO_2	44 %
Al_2O_3	16 %
B_2O_3	8 %
MgO	16 %
K_2O	10 %
F	6 %

Some properties of similar products are listed in Table 13.

TABLE   13

PROPERTIES OF MACHINABLE GLASS CERAMICS

Properties	Glass ceramic type (Corning)			
	9650	9652	9654	9656
Linear expansion coefficient α 10^{-7} K^{-1} (25°...400°C)	97	74	64	63
Thermal load capacity (max.) in °C	800	800	800	800
Tensile strength in Pa 10^6 at 25°C 400°C 600°C	86.3 59.8 –	92.2 78 70.6	49.1 46 45.1	60 47.1 45.1
Bending stiffness in Pa 10^6	255.06	343.35	–	196.2
Young's modulus 10^3 N mm^{-2}	60.82	63.16	56.90	58.86
Volume resistivity in Ω cm at 500°C	17^7	10^{11}	10^{12}	10^{11}
Dielectric constant 10 kHz, 25°C	6.1	5.7	5.6	5.6
Dielectric losses tan δ 10^{-4}, 10 kHz, 25°C	30	60	60	20

In the region between glass and ceramic there are also a few photosensitive types of glass, where crystallization is initiated by uv irradiation and commences during subsequent heating. For this purpose, a few percent alkalifluoride, Zn and Al oxide must be present and a small quantity of silver compounds ($<0.2\%$) and ceroxide ($<0.05\%$) in the silicate glass matrix. Electrons are separated from the Ce-ions by uv irradiation and trapped by the silver ions during heating up so that metal atoms form which immediately deposit on one another as metal colloid particles. These act as nuclei for devitrification, which transforms the system to a yellow to brown coloured glass ceramic during subsequent annealing. In this way, with uv optics, photographic pictures of patterns can be recorded and fixed in the glass.

If other halides are present (bromine, chlorine), by repeated uv irradiation and heating, according to the degree of the latter any other colour tones can also be achieved based on various crystallite forms. These new «polychromatic» glasses (Corning) will presumably be applied for both technical and decorative purposes.

It is also significant that the little crystals formed can be etched by hydrofluoric acid at least 10 times faster than the surrounding glass matrix. Due to these properties, it is possible to manufacture mould etch parts of high precision, such as perforated panels for displays, printing stencils and such.

REFERENCES

[1]   G. W. Morey, The Properties of Glass, Reinhold, New York, 1945.
[2]   H. J. Güntherodt and H. Beck, (Eds.), Glassy Metals, (Topics in Appl. Physics, Vol. 46) Springer, Berlin, 1981.
[3]   I. I. Kitaigorodski, Technologie des Glases, Oldenbourg Verlag, Munich, 1957.
[4]   W. Vogel, Glaschemie, VEB Deutscher Verlag Grundstoffindustrie, Leipzig, 1979.
[5]   M. Faulstich, in Beiträge zur angewandten Glasforschung, E. Schott, (Ed.), Wissenschaftl. Verlagsges. m.b.H., Stuttgart, 1959, p. 269.
[6]   K. H. Sun, J. Am. Ceram. Soc., 22 (1947) 277.
[7]   G. M. Bartenev, (translated by P. and F. F. Jaray), The Structure and Mechanical Properties of Inorganic Glasses, Wolters-Noordhoff, Groningen, 1970.
[8]   W. H. Zachariasen, J. Am. Chem. Soc., 54 (1932) 3841.
[9]   a) B. E. Warren, Z. Krist, 86 (1933) 349 and J. Appl. Phys., 8 (1937) 645.
      b) P. Chaudhari, P. Beardmore and M. B. Bever, Phys. Chem. Glass, 7 (1966) 157, and R. Blachnik and A. Hoppe, J. Non-Cryst. Solids 34 (1971) 191.
      c) J. Zarzycki and R. Meznard, Phys. Chem. Glass 3 (1962) 163.
[10] J. C. Phillips, J. Non-Cryst. Solids, 34 (1979) 153, and Physics Today, 35 (1982) 27.
[11] N. Valenkov and E. A. Porai-Koshits, Z. Krist, 95 (1936) 195; Nature, 137 (1936) 273.
[12] E. A. Porai-Koshits, DAN SSSR, 40 (1943); Uspekhi Khimii, 13 (1944) 115; The Structure of Glass, Izd. AN SSSR (1955) pp. 30-43, The Glass-Like State, Izd. AN SSSR (1960) pp. 14-23.
[13] E. A. Porai-Koshits and N. S. Andreev, Nature, 182 (No. 4631) (1958) 336; DAN SSSR, 118 (1958) 735.
[14] O. K. Botvinkin, Physical Chemistry of Silicates, Promstroiizdat, Moscow, 1955; Introduction to the Physical Chemistry of Silicates, Goskhimizdat, 1938; The Glass-Like State, Izd. AN SSSR (1960) pp. 120-123.
[15] M. S. Aslanova, Glass Ceramics (USSR), Vol 17, (1960) 10; The Glass-Like State, Izd. AN SSSR (1960) pp.391-396.
[16] K. S. Evstrp'ev and N. A. Toropov, Chemistry of Silicon and Physical Chemistry of Silicates, Promstroiizdat, Moscow, 1956.
[17] E. J. Muchin and N. G. Gutkina, Crystallization of Glasses, Oborongiz., Moscow, 1960.
[18] W. Holzmüller and K. Altenburg, Physik der Kunststoffe, Akademie Verlag, Berlin 1961.
[19] U. S. Precision Lens, The Handbook of Plastic Optics, 1973.
[20] B. Welham, in Applied Optics and Optical Engineering, Vol. 7 (1979) 79, R. R. Shannon, J. C. Wyant, (Eds.), R. Kingslake (ser. Ed.) Academic Press, New York.
[21] G. M. Bartenev, DAN SSSR, 76 (1951) 227.
[22] G. M. Bartenev and I. A. Lukyanov, Zh. Fiz. Khim., 29 (1955) 1486.
[23] G. M. Bartenev and Yu. A. Gorbatkina, High Molecular Combinations, 1 (1959) 769.
[24] A. Winkelmann and O. Schott, Ann Phys. Chem., 51 (1894) 730.

[25] H. Salmang, Die Glasfabrikation: Physikalische and chemische Grundlagen, Springer, Berlin, 1957.
[26] W. A. Weyl, J. Soc. Glass. Technol., 33 (1948) 233.
[27] F. Rizzi, Rendiconti Acad. Sci. Fis., Napoli, (3) 30 p. 174, rev. in Glastechn. Ber., 5 (1927) 177.
[28] W. L. Wolfe, Properties of optical materials. In Handbook of Optics, W. G. Driscoll, (Ed.), McGraw-Hill, New York, 1978.
[29] H. Schröder and N. Neuroth, Schott Inform., 4 (1967) 1.
[30] F. C. Lin, Infrared transmission of several non-silicate glass-systems, Glass Industry, 1963.
[31] A. Smakula, Opt. Acta., 9 (1962) 205.
[32] W. L. Wolfe, Optical Materials for Infrared Instrumentation, Rep. 2389-11-S (1959) Supplement 2389-11-S1 (1961). The University of Michigan, Ann Arbor, Mich.
[33] S. S. Ballard and J. S. Browder, Appl. Optics, 5 (1966) 1873.
[34] P. Billard, Acta Electronica, 6 (1962) 7.
[35] a) Landolt Börnstein, Physical. Chem. Tabellen, 5. Aufl., 2. Band, Berlin, 1923, pp. 893-998.
      b) Landolt Börnstein, 6. Aufl., Vol. 2, Part. 8, Berlin 1962 p. 2-43, 2-397 (Koritnik), 2-405 to 2-433 (H. Pick), 3-465 to 3-542 (O. Lindig).
[36] M. Robinson, R. Pastor, R. Turk, D. Devor and M. Braunstein, Mat. Res. Bull. 15 (1980) 735.
[37] S. Musikant, (Ed.), Proc. SPIE, Vol. 297, Emerging Optical Materials, (1981) p. a.) 2, b.) 13, c.) 80, d.) 204.
[38] A. J. Cohen and H. L. Smith, Science 134 (1962) 981.
[39] W. H. Arminstead and S. D. Stookey, FRG Pat. No 809 847 (Aug. 1951).
[40] S. D. Stookey and F. W. Schuler, Proc. 4th Internat. Congr. on Glass, Paris, 1956.
[41] W. H. Arminstead and S. D. Stookey, Science 144 (1964) 150.
[42] K. Gerth and A. Rehfeld, FRG Pat. No 2 149 568; GDR Pat. No 97 121.
[43] A. Rehfeld and J. Rentsch, FRG Pat. No 2 260 879; FRG Pat. No 2 256 775; GDR Pat. No 10 13 75.
[44] VEB Jenaer Glaswerke, Schott und Gen., Jena, Druckschrift No 0107/1.
[45] L. G. Sawchuk and S. D. Stookey, US Pat. No 3 293 052 (Mar. 1963);FRG Pat. No 1 496 082 (Jan. 1964).
[46] R. J. Araujo, US Pat. No 3 325 299;FRG Pat. No 1 494 093.
[47] R. J. Araujo, FRG Pat. No 1 496 091.
[48] S. D. Stookey, US Pat. No 3 205 079.
[49] S. D. Stookey, J. Am. Ceram. Soc. 43 (1960) 190.
[50] S. D. Stookey, Glastechn. Ber. 32K (1959) 1.
[51] Schott Zerodur-Leaflet, 1981.
[52] D. G. Grossmann, US Pat. No 3 839 055.
[53] G. H. Beall, C. K. Chyung and H. J. Watkins, US Pat. No 3 801 295.
[54] G. H. Beall, FRG-OS Nos 2 224 990 and 3 756 838 and 2 133 652.

CHAPTER 3

3. NATURE OF A SURFACE

The surface of a solid material is the place where the material is in contact with the surrounding medium. The term "surface" is used if the medium is vacuum or a gaseous atmosphere. The contact with liquids or other solids is termed an "interface". Surfaces and interfaces are a universal phenomenon of the real structure of solid materials. The open arrangement of surface atoms or molecules compared with those of the bulk leads to special surface properties. Generally, the changes in the physical and chemical properties of surfaces are not limited to their topmost monolayer but to a more or less extended transition region.

Surfaces have received the attention of many scientists, and through the centuries many investigations of their properties have been undertaken. Early experiments concentrated mainly on an extensive study of their attractive forces. Thus, for example, Newton [1,2] was familiar with the phenomenon of two flat glass plates which, after being pressed together, could hardly be separated. He also knew that the presence of a liquid between the plates made the separation much more difficult. A number of experiments to study the rise of liquid columns in capillaries of varying diameter were carried out later [3,4,5,6]. From investigations of crystal growth, it follows that growth takes place predominantly by stepwise addition of atoms on surfaces or interfaces [7]. Degrading effects such as corrosion, abrasion and wear also occur on the surface or in regions near the surface.

Today, it is generally accepted that many technical material problems are in reality surface problems. This apparently is the reason for the great effort being invested in studying surface properties and for the development of various surface-treatment technologies.

3.1 CHARACTERIZATION OF A SURFACE

A surface is characterized by various specific physical, mechanical and chemical properties. The surface structure can be crystalline, amorphous or of a mixed type and be very different from the bulk. The atoms in the surface have a higher potential energy. The properties of surfaces are determined by surface forces and surface energies.

Depending on material and treatment, the surface can be an insulator, a semiconductor or a conductor. Technically produced surfaces are flat or curved. A distinction is made between flatness, waviness and roughness. The surfaces contain defects on an atomic and on a macroscopic scale. Such atomic defects are: point defects, dislocation lines, monoatomic ledges on cleavage planes.

Macroscopic defects are: polishing scratches, glass-drawing asperities, pores due to less than theoretical density of the material, grinding scratches, crystallite boundaries, surface warp and fused particles. The different nature of these defects requires a variety of methods to characterize quantitatively the condition of a surface. Finally, surfaces are characterized by a different chemical composition from that of the bulk material. This phenomenon depends on the generation and treatment of a surface, adsorption/desorption processes, surface swelling, leaching, surface reactions and segregation effects.

3.1.1 STRUCTURE OF A SURFACE

For the investigation of surface structures, electron-diffraction techniques such as LEED [8] and HEED [9] are mainly used. Recently, an effort has also been made to obtain structural information by spectroscopic techniques, i.e. SIMS [10]. The clean surface of a single or polycrystalline material is structurally heterogeneous.

There are several different atomic positions that are distinguishable by their number of nearest neighbours. There are atoms in terraces, where they are surrounded by the largest number of neighbours. There are also atoms in steps and at kinks in the steps, however. These sites have fewer neighbours than the atoms in the terraces. Adatoms that are located on top of the terrace have, of course, the smallest coordination number. This simple model describes the real situation very well. It is interesting to note that each of the surface sites may have a different chemical reactivity [11,12,13]. This explains much of the complexity revealed in studies of heterogeneous catalysis or corrosion. Surface atoms in any crystal face are in an anisotropic environment, which is different from that around the bulk atoms. Each bulk atom has a markedly higher crystal symmetry than atoms placed on the surface. The changed symmetry and the lack of neighbours perpendicular to the surface favours displacement of surface atoms in ways that are not allowed in the bulk.

Surface relaxation can give rise to a multitude of surface structures, depending on the electronic structure of a given material. These displacement reactions are called surface reconstructions. The surfaces of many metal, semiconductor and compound materials have atomic structures that are different from those expected from the projection of the X-ray bulk unit cell. The phenomenon is not fully interpreted. Further studies are required to understand the mechanism and nature of surface restructuring.

The energy released by the creation of a surface can also induce chemical changes. Surfaces of aluminium oxide and vanadium pentoxide exhibit both non-stoichiometry and surface reconstruction [14]. Both oxide surfaces show a deficit in oxygen. Similar effects were also observed with alkali halides.

Technical surfaces are almost always exposed to the atmosphere, in practice, and are therefore, depending on the nature of the surface, covered with an oxide film and/or adsorbed gases and water vapour. Adsorption and compound formation are not limited to the geometrical surface but can also occur at the grain boundaries.

Thicker compound films can be amorphous. In contrast to crystalline materials, the polymeric structure of glasses shows a lack of symmetry and periodicity. The constituent atoms have a statistical distribution and hence processes like gas sorption and corrosion, occurring on a fresh surface made by fracturing, should proceed almost uniformly over the surface. Experiments with certain technical multicomponent glasses, however, showed that glass may contain small domains of varying size in the range between 0.01 and 0.1 μm. The domains are regions of differing chemical composition and may also be crystalline. That means, for example in the case of water-vapour adsorption, higher H_2O concentrations in regions containing more hygroscopic oxides. Glass surfaces are often chemically cleaned or annealed. During such treatments they lose soluble and volatile components leaving behind a silica-enriched layer consisting, in the case of soda lime glass, of silicon and oxygen in a random network structure as in fused silica [15]. The porous films that form on weathered glass are also non-crystalline silica.

Particular types of macromolecules, production technologies used, or later treatments may cause organic glasses also to contain small crystalline domains. Although these domains are small compared with the wavelength of light, they cause a more or less milky appearance in such heterogeneous materials. It is therefore likely that super lattices are formed. There have been no special investigations on the surface structure of organic glasses. It can be assumed, however, that some crystalline areas may also exist on or near the surface.

3.1.2 CHEMICAL COMPOSITION OF A SURFACE

Until recently, analytical investigations of surfaces were handicapped by the lack of suitable methods and instrumentation capable of supplying reliable and relevant information. Electron diffraction is an excellent way to determine the geometric arrangement of the atoms on a surface, but it does not answer the question as to the chemical composition of the upper atomic layer. The use of the electron microprobe (EMP), a powerful instrument for chemical analyses, is unfortunately limited because of its extended information depth. The first real success in the analysis of a surface layer was achieved by Auger electron spectroscopy (AES) [16,17], followed a little later by other techniques such as electron spectroscopy for chemical analysis (ESCA) and secondary-ion mass spectrometry (SIMS), etc. [18-23].All these techniques use some type of emission (photons, electrons, atoms, molecules, ions) caused by excitation of the surface state. Each of these techniques provides a substantial amount of information. To obtain the optimum information it is, however, often beneficial to combine several techniques.

In considering the possible surface compositions, we must distinguish between single and multicomponent materials. The composition of a given material in the topmost layer is in both cases generally very different from the bulk composition, for several reasons. Technical surfaces have modified surface compositions owing to the method used to create the surface. Cleaning procedures and finally adsorption of gases and vapours, possibly covered up as a result of surface reactions, modify the surface composition. In the case of gas adsorption in a multicomponent material with sufficiently mobile atoms, surface enrichment of one constituent can occur by the formation of preferred strong adsorption bonds (chemisorption, chemical reaction) at the expense of other constituents that form weaker bonds. Materials with mutual regular solution behaviour of their constituents minimize their surface free energy by a segregation of that component with the lower surface tension. Embedded or solute atoms and molecules with space requirements different from those of the atoms of the host-lattice cause excessive strain in the material, so that generally segregation of the component that is a misfit in the lattice occurs. In the case of the formation of stable compounds of high lattice energy, the possible segregation induced by the different surface tensions of the constituents ceases if the necessary exchange energy exceeds the influence of surface forces. If no other effects occur, the surface composition under those conditions is the same as the bulk composition [24].

With multicomponent amorphous polymeric materials, prediction of the occurrence of a surface enrichment effect is very difficult because of many uncertain parameters. With worked multicomponent silicate glasses, the wet grinding and polishing process, and the cleaning with liquids, leaches out the soluble constituents, e.g. the alkali and partially also the alkaline earth metal oxides. The silica network is modified by chemical and physical water uptake and varying amounts of the oxidic polishing agents which are incorporated into and near the surface. A glass surface made by fire polishing or by casting has a composition that depends on the extent to which the more volatile constituents have evaporated and been replaced by diffusion from the bulk material. In addition to the lowering of mainly the alkali surface concentration, reaction or corrosion products may also be deposited. Treatment with acids extracts all the basic oxides, leaving behind a silica-enriched surface layer. That means that, after acid cleaning, different types of glass of evidently varying volume composition tend to have almost the same surface composition. The thickness of surface layers with modified composition on glass is, depending on the type and duration of treatment, generally in the range between 1 and 100 nm.

The surface composition of clean plastics is generally homogenous. Surfaces produced by injection moulding may be covered with mould-release agents or other processing impurities that are hard to remove. To eliminate dust-collecting electrical charges generated during processing, the surfaces are often provided with antistatic layers, which can be removed by washing with water. Surface swelling occurs in contact with many organic solvents and with water. Careful thermal treatments displace the absorbed liquids without degradation of the plastic, and the swelling disappears. Plastic substrate surfaces are sometimes treated with a non-gassing base coat to improve surface smoothness and to promote adhesion of a later metal coating. The base coat can also play a significant role in matching a given coating-substrate combination.

3.1.3 ENERGY OF A SURFACE

The creation of a surface requires work [25]. This process is always accompanied by a positive change in free energy S. The surface tension γ is the reversible work required to create a unit area of surface at constant temperature, volume, and chemical potential. The surface area may be increased by adding more atoms or molecules to the surface or by stretching the existing surface. Surface-tension investigations of solids result, depending on the experimental conditions, in values that are a combination of values for both surface tension and surface stress, and it is very difficult to distinguish one from the other. With liquids, there is no difficulty in distinguishing between surface stress and surface tension because the diffusion of atoms in the liquid is fast enough to prevent or to remove the stress.

Another problem with solid surfaces is the lack of data on surface tension as a function of temperature. Most data available are for higher temperatures. Although it can often be assumed that the surface tension decreases with increasing temperature, the behaviour over larger temperature ranges is usually unknown. The surface tension of a solid also depends on the crystallographic orientation and on the crystallite size, which determine the influence of curvature on surface tension. Both parameters further increase the difficulty in obtaining reliable surface-tension data.

TABLE 1

SURFACE FREE ENERGY OF OXIDES

Material	Surface free energy S (erg cm^{-2})	Temperature T (K)	Reference
Al_2O_3	928	300	26 – 28
B_2O_3	96	solid	29
BaO	290	1373	30
CaO	820	298	31
FeO	854	300	32
MnO_2	620	2123	33
PbO	132	1173	34
SiO_2	605	298	31
TiO_2	380	MP	35
ZnO	90		36
ZrO_2	1130 – 770	< 1423	37, 38

In Table 1, the surface free energies of several important oxides are listed. As may be seen in the table, the differences between the various materials are remarkable. In single and multicomponent materials, the tendency towards the lowest possible value of surface free energy is the reason for recrystallization effects and for sorption, phase separation, and segregation phenomena. Surface enrichment of the component with the lowest surface-tension value is generally observed. Calculations of the surface-tension and surface-enrichment effects were made for alloys (real solutions) using the values of the pure elements [39]. The surface energies of multicomponent silicate glass at 1000° C are between 200 and 300 erg.cm^{-2}. With alkali glass, the surface free energy may be reduced by sodium migrating to the surface. It can be expected therefore that the surface free energy of freshly fractured surfaces will be higher than that of a melted sample. It is interesting to note that water vapour reacts with glass surfaces and decreases the surface tension. Other atmospheric gases, however, have no significant influence [40]. Later results in data collection of published values for the surface tension of glass [41] are listed in Table 2. An increase of temperature of the molten glass often decreases its surface tension. With the same glasses, a sharp decrease of surface tension was observed between 600° C and 750° C [42], the range in which the glass softens.

Temperature coefficients of surface tensions between − 0.4 and − 0.02 dyn cm^{-1}·°C^{-1} [42-44] were measured and, depending on the glass composition, positive coefficients [45,46] as well as the negative TC values were observed. The influence of glass composition on surface tension was studied by substituting one oxide component for another [44]. An increase in γ was observed by adding oxides in the sequence Na$_2$O, BaO, CaO and MgO. It is assumed that the surface tension of glass can be treated as an additive function of composition [47]. No data are available for organic glasses. It is very likely, however, that their surface free energy values are below 80 erg cm^{-2}. Surface energy can be reduced by segregation products of polymer impurities and low-molecular-weight species.

TABLE 2

SURFACE FREE ENERGY OF SILICATE GLASSES

Material	Surface free energy S (erg cm^{-2})	Temperature T (°C)	Reference
Soda-lime silicate glasses	270 − 330	1100 − 1400	41
Soda-lime borosilicate glasses	244 − 278	1000 − 1400	41
Lead silicate glasses	142 − 235	1000 − 1300	41

A knowledge of surface energy is important because it influences the adsorption of gases, wetting by liquids, adhesion to solids, and frictional resistance.

3.1.4 MORPHOLOGY OF A SURFACE

The morphology of a surface is a very important property since it influences physical and chemical behaviour of a material (e.g. light scattering and heterogeneous catalysis). In the case of a coating process, the surface is also the place where the film-substrate interaction influencing many properties of the deposited film (e.g. adherence, microstructure, topography) occurs. One of the most common methods of forming surfaces on glass is by optical working, but it is difficult in this case to formulate generalizations about the surface finish. When in a fire-polishing process glass is melted and cooled down, the surface is as smooth as that of a liquid. Once in contact with the atmosphere, however, the surface may be rapidly attacked by various corrosive agents, developing fissures and making the material porous, although glass is generally nonporous. A less perfect surface finish is often obtained by the technical wet-grinding and polishing processes. The topography and the nature of surfaces on optically worked glass depend on the type and mechanism of the applied polishing process.

The irregularities of a surface can be classified into three components of different periodicities. As may be seen in Fig. 1, "flatness" has the longest periodicity of several centimetres or even more. A medium value of periodicity of the order of the wavelength of light is referred to as "waviness" and the shortest values in the nanometer region are designated "roughness". The flatness or planarity of substrates is of main importance in the production of high quality mirrors and filters. It is inspected by various interferometric techniques utilizing a reference glass (mirror). The entire substrate is viewed and the interference fringes form a contour map of the surface. Atomically-flat surfaces over large substrate areas are almost unattainable. Even the best single-crystal substrate acquires defects in the fabrication and manufacturing process.

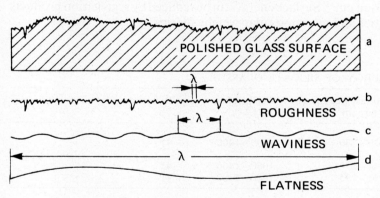

Fig. 1
Schematic representation of the surface topography of polished glass (a) and the various components of topography (b, c, d).

A very good technique for the characterization of the smoothness of a substrate is the well known stylus technique. With this technique, the surface waviness and surface roughness can be determined very rapidly, and a quantitative measure of the surface quality is also obtained. Polished crystal plates such as sapphire and polished vitreous materials as, for example, fused silica, silicate glasses and organic polymers show very uniform surfaces and yield smooth stylus traces. Drawn glass also has smooth surfaces, with occasional surface irregularities of approximately 100 nm in height that stem from the forming operation. Drawn borosilicate glass, however, tends to be wavy because of its relative high drawing temperatures and the short working ranges. The deviations from a flat plane, however, are generally negligible. Roughness and scratches are more critical. These defects are replicated by the deposited layers or propagate into the films as, for example, dislocations continue into epitaxial layers and even fine scratches in the surface generate irregularities in the films.

3.1.5 INTERACTIONS SOLID/GAS AND SOLID/SOLID

Surfaces always differ in behaviour from the bulk of a material because of the abrupt changes that occur at and near phase boundaries. Surface atoms and molecules are not in equilibrium states, since they are neither in one phase nor in the other. Unsaturated bonds abound. This leads to an excess energy associated with the surface, the so-called surface free energy which has different values for different crystallographic orientations. There are different ways to minimize surface energy. A simple way would be to reduce the surface area under the influence of surface tension. But this is not realistic with solid materials. Surface free energy, however, can also be lowered by adsorption and segregation phenomena.

One of the most characteristic properties of a solid surface is its ability to adsorb gases and vapours. Adsorption occurs as a result of the interactions between the field of force at the solid surface and that existing around the molecule of the gas or the vapour which is to become adsorbed. For vapours at pressures near the saturation pressure and for gases showing a specific chemical affinity for the solid, the amount of adsorption can be substantially larger and may approach or even exceed the point of monolayer formation. Adsorption is classified as physical if it involves only Van der Waals' forces and chemical if an exchange or sharing of electrons also take place. Both types can be distinguished by the heat of adsorption. In chemisorption the heat of adsorption is of the order of the heat of chemical reactions (e.g. $- \Delta H \simeq$ 10-30 kcal$\cdot$mole^{-1}), and so it is, in general, much higher than for physical adsorption (e.g. $- \Delta H \simeq$ 2-6 kcal$\cdot$mole^{-1}). However, much smaller values can also be obtained for true chemisorption. Thus, for example, a value of 3 kcal$\cdot$mole^{-1} was reported for chemisorbed hydrogen on glass [48,49]. In principle, chemisorption can even be endothermic, like certain bulk chemical reactions. A further distinction can be made by the rate of adsorption. Like other chemical reactions, chemisorption is usually an activated process and so it proceeds at a finite rate that increases with rising temperature. The non-activated physical adsorption is fast at all temperatures, even at the very low ones. A distinction can also be made by the thickness of the adsorbed layer. Chemisorption is always confined to a single atomic or molecular layer whereas in physical adsorption the layer, though monomolecular at lower pressure, will become multimolecular at higher relative pressures.

Sorption phenomena have been extensively studied and there are many books and review articles available [50-59]. Silicate glass usually has a great affinity for water vapour. Besides containing smaller amounts of CO_2, SO_2, O_2, N_2 glass invariably contains water bound in the bulk structure and on the surface as hydroxyl groups and, mainly on the surface, also as adsorbed H_2O molecules. A schematic presentation of adsorbed water on glass is shown in Fig. 2. It seems that physical

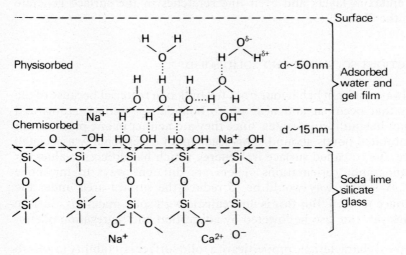

Fig. 2
Schematic representation of the formation of a water film on glass by adsorption of water vapour.
The typical reactions are:

$$-Si-O-Si- \ + \ H_2O \ = \ -Si-OH \ + \ HO-Si-$$

$$-Si-O\text{-}Na^+ \ + \ H_2O \ = \ -Si-OH \ + \ Na^+ \ + \ OH^-$$

$$-O-Si-O-Si- \ + \ NaOH \ = \ -O-Si-O\text{-}Na^+ \ + \ HO-Si-$$

In the physisorbed film mainly Van der Waals forces are responsible for bonding:

$$
\begin{array}{c}
H^{\delta+} \\
\diagdown \\
O^{\delta-} \cdots H^{\delta+} - O^{\delta-} \diagdown \\
\diagup H^{\delta+} \\
H^{\delta+}
\end{array}
$$

adsorption occurs only on sites where hydroxyl groups are chemisorbed [60]. A fresh surface of, for example, soda-lime silicate glass in contact with an atmosphere of unsaturated water vapour reacts (chemisorption) on active surface sites and forms OH^- ions and it is also partially covered by adsorbed (physisorption) H_2O molecules. During longer exposure, the adsorbed water reacts with the alkali component in the glass, yielding alkaline solutions on the surface, which attack the silica network, forming silica gel. Contact with various gases, e.g. CO_2, may lead to surface absorption by the alkaline aqueous layer. It is obvious that adsorption phenomena on multicomponent silica glass are rather complex. When glass is heated in a vacuum, gas may be released both from the surface and from the bulk, depending on the applied temperature. The nature and quantity of gases evolved during baking in a vacuum was extensively studied [61-65] and was shown to depend on glass type, manufacture technique, and surface treatment. A large collection of data appears in [66]. For ordinary glass, the surface-water is removed by baking at $450 - 500°C$ or below. The ratio of the liberated gases $H_2O: CO_2: N_2$ is about 100:10:1. In temperature treatments above 550°C also, and mainly in the temperature range at which glass softens and melts, the chemically-bound water in the interior is dissociated and is released into the atmosphere [67,68]. Organic glasses also adsorb water vapour. They also show permeation. For polystyrene, the total water sorption is about 0.3% and the coefficient of permeation is 7.8×10^{-6} $cm^2 \cdot s^{-1}$ [69]. Larger differences in the permeation property are caused by different solubilities. In general, there is no chemical reaction between the plastic-polymer surface and the water vapour.

The solid/solid interaction is concerned with the nature and properties of the adhesion between glass surfaces and the surfaces of other solid materials. The ASTM* defines adhesion as "a condition in which two surfaces are held together by either valence forces or by mechanical anchoring or by both together" [70]. The bonding forces can be Van der Waals' forces, electrostatic forces (charging effects) [71] and/or chemical-bonding forces which are effective across the interface. The resulting adhesion energies are found to be between 0.1 and 10 eV. The physisorption contributes up to approximately 0.5 eV of energy to the adhesion. The force lies between 10^4 and 10^8 dyn·cm^{-2}. The energy contributed from chemical bonds ranges between 0.5 and approximately 10 eV. The corresponding forces are greater or equal to 10^{11} dyn·cm^{-2}.

The quality of the adhesion depends to a large degree also on the microstructure of the interface layer that is forming.

A distinction is made between five types of interface layers [72,73,74] based on effects such as mechanical anchoring, more or less abrupt material transition, chemical bonding, and diffusion as well as pseudo-diffusion. More details on this topic are given in Chapter 5.

This seemingly clear picture of differing interface layers and types of interaction is clouded owing to the fact that surfaces do not behave uniformly because they are under the influence of so-called active and passive centres. Active centres include, for example, grain boundaries, dislocations, vacancies, chemical inhomogeneities or crystallite faces with varying chemisorption free energy and activation energies of the chemisorption. Passive centres also exist analogously to the active centres. Passive centres are surface areas already covered with foreign material at which no chemisorption can take place to any great extent. The quality of adhesion improves with time in some cases, e.g. Ag on glass; this phenomenon is explained by the slow formation (diffusion) of an oxygen-bonded interface layer.

---

* ASTM = American Society for Testing and Materials.

Adhesion is determined in a complex way by the sum of various intermolecular atomic interactions and may be influenced by various parameters; these will also be discussed later in Chapter 5.

3.2 PRODUCTION OF GLASS SURFACES

In the industrial production of glass, a distinction must be made between mass production for container glass and flat or sheet glass, and the production of more special products like fibre glass for technical purposes and, in the highest qualities, for optical communications. Finally, there are high quality products like optical glasses and glasses for electronics, space technology, and biomedical applications. Although glass containers are also coated with various films [70-73], the interest of this monograph is directed more towards coatings on flat glass and on miscellaneous optical glasses.

The evolution of glass-production technology has been steady but rather slow. It is interesting that this is in sharp contrast to the very rapid evolution of thin-film technology. The soda-lime silicate glass known since early times is still the basic glass material, and this will not change in near future. Flat-glass manufacturing, however, has made significant progress with the introduction of the float-glass process. Flat-glass consumption by the end-use sector in the EEC in 1978 amounted to a total of 3 million metric tons. Of this quantity, about 56% was used for glazing in buildings, 23% in the automobile industry, and only 21% in other branches. Although there are more than 250 different types of optical glasses in use, the total quantity is small compared with that of flat glass. The tendency to reduce the weight and to achieve an acceptable level of safety led to the fabrication of thin laminated glass structures and also, because of the lower costs, to an increase in the use of organic-polymer glasses. The numerous scientific and technical applications of glass require quite different surface qualities. In the field of thin-film physics, the need for well defined surfaces is widely recognized. It is recognized that the quality of a surface finish depends strongly on the applied production technology. The most important surface manufacturing methods are therefore briefly reviewed here.

3.2.1 DRAWING AND CASTING

The fabrication of sheet and plate glass by the well-established Fourcault, Libbey-Owens and Pittsburgh methods [74,75] uses drawing processes. In a vertical sheet-glass machine, for example, the molten glass passes a slot of debiteuse which also forms because of cooling a highly viscous glass ribbon. During further cooling, this ribbon runs through transporting equipment consisting of many shaping rolls of polished steel. The glass ribbon has a width of 2-3 meters and its adjustable thickness is between 0.9 and 6.5 mm. The composition of the surface of flame-polished glass is variable and depends on the extent to which the more volatile elements have evaporated and been replaced by diffusion from the bulk. During cooling, a rather smooth surface is created under the influence of the surface tension. Most applications of plate glass manufactured this way, apart from mirror glass, demand no subsequent machining. The plane parallelity of the two surfaces, however, is disturbed to a variable degree by the waviness of the glass.

Since the early 1960's, the classical manufactured sheet and plate glasses have been increasingly replaced by a modern flat glass product called "float glass". This product has the quality of the best mirror glass and can be fabricated in one run. The float-glass process is the result of intensive experiments carried out in the laboratories of Pilkington Brothers Ltd. [75]. The glass type and melting procedures are similar to those used in the other flat-glass manufacturing methods. Instead of being drawn, however, the melt is cast on a float band which consists of molten tin. The tin bath has a width of 4 to 8 meters and is about 60 meters long. At the entrance of the float band, where the molten glass first comes into contact with the tin surface, the temperature of the liquid metal is 1000 °C. At the end, it is only 600 °C. Special rolls take the glass off and transport the ribbon through a cooling tunnel. During this treatment, the temperature decreases to about 200 °C. Further cooling to room temperature is performed in open transporting equipment. Finally, the flat glass ribbon, which can be adjusted to thicknesses between 1.5 mm to 20 mm, is cut up in different-sized pieces. Float glass is produced fully automatically and in outputs of up to about 3000 $m^2 \cdot h^{-1}$. The quality of the surface, which is generally very smooth, is mainly determined by the interface between the molten tin and the glass melt. The cleanliness of the liquid-tin surface is therefore very important. To prevent oxidation, the hot tin bath is kept under a slightly reducing atmosphere.

3.2.2 PRESSING AND MOULDING

Amongst the other available techniques hot pressing constitutes a frequently required form of raw glass treatment for precision optical lenses and for lenses in ophthalmic applications. This initially involves shearing a continuous flow of temperature-controlled glass into gobs of a certain weight and then pressing them into the required shape. The convex and the concave surfaces are moulded as the glass comes into contact with the top and bottom cast iron plungers under pressure. An annealing process is then essential, over a number of hours, to restore homogeneity. Mouldings thus produced are consistent in diameter and thickness within tight tolerances.

Accurate surface shapes are produced in modern curve-generating machines by milling and lapping operations starting with sintered diamond tools. At least three grinding operations are required for rough machining with increasingly fine grinding grain. These are then followed by one or two fine-grain operations to achieve a finish suitable for starting the polishing process.

Hot pressing and moulding of glass without further surface finishing operations is used to produce, for example, mirror substrates for various small projector lamps, signal lamps, and automobile reflectors. The surface of this type of glass is quite satisfactory. Optical parts made of organic-polymer glasses are cast, fabricated or injection moulded. Cast optics are used mostly in the ophthalmic field. Fabrication techniques are economical for producing the larger optical elements with diameters of 100 mm or more. These elements are usually machined and then polished from cast blocks. With injection moulding, however, a larger volume of plastic optical parts can be produced at relatively low cost. For producing quality optical elements, injection moulding is performed in an ultra-clean environment that is temperature and humidity controlled. This process, like many other optical procedures, is highly proprietary among lens producers, and details are not revealed. In general, however, the process uses steel moulds containing cavities that are machined to the desired optical configuration. A polishing process refines the cavity surface to an optically finished quality. The plastic is supplied as extruded and cut pellets. After being heated to plasticity, it is forced under pressure into the temperature-controlled mould. The melt in the mould is cooled down and then the formed optical elements are removed. Finally each element receives an antistatic coating to eliminate dust collecting due to electrical charges generated during processing. No additional machining or polishing is required, since the mould-cavity configuration conforms to the finished optic requirement, and because the quality of the mould surface was imparted to the melt. The surface smoothness of injection-moulded lenses can be extraordinarily good if the mould surface is excellent. There is also a high lens-to-lens reproducibility.

3.2.3 GRINDING AND POLISHING

Many special types of glass, e.g. those for optical applications, are available in the form of strips, blanks and blocks of various size. Depending on the product dimensions, most of the tank-melted glass is directly rolled, cast or pressed. Glass-forming characteristics and product geometry determine which glass types can be produced in this way. Optical glass is also supplied in restricted gob form having a shallow convex curve on the bottom surface and a fire-polished surface.

Grinding and polishing are surface-treatment operations used for shaping and finishing materials like single crystals as well as inorganic and organic glasses. First of all, a glass block, for example, is cut up with a diamond saw and brought to the rough form by milling. Subsequently, the object is shaped by grinding. This is a predominantly physical operation. In the grinding process, glass is removed by breaking out small glass fragments under the action of loose or bound abrasive grains of silicon carbide, electrocorundum or diamond on discs of iron or brass. Water is used to distribute the grinding grains evenly and to prevent an undesired rise in surface temperature, which would possibly generate plastic deformation. The grinding material grains must be harder than glass and the particle size broken out during the grinding operations depends on the size of the abrasive grain. The ground surface, however, not only appears rough but is also covered with flaws and fissures penetrating the glass. It is assumed that the combined effect of the load of the abrasive particles and the additional tensile stress in the surface which results from their frictional pull is responsible for breaking and flaw formation. Generally, the weight of glass removed during grinding is directly proportional to the weight on the shaping tool or to the relative speed between tool and glass.

The purpose of polishing is to produce regular transparent surfaces on glass or other materials and these surfaces are usually required to be flat, spherical, aspherical or of a complex shape. The method by which ground glass surfaces are smoothed by polishing with finely divided metal oxides (αFe_2O_3, CeO_2, ZrO_2, TiO_2, Al_2O_3, etc.), on cloth or pitch polishers in the presence of water has been the subject of conjecture for more than three hundred years and even modern investigations with polyethylene polishers have not yet produced a generally accepted qualitative theory. The polishing speed, pressure, type of glass, surface finish from grinding, type of polishing material and polishing compound, all have an important influence on what happens during the process. It is generally accepted, however, that polishing is a physical and chemical process. There is an excellent review on this topic [66] according to which the different theories can be roughly classified into three groups. In the wear theory, polishing is viewed as an extension of grinding involving the mechanical removal of asperities . In the flow theory, it is assumed that the asperities flow plastically into surface cavities. The required surface liquification is produced by pressure, or stems from frictional heating. Finally, according to the chemical theory, the asperities are removed by transformation into compounds produced in chemical reactions with the polishing agents that can be easily removed mechanically. There are various ways to polish a surface so that, depending on the (special) polishing conditions chosen, one or other of the proposed mechanisms is dominant.

In technical applications, polishing is performed in at least two steps. In initial polishing, the ground grey surface is smoothed by the action of an oxide/water slurry on felt or pitch. The rough surface is worn down nearly to the bottom of the deepest cavities. Silica gel is produced as a reaction product of glass and the polishing agents. It smears and glazes the smoothed surface areas and fills, to a certain extent, the remaining cavities. With felt instead of pitch, the surface created is rather smooth but wavy. In final polishing, the conditions are adjusted to favour smearing and enhance polishing. The latter is important to remove the last cavities. The surface flow is increased by the application of higher polishing pressures (about 1 kg·cm^{-2}) and higher spindle rotation to obtain frictional heating, which may also promote silica gel formation. At the end of the polishing process, the polishing agents, which are mainly CeO_2 and ZrO_2 for inorganic glasses today, may be worked out of the tool and the water is allowed to dry up. The remaining polish layer consists mainly of silica gel and therefore differs appreciably from the bulk glass.

It is known [76] that the thickness of the polish layer is quite different for different types of glass and polishing tools. The polish layer of felt-polished soda-lime plate glass has a thickness of 0.01 μm; for the same glass with pitch polishing, it has a thickness of 0.03 μm; and for pitch-polished borosilicate crown (BK7) the layer is even 0.3 μm thick. Grinding and polishing operations are used to produce glass surfaces with the best surface quality. In these cases, optimal polishing is required which removes the grinding topography to below the deepest

48

cavities. Plastic polishing tools have been developed, with a better performance than pitch, for use where the highest glass surface accuracy is not required.

The question of the mechanism of polishing is more complex when the polishing of organic polymer glass is considered. Although small optical parts of, for example, acrylic are made by injection moulding, because of the shrinkage the larger parts (e.g. $\varnothing > 100$ mm) are fabricated by machining and polishing. The raw material used for many applications is cast acrylic. The maximum standard size of acrylic in sheet form is 1.8 m in length with a thickness of about 100 mm. Manufactured acrylic optics pieces, finished by proprietary machining and polishing techniques mainly with Al_2O_3 and SnO_2, are equal or even superior to a really good «medium quality» optical glass.

Inspection and test procedures are required to control and to guarantee a quality standard. Light and electron microscopy are valuable tools to study the polishing process [77] as well as to control the quality of the finished product [78,

Fig. 3
Electron micrographs (Pt/C surface replicas) of surfaces of:
a) a high quality pitch polished BK 7 (Schott) glass,
b) a medium quality polished BK 7 glass and
c) a carefully polished CaF₂ synthetic crystal.

79] to avoid manufacturing defects. Various interferometric techniques can also be used for this purpose, [80]. A further powerful technique to control the remaining surface roughness of high quality glass substrates is the measurement of light scattering. It was shown, using this technique [81], that the polishing time is an important parameter, with inorganic glass. Excellent surface finishes are obtained only after a polishing time of longer than 5 hours.

Figures 3 and 4 show electron micrographs of various surfaces.

Fig. 4
Electron micrograpns (Pt/C surface replicas) of the surfaces of polymethyl metha-crylate lenses produced by casting into optically worked moulds: a) low quality, b) high quality.

A detailed description of modern optical production technology is found in refs. [82] and [83].

REFERENCES

[1] I. Newton, Principia, London, (1686), Prop. LXXXV, Theorema XXXV; Prop. LXXXVI.
[2] I. Newton, Optics, London, (1717), Query 31.
[3] B. Taylor, Phil., Trans., 27 (1711) 538.
[4] F. Hawksbee, Phil. Trans., 27 (1711) 395, 473, 539.
[5] F. Hawksbee, Phil. Trans., 28 (1712) 151.
[6] J. Jurin, Phil. Trans., 30 (1719) 1083.
[7] W. Kossel and I. N. Stranski, see:Grundlagen der Festkörperphysik, Ch. Weissmantel and C. Hamann (Eds.), Springer, Berlin, 1979.
[8] G. A. Somorjai, LEED and Surface Topography, in Surface Science, Vol. I, Lectures at Internat. Centre of Theoret. Physics, Trieste, 1974, Published by Internat. Atomic Energy Agency, Vienna, 1975.
[9] R. Brill and R. Mason, (Eds.), Advances in Structure Research by Diffraction Methods III, Viehweg, Braunschweig, 1970.

50

[10]   R. Buhl and A. Preisinger, Surf. Sci., 47 (1975) 344.
[11]   G. A. Somorjai, Advan. Catalysis, 1976.
[12]   D. W. Blakely and G. A. Somorjai, Nature (London), 258 (1975) 580.
[13]   D. W. Blakely and G. A. Somorjai, J. Catalysis, 42 (1976) 181.
[14]   M. A. Chesters and G. A. Somorjai, Ann. Rev. Mat. Sci., 5 (1975) 99.
[15]   J. J. Antal and A. H. Weber, Phys. Rev., 89 (1953) 900.
[16]   J. J. Lander, Phys. Rev., 91 (1953) 1382.
[17]   P. W. Palmberg, G. K. Bohn and J. C. Tracy, Appl. Phys. Lett., 15 (1969) 254.
[18]   A. Benninghoven, Appl. Phys., 1 (1973) 3.
[19]   C. T. Hovland, W. C. Johnson and W. M. Riggs, Ind. Res./Developm., Sept. (1978) 124.
[20]   N. C. MacDonald, G. E. Riach and R. L. Gerlach, Ind. Res., Aug. (1976) 12.
[21]   W. M. Riggs and R. G. Beimer, Chem. Technol., 5 (1975) 652.
[22]   B. F. Philipps, Standardiz. News, 6 Feb. (1978) 23.
[23]   M. V. Zeller and B. F. Phillips, Ind. Res./Developm., March (1981) 136.
[24]   G. A. Somorjai, J. Coll. Interf. Sci., 5 (1977) 150.
[25]   F. Garcia-Moliner, Surface Thermodynamics, in Surface Science, Vol. I, Lectures at Internat.
       Centre for Theoret. Physics, Trieste, 1974, Published by Internat. Atomic Energy Agency, Vienna,
       1975.
[26]   A. Portevin and P. Bastien, C. R. Acad. Sci., 202 (1937) 1072.
[27]   S. K. Rhee, J. Am. Ceram. Soc., 55 (6) (1972) 300.
[28]   P. P. Budnikov and F. Y. Xaritonov. Izv. Akad. Nauk SSSR, Neorg. Mat., 3 (1967) 496.
[29]   S. I. Popel and O. A. Esin, Zh. Neorg. Khim., 2 (1957) 632.
[30]   W. D. Kingery, Metallurg. (Moscow) (1963) 446.
[31]   R. Panpuch, Silic. Ind., 23 (1958) 191.
[32]   B. Sikora, Pr. Inst. Hutn., 20 (6) (1968) 375.
[33]   S. I. Popel, V. I. Sokolov and O. A. Esin, Zh. Fiz. Khim., 43 (1969) 3175.
[34]   L. Shartsis, S. Spinner and A. W. Smock, J. Am. Ceram. Soc., 31 (1948) 23.
[35]   V. I. Kostikov, M. A. Maurakh, B. S. Mitin, I. A. Pen'kov and G. M. Sverdlov,
       Sb. Mosk. Inst. Stali. Splavov., No. 49 (1968) 106.
[36]   P. C. Bonsall, D. Dollimore and J. Dollimore, Proc. Brit. Ceram. Soc., 6 (1966) 61.
[37]   R. C. Garvie, J. Phys. Chem., 69 (1965) 1238.
[38]   Yu. M. Polezhaev, Zh. Fiz. Khim., 41 (1967) 2958.
[39]   S. H. Overbury, P. A. Bertrand and G. A. Somorjai, Chem. Rev., 75 (1975) 547.
[40]   N. M. Parikh, J. Am. Ceram. Soc., 41 (1958) 18.
[41]   D. W. Mitchell, S. P. Mitoff, V. F. Zackay and J. A. Pask, Glass Ind., 33 (1952)
       453 and 33 (1952) 515.
[42]   W. B. Pietenpol, Physics, 7 (1936) 26.
[43]   G. W. Pamelee and K. C. Lyon, J. Soc. Glass. Technol., 21 (1937) 44T.
[44]   C. L. Babcock, J. Am. Ceram. Soc., 33 (1940) 12.
[45]   L. Shartsis and R. Canga, J. Res. Nat. Bur. Stand., 43 (1949) 221.
[46]   L. Shartsis and A. W. Smock, J. Res. Nat. Bur. Stand., 38 (1947) 241.
[47]   K. C. Lyon, J. Am. Ceram. Soc., 27 (1944) 186.
[48]   P. M. Grundy and F. C. Tomkins, Quart. Rev. Chem. Soc., 14 (1960) 257.
[49]   B. M. W. Trapnell, Proc. Roy. Soc., 206 A (1951) 39.
[50]   B. M. W. Trapnell, Chemisorption, Butterworth, London, 1955.
[51]   D. M. Young and A. D. Crowell, Physical Adsorption of Gases, Butterworth, London, 1962.
[52]   S. Ross and J. P. Olivier, On Physical Adsorption, Intersci. Publ., New York, 1964.
[53]   S. Gal, Methodik der Wasserdampfsorptionsmessungen, Springer, Berlin, 1967.
[54]   G. Wedler, Adsorption, Verlag Chemie, Weinheim, 1970.
[55]   J. H. de Boer, The Dynamical Character of Adsorption, Clarendon Press, Oxford, 1953.
[56]   R. Gomer, Interactions on Metal Surfaces, Springer, Berlin, 1975.
[57]   J. G. Dash, Films on Solid Surfaces, The Physics and Chemistry of Physical Adsorption, Academic
       Press., New York, 1975.
[58]   H. H. Dunken and V. I. Lygin, Quantenchemie der Adsorption an Festkörperoberflächen,
       Deutscher Verlag f. Grundstoffindustrie, Leipzig, 1978.
[59]   G. A. Somorjai and M. A. van Hove, Adsorbed Monolayers on Solid Surfaces, Springer, Berlin,
       1979.
[60]   G. J. Young, J. Coll. Sci., 13 (1958) 67.
[61]   I. Langmuir, Trans. Am. Inst. Electr. Eng., 32 (1913) 1921.
[62]   I. Langmuir, J. Am. Chem. Soc., 38 (1916) 2221.

[63] I. Langmuir, J. Am. Chem. Soc., 40 (1918) 1361.
[64] B. Johnson Todd, J. L. Lineweaver and J. T. Kerr, J. Appl. Phys., 31 (1960) 51.
[65] Cartwright, in S. Dushman, (Ed.), Scientific Foundations of Vacuum Technique, 2nd edn., J. Wiley and Sons Inc., London, 1962.
[66] L. Holland, The Properties of Glass Surfaces, Chapman & Hall, London, 1964.
[67] S. Barbe and K. Christians, Vak. Techn., 11 (1962) 9.
[68] J. J. Naughton, J. Appl. Phys., 24 (1953) 499.
[69] H. F. Müller, in R. Houwink, (Ed.), Chemie und Technologie der Kunststoffe, 3. Aufl. Bd. 1, Leipzig 1954.
[70] J. D. J. Jackson, B. Rand and H. Rawson, Thin Solid Films, 77 (1981) 5.
[71] S. M. Budd, Thin Solid Films, 77 (1981) 13.
[72] N. Jackson and J. Ford, Thin Solid Films, 77 (1981) 23.
[73] R. D. Southwick, et al., Thin Solid Films, 77 (1981) 41.
[74] I. I. Kitaigorodski, Technologie des Glases, Oldenbourg, München, 1957.
[75] H. G. Pfaender and H. Schröder, Schott Glaslexikon, Moderne Verlags GmbH, München, 1980.
[76] H. Schröder, Glastechn. Ber., 22 (1949) 424.
[77] H. Poppa, Glastechn. Ber., 30 (1957) 387.
[78] H. K. Pulker, Glastechn. Ber., 38 (1965) 61.
[79] H. K. Pulker, 12. Nat. Vac. Symp. New York 1965, Extended. Abstract Print (1965) 1-7.
[80] W. F. Köhler, J. Opt. Soc. Am., 43 (1953) 743.
[81] F. Flory, Thesis, 1978, University D' AIX-MARSEILLE, France.
[82] D. F. Horne, Optical Production Technology, Adam Hilger, London, 1972.
[83] D. F. Horne, Spectacle Lens Technology, Adam Hilger, Bristol, 1978.

CHAPTER 4

4. CLEANING OF SUBSTRATE SURFACES

Surfaces exposed to the atmosphere are generally contaminated. Any unwanted material and/or energy on a surface is regarded as a contaminant. Any manipulation at all contributes to the generation of contaminations. Surface contamination can be gaseous, liquid, or solid in its physical state and may be present as a film or in particulate form. Furthermore, it can be ionic or covalent and inorganic or organic in its chemical character. It can originate from a number of sources, and the first contamination is often a part of the process creating the surface itself. Sorption phenomena, chemical reactions, leaching and drying procedures, mechanical treatment, as well as diffusion and segregation processes can give rise to various compositions of surface contamination. Most scientific and technical investigations and applications, however, require clean surfaces. So, for example, before a surface can be coated with thin films, it must be clean; if it is not clean, then the film will not adhere well or may not adhere at all. In the case of coating clean often means absence of detrimental impurities. In optical coating applications, even very minute contaminations of special materials can interfere with the functioning of sensitive devices whereas other materials may have practically no influence.

Therefore, the cleaning of surfaces is a very important but also a very difficult and delicate operation. The adhesive forces holding small particles onto surfaces may be quite strong in terms of bond strength per unit area. The mechanical energy available to break such bonds is relatively small in most cleaning processes. Consequently, in a given cleaning process, a strong effort is necessary to reduce the adhesive bonds between contaminant and surface as much as possible if the cleaning process is to be at all effective. The reduction of these adhesive bonds is affected by the choice of the cleaning fluid and the conditions under which it is used, particularly the temperature. Pure water, for instance, may be relatively ineffective in removing the contaminants from a given surface, whereas hot water with detergent is quite effective in the same process.

There are a number of diverse cleaning techniques used in various scientific and industrial cleaning problems. Probably the universally used basic cleaning method is scrubbing or polishing a surface. This is used with various cleaning fluids, and for all degrees of cleaning abrasion, ranging from the lightest touch of a non-abrasive brush to rigorous brushing. The latter procedure is a relatively harsh treatment used only to remove gross contaminants from unfinished parts. It is wrong, however, to assume that all the contaminants can be removed in a single cleaning operation. In many cases, particulate material can tenaciously adhere to a surface, and suprising forces can be developed between a surface and a small particle. The cleaning techniques will only be partially effective and, in general, will probably be less and less effective as particle size decreases. In the case of glass and crystal surfaces, the undesired materials on the surface may be mainly water-containing oxidic polishing residues, inorganic and organic dust particles, or films consisting of oil and grease or a combination of these.

4.1 CLEANING PROCEDURES

Cleaned surfaces can be classified into two categories: atomically clean surfaces and technologically clean surfaces. Surfaces of the first category are required for special scientific purposes and these can only be realized in ultra high vacuum.

With the exception of these very sophisticated products, practical coating applications require only technologically clean or slightly better qualities of surface. In accordance with a principle described in ref [1], the degree of surface cleanliness must meet the following two criteria: it must be good enough for subsequent processing and it must be sufficient to ensure the future reliability of the product for which that surface will be used.

A further distinction must be made between cleaning methods that are applicable in the atmosphere and those that are applicable only in vacuum. In all cases where handling of the parts and the use of solvents is required, cleaning cannot be performed in a vacuum. If cleaning in vacuum, e. g. by heating operations and particle bombardment is used, then this is generally conducted inside the deposition system.

4.1.1 CLEANING WITH SOLVENTS

Cleaning with solvents is a widespread procedure that is always included whenever cleaning of glass surfaces is discussed. In this process, various cleaning fluids are used. A distinction must be made between demineralized water or aqueous systems such as water with detergents, diluted acids or bases and non-aqueous solvents such as alcohols, ketones, petroleum fractions and chlorinated or fluorinated hydrocarbons. Emulsions and solvent vapours are also used. The type of solvent used depends on the nature of the contaminants. There are many papers on the cleaning of glass by solvents; some examples are given in ref. [2-10].

4.1.1.1 RUBBING AND IMMERSION CLEANING

Perhaps the simplest method of removing superficial dirt from glass is to rub the surface with cotton wool dipped in a mixture of precipitated chalk and alcohol or ammonia. There is evidence, however, that traces of chalk can be left behind on such surfaces so that after that treatment the parts must carefully be washed off in pure water or alcohol. This method is best suited as a precleaning operation, i.e. as the first step in a cleaning sequence. Rubbing a lens or a mirror substrate with a lens tissue saturated in solvent is almost a standard cleaning procedure. It takes advantage of solvent extraction and imparts a high liquid shearing force to attached particles when the fibres of the tissue pass over the surface. The resulting cleanliness is dependent on the presence of contaminants in the solvent and in the lens tissue. Recontamination is avoided by discarding each tissue after one pass over the surface. A very high level of surface cleanliness is achievable with this cleaning operation.

A simple and often used cleaning technique is that of immersion or dip cleaning. The basic equipment employed in dip cleaning is easy to construct and inexpensive. An open tank of glass, plastic or stainless steel is filled with a cleaning fluid and the glass parts, which are clamped with a pair of tweezers or are inserted in a special holder, are then dipped in the fluid. Agitation may or may not be employed. The wet parts are taken out of the tank after a short time and may then be dried by rubbing with a towel of pure cotton free of washing-powder contaminants. The parts are subsequently inspected by dark-field illumination. If the degree of cleaning is not sufficient, the operation can be repeated by further dipping in the same fluid or other cleaning fluids.

In addition to physical cleaning, chemical reactions can also be exploited for cleaning purposes. Various acids with strengths ranging from weak to strong, as well as mixtures such as chromic and sulphuric acid are used. All acids, with the exception of hydrofluoric acid, must be used hot, i.e. between 60 and 85°C, to produce a clean glass surface. Silica is not readily dissolved by acids, apart from hydrofluoric acid, and the surface layer on aged glass is invariably finely divided silica. Higher temperatures may aid the dissolution of silica so that a new surface is created on the treated glass. Acid cleaning cannot be used for all types of glass. This is especially true for glasses having a high barium or lead oxide content, such as some optical glasses. These are leached by even mild acids, producing a loose silica surface film. According to a report in [10], a cold diluted mixture of 5% HF, 33% HNO_3, 2% Teepol® and 60% H_2O should be an excellent universal fluid for cleaning glass and silica.

Caustic solutions exhibit a detergency and the ability to remove oils and greases. The lipids and fatty materials are saponified by the bases to soaps. These water-soluble reaction products can readily be rinsed off the clean surface. It is generally desirable to limit the removal process to the contaminant layer, but a mild attack of the substrate material itself is often tolerable and ensures that the cleaning process is complete. Attention must be paid to unwanted, stronger etching and leaching effects. Such processes may destroy the surface quality and should therefore be avoided. The chemical resistance of inorganic and organic glass can be found in the glass catalogues. Single and combined immersion cleaning processes are mainly used to clean small pieces.

4.1.1.2 VAPOUR DEGREASING

Vapour degreasing is a process that is primarily useful for removing grease and oil films from surfaces. In glass cleaning, it is often used as the last step in a sequence of various cleaning operations. A vapour-degreasing apparatus consists essentially of an open tank with heating elements in the bottom and water-cooled condensing coils running around the top perimeter. The cleaning fluid may be isopropyl alcohol or one of the chlorinated and fluorinated hydrocarbons. The solvent is vaporized and forms a hot, high-density vapour, which remains in the equipment because the condenser coils prevent vapour loss. The precleaned cold glass pieces, in special holders, are immersed in the dense vapour for periods ranging from 15 sec to several minutes. Pure cleaning fluid vapour has a high solvency for fatty substances, and when it condenses on cold glass a solution is formed with the contaminant, which drips off and is replaced by more pure condensing solvent. The process runs until the glass is so hot that condensation ceases. The greater the thermal capacity of the glass, the longer the vapour will continue to condense, washing the immersed surface.

A glass cleaned in this way shows static electrification. This charge must be eliminated by a treatment in ionized clean air to prevent the attraction of dust particles from the atmosphere which adhere very strongly because of the electric forces.

Vapour degreasing is an excellent way to obtain highly clean surfaces. The efficiency of cleaning can be checked by determining the coefficient of friction, in addition to dark-field-inspection, contact-angle and thin-film-adhesion measurements methods. High values are typical for clean surfaces [7].

4.1.1.3 ULTRASONIC CLEANING

Ultrasonic cleaning provides a valuable method of removing stronger adherent contaminants. This comparatively recent process produces an intense physical cleaning action and is therefore a very effective technique for breaking loose contaminants that are strongly bonded to a surface. Inorganic acidic, basic and neutral cleaning fluids are used as well as organic liquids. The cleaning is performed in a stainless steel tank containing the cleaning fluid and equipped with transducers on the bottom or at the side walls. These transducers convert an oscillating electrical input into a vibratory mechanical output. Glass is chiefly cleaned at frequencies between 20 and 40 kHz. The action of these sound waves gives rise to cavitation at the glass surface / cleaning liquid interface. The instantaneous pressure generated by small imploding bubbles may reach about 1000 atm. It is obvious, therefore, that cavitation is the prime mechanism of cleaning in such a system, although detergents are sometimes used to assist in emulsification or dispersion of released particles. In addition to other factors, an increase in power input will provide a higher cavitation density at the surface, which in turn increases the cleaning efficiency. It is also a very fast process: cleaning cycle times are between a few seconds and a few minutes. Ultrasonic cleaning is used to remove pitch and polishing-agent residues from optically worked glass. Since it is often also used in cleaning sequences to produce surfaces with very low residual contamination levels ready for thin-film deposition, cleaning facilities are often located in a clean room rather than in a manufacturing area.

4.1.1.4 SPRAY CLEANING

The spray-cleaning process uses the shearing forces exerted by a moving fluid on small particles to break the adhesion forces holding the particles to the surface [11, 12]. The particles will be suspended in the turbulent fluid and carried away from the surface. In general, the same types of liquids that are used for immersion cleaning can also be used for spray cleaning. The more viscous and dense the cleaning liquid, the more momentum will be imparted to the attached contamination particle, assuming a constant stream velocity of the liquid. Increasing the pressure and the corresponding liquid velocity results in an increase in cleaning efficiency. Pressures of about 350 kPa are used. With a narrow fan-spray nozzle, the nozzle-to-surface distance should not exceed one hundred nozzle diameters to obtain optimal results. High-pressure spraying of organic liquids causes problems with surface cooling followed by unwanted water-vapour condensation which can leave surface spots. This can be prevented either by using a surrounding nitrogen atmosphere or using a water spray, which shows no spotting, instead of an organic liquid. High-pressure liquid spraying is a very powerful and effective method to remove particles as small as 5 μm. In some cases also, high-pressure air or gas jets are very effective.

4.1.2 CLEANING BY HEATING AND IRRADIATION

Placing substrates in a vacuum causes evaporation of volatile impurities. The effectiveness of this process also depends on how long the substrates remain in the vacuum and on the temperature as well as on the type of contaminant and on the substrate material. Under high-vacuum conditions at ambient temperature, the influence of partial pressure on desorption is negligible. Therefore, desorption is produced here by heating. Heating the glass surfaces causes a more or less strong desorption of adsorbed water and various hydrocarbon molecules, depending on the temperature. The applied temperatures are in the range between 100 and 350° C and the required heating time is between 10 and about 60 minutes. Only in the case of ultra-high-vacuum applications is it necessary to use heating temperatures higher than 450° C in order to obtain atomically clean surfaces. Cleaning by heating is particularly advantageous in all those cases where, because of desired special film properties, film deposition is performed at higher substrate temperatures. But, as a consequence of heating, polymerization of some hydrocarbons to larger aggregates and decomposition to carbonaceous residues may also occur. This sometimes makes such heat treatments problematical. However, treatment with high-temperature flames, for instance a hydrogen-air flame, shows good results, although the surface temperature in such a process reaches only about 100° C. In a flame, various kinds of ions as well as impurities and molecules of high thermal energy [13, 14] are present. It is assumed [7] that the cleaning action of a flame is similar to that of a glow discharge in which highly energetic, ionized particles strike the surface of the parts to be cleaned. According to this model, removal of material from a glass surface in a flame may occur because of the high-energy particles which impart their energy to the adsorbed contaminants. Particle bombardment and surface recombination of ions will liberate heat and may in this way also help to desorb contaminant molecules.

A relatively new technique for cleaning surfaces is the use of ultraviolet radiation to decompose hydrocarbons. Exposure times of about 15 hours in air produced clean glass surfaces [15]. If properly precleaned surfaces are placed within a few millimeters of an ozone-producing uv source, clean surfaces are produced in even less than one minute [16]. This clearly demonstrates that the presence of ozone increases the cleaning speed. As for the cleaning mechanism, it is known that the contaminant molecules are excited and/or dissociated under the influence of uv. Furthermore, it is also known that the production and the presence of ozone produces highly reactive atomic oxygen. It is now assumed that the excited contaminant molecules and the free radicals, produced by the dissociation of the contaminant, react with atomic oxygen and form simpler and volatile molecules like H_2O, CO_2 and N_2. An increase in temperature was found to increase the reaction rates [17].

4.1.3 CLEANING BY STRIPPING LACQUER COATINGS

The use of strippable adhesive or lacquer coatings to remove dust particles from a surface is a very special and somewhat unconventional cleaning technique. It is preferably used to clean small pieces such as, for example, laser-mirror substrates. It can be concluded from published results [18] that even very small dust particles that have become embedded in the adhesive coating can be effectively removed from the surface. It was found that among various commercially available strip coatings, nitrocellulose in amyl acetate is best suited for stripping dust without leaving a residue. However, probably dependent on the type of coating used, sometimes small amounts of organic residue remain on the surface after stripping. If this happens, the stripping operation may be repeated or the surface may be cleaned with an organic solvent, possibly in a vapour degreaser.

The basic cleaning procedure is quite simple. The thick lacquer coating is applied to the precleaned surface with a brush or by dipping. The parts are then allowed to dry completely. In a subsequent operation, performed in a laminar flow box to prevent recontamination, the lacquer film is stripped off. Stripping is easier if a wire loop is embedded in the coating. Attempts to strip off the film in vacuum prior to thin-film deposition were only partly successful because of the difficulty in detecting surface residues inside the evacuated system.

4.1.4 CLEANING IN AN ELECTRICAL DISCHARGE

This type of cleaning is the one most widely used in practice. It is performed in the coating plant at reduced pressure immediately prior to film deposition. There are various experimental arrangements in use to sustain a glow discharge for cleaning [7, 19, 20]. Generally, the discharge burns between two, only negligibly sputtered, aluminium electrodes, which are positioned near the substrates. Oxygen and sometimes argon are normally used to form the necessary gas atmosphere. It seems, however, that mild cleaning is effective only when oxygen is present. Typical discharge voltages are in the range between 500 and 5000 Volts. The substrates are immersed into the plasma, without being a part of the glow-discharge circuit. Only precleaned substrates are treated. A glass surface immersed in the plasma of a glow discharge is bombarded by electrons, mainly positive ions and activated atoms and molecules. Therefore, the cleaning action of a glow discharge is very complex. The contributions of the individual phenomena, which have been quite well investigated, depend strongly on the various electrical and geometrical parameters and on the discharge conditions [21].

A number of processes are responsible for the beneficial action of glow-discharge treatment of substrates before film deposition, as now considered. Particle bombardment and surface recombination of ions with electrons transfer energy to the substrate and cause heating. It is possible to obtain temperatures up to 200°C. Heat and the bombardment with electrons as well as with low-energy ions and neutral atoms favour the desorption of adsorbed water and some organic contaminants. The impact of activated oxygen leads to chemical reactions with organic contaminants, resulting in the formation of low-molecular-weight and therefore volatile compounds. Furthermore, the surface is modified chemically through the addition of oxygen and/or the sputtering of easily removable glass components such as alkali atoms. Physical modifications may occur by bombardment-induced surface disorder, prenucleation by sputtered and deposited foreign material [22] and deposited polymerized or even carbonized hydrocarbon from residual-gas components.

The most important parameters in glow-discharge cleaning are the type of applied voltage (ac or dc), the value of discharge voltage, the current density, type of gas and gas pressure, duration of treatment, type of material to be cleaned, shape and arrangement of the electrodes and position of the parts to be cleaned. It is easy to see that it is impossible to establish universally approved data for an optimal cleaning process for all types of deposition plants and substrate materials. In commercially available coating systems, there is often only a simple rod-shaped aluminium glow discharge electrode. In a dc-discharge, this electrode acts as the cathode. The walls of the plant represent the anode and are grounded. The insulating substrates are placed near the anode. In such an arrangement, the walls of the vacuum chamber receive the greatest amount of discharge current and only a smaller portion bombards the substrates. Gas desorption of the walls is thus improved, decreasing the time required for evacuation. Deposition of a minute quantity of material sputtered from the walls may contaminate the cleaned substrates. To obtain a controlled gas atmosphere, the plant is first pumped down to high vacuum and then filled up with oxygen to the desired discharge pressure. To prevent recontamination and to improve the cleaning efficiency, special arrangements have been developed and carefully tested. These are described mainly in refs. [19] and [21]. In the region of the positive column, there are equal numbers of lower energy ions and electrons with a Maxwellian velocity distribution. Immersion of insulating substrates in this zone of the glow discharge plasma enables a careful cleaning with low-energy electrons. However, because of the water vapour content of the residual gas, regeneration of adsorbed water layers can often not be avoided.

Substrate cleaning in high vacuum by bombardment with ions generated in special ion guns is seldom used for industrial coating processes. Ion-beam technologies, however, offer interesting possibilities not only in cleaning but also in polishing and machining of optical surfaces, as well as for modifying the optical constants of sub-surface layers [23, 24]. This may be interesting for the production of integrated optical devices [25, 26].

4.1.5 CLEANING CYCLES

Surface cleaning is performed by various methods, such as washing with solvents, heating, stripping and plasma treatment. Each has its range of applicability. Solvent cleaning has the greatest range of utility but is inadequate in many cases, particularly where the solvents themselves are contaminants. Heating is useful up to the temperature limits of the surfaces to be cleaned. Plasma treatment provides a cleaning method where contaminant bond strengths exceed the temperature limits of the system. The plasma energy can be much higher than that achieved thermally and still not damage surfaces because of the low thermal flux.

No one method, however, has all the desired features of simplicity, low cost and effectiveness. To achieve optimum cleanliness of substrate surfaces, combinations of the various cleaning methods must be used. As an example, a part is frequently vapour degreased before spray cleaning. The vapour degreaser removes the oil film but is not effective on the particulate materials. The spray cleaner, however, is quite effective in removing these materials, but might not be if a residual oil film had been left on the surface. Only after removing this oil film, can the maximum effectiveness of the spray cleaning be realized. Cleaning fluids are frequently incompatible with one another and it is necessary to completely remove one cleaning fluid from the surface before proceeding to another cleaning fluid. These few examples show the necessity of using sequences of cleaning operations. There is no universal approach to cleaning cycles. They are quite varied, and are specifically tailored to the particular requirements of the surfaces being cleaned and the contaminants that exist upon them. There are, however, some general guidelines to follow when establishing a cleaning sequence.

Precleaning of glass parts usually starts with immersion cleaning in detergent solutions, assisted by rubbing, wiping or ultrasonic agitation, followed by rinsing in demineralized water and/or alcohol. It is important to get the parts dry without allowing solution sediment to remain on the surface because it is often hard to remove later [27].

In a cleaning operation, the sequence of cleaning liquids must be chemically compatible and mixable without precipitation at all stages. A change from acidic solutions to caustic solutions requires rinsing with plain water in between. The change from aqueous solutions to organic fluids always requires an intermediate treatment with a mixable co-solvent such as alcohol or special dewatering fluids. Corrosive chemical agents from the fabrication process as well as corrosive cleaning agents may remain on the surfaces for only a very short time. The last steps in a cleaning cycle must be performed extremely carefully. In a wet operation, the final rinsing fluid used must be as pure as possible and, generally, it should be as volatile as is practical. The choice of the best cleaning cycle often requires an empirical approach, e.g. [34]. Finally, it is important that cleaned surfaces are not left unprotected. Proper storage and handling before further treatment by film deposition are stringent requirements.

4.1.6 CLEANING OF ORGANIC GLASS

Cleaning of organic glass and plastic materials requires special techniques and handling because of their low thermal and mechanical stability. Organic glass surfaces may be covered with low molecular weight fractions, surface oils, antistatic films, finger prints, etc.. Most of the contaminants can be removed by an aqueous detergent wash or by other solvent cleaning possibly associated with mild liquid etching [36]. However, care must be taken with cleaning fluids because they may be absorbed into the polymer structure causing it to swell and possibly to craze on drying.

With organic polymers cleaning means that the surface must be modified in such a way that it does not represent an area of insufficient adhesion at the interface substrate surface / film formed after film deposition. Therefore frequently cleaning of plastics may mean simply the modification of the surface so that the contaminant initially present is afterwards no longer considered as a contaminant. The proper treatment in a glow discharge plasma is very effective for that purpose since, in addition to micro roughening, it also causes chemical activation and cross-linking [37]. In particular cross-linking is advantageous because it increases the surface strength of the polymer and reduces the amount of undesirable low molecular weight components. Proper cleaning fluids and short cleaning time as well as carefully established energy limits and proper doses in particle bombardment or in radiation treatment are important for optimum results.

4.2 METHODS FOR CONTROL OF SURFACE CLEANLINESS

The methods suitable for control of surface cleanliness are mainly those discussed in Section 3.1.2. Compositional changes on and near the surface of glass (i.e. to a depth of about 20 Å) can be measured with Auger electron spectroscopy (AES), electron spectroscopy for chemical analysis (ESCA), ion-scattering spectroscopy (ISS) or secondary-ion-mass spectrometry (SIMS). Coupling these methods with sputteretching, yields highly detailed compositional profiles of the intermediate glass surface in the thickness range from 20 to ~ 2000 Å [35]. Measurement of the average composition through to the far surface, that is to about 10,000 Å, is now routinely available with electron microprobe analysis (EMPA), energy-dispersive X-ray analysis (EDX) in the scanning electron microscope (SEM), or with infrared-reflection spectrometry (IRRS). The related merits of these glass characterization techniques have been recently reviewed [28]. These highly specialized techniques are mainly used to check the efficiency of a newly established cleaning sequence and to troubleshoot when sudden problems arise in a cleaning cycle. In daily practice, however, the use of such excellent inspection methods is much too time-consuming and expensive.

There are also many other cheaper cleanliness tests such as inspection of the breath figure, the atomizer test [29] and the water-break test [30]. All of these qualitative tests are based on the wettability, which is generally high on clean glass surfaces. Oil- and grease-contaminated, and therefore hydrophobic, glass shows a grey breath figure consisting of individual, relatively large water droplets with large contact angles to the surface. Clean glass shows a black breath figure and the water film consists of droplets with contact angles approaching zero. Exposure of a dry and clean glass to a water spray from an atomizer yields a formation of very fine water droplets, which in areas of surface contamination coalesce to larger aggregates. Similar behaviour is responsible for the break of a continuous water film in contaminated surface regions.

The contact angle of water droplets and droplets of other materials on a surface can even be a quantitative measure of wettability [31].

The frictional resistance to solids rubbed on glass is a further sensitive measurement of the cleanliness of the surface. Clean glass has an abnormally high friction coefficient, which is near $\mu_s = 1$ [7]. The presence of an adsorbed monomolecular layer of a fatty acid, such as stearic acid, however, is sufficient to have a marked lubricating effect [7]. The corresponding coefficient of friction is then only $\mu_s \simeq 0.3$.

To detect residual particles (organic and inorganic dust) on a glass surface, light scattering is a valuable tool. Inspection of a cleaned surface with a simple dark-field lamp or in a microscope with dark-field illumination is often used to assess the result of cleaning. The light-scattering effect can also be used for a quantitative measurement, as can be seen in ref. [81] of Chapter 3.

4.3 MAINTENANCE OF CLEAN SURFACES

The stability and maintenance of a cleaned surface is often more critical than the final surface state which is achieved after the cleaning process. Storage in an ultraclean, controlled environment, a very expensive but most effective measure, is usually seldom required. Instead of using a universal protection device, it is often easier and cheaper to identify the undesirable contaminants and to eliminate them from the storage environment. Such contaminants are usually the airborne ones, including various types of dust particles, atmospheric condensates of chemical vapours, and, last but not least, water vapour. The use of some preventive measures is, therefore, well worth considering. Contact with dust may be reduced drastically by storing the parts in a closed container or in a clean bench.

Adsorbance of hydrocarbon vapours can be minimized by storing the substrates in freshly oxidized aluminium containers, which preferentially absorb the hydrocarbons. A disadvantage of this technique is the extra steps involved in periodically stripping and reoxidizing the metal surface of the container. The ultraviolet/ozone technique, mentioned above, may also be used to keep oxidic surfaces clean in an ambient environment. In general, however, cleaned surfaces should at least be stored in clean glass or plastic containers.

Cleaning and maintenance of glass cleanliness also depends strongly on the microstructure of the surface. Contaminants from polishing and subsequent operations may be hidden in micropores and fine flaws. These may then, in the presence of adsorbed water vapour, attack the glass network. Water vapour favours corrosion of the surface and advanced corrosion requires repolishing of the glass surfaces. Such effects are strong with some special types of glass such as those with a high content of lanthanum oxide [32, 33]. Even glasses with a very high chemical durability, such as soda lime borosilicate compositions, exhibit the problem of maintenance of a cleaned surface state to some degree. Often glass surfaces are sealed in a hermetic device. In some cases, these surfaces are contaminated by the sealing operation, which frequently involves heating of other non clean surfaces. Before sealing, all interior surfaces should be as clean as possible. Highly unstable, water-vapour-sensitive surfaces are generally stored in vacuum desiccators.

However, the best precaution for preventing recontamination, beside cleanliness of operator and environment, is to avoid long intervals between cleaning and film deposition. The cleaned parts should be mounted into the substrate holders immediately after cleaning and the whole device should be placed carefully and quickly in the coating plant.

REFERENCES

[1]   K. L. Mittal, Surface Contamination, Vols. 1 and 2, Plenum, New York, 1979.
[2]   L. L. Hench and E. C. Ethridge, ref. [1], Vol. 1, p. 313.
[3]   P. B. Adams, ref. [1], Vol. 1, p. 327.
[4]   S. Tsuchihashi, Kagaku, 33 (7) (1978) 545.

[5]   P. B. Adams, J. Testing Evaluation, 5 (1977) 53.
[6]   C. G. Patano and L. L. Hench, J. Testing Evaluation, 5 (1977) 66.
[7]   L. Holland, Properties of Glass Surfaces, Chapman and Hall, London, 1964.
[8]   W. W. Fletcher, E. S. Keir, P. G. Johnson and B. Slingsby, Glass Technol., 3 (1962) 195.
[9]   T. Putner, Brit. J. Appl. Phys. 10 (1959) 332.
[10] R. H. A. Crawley, Chem. Ind., 45 (1953) 1205.
[11] J. M. Corn, J. Air Pollution Control Assoc., 11 (11) (1961) 523; 11 (12) (1961) 566.
[12] A. D. Zimon, Adhesion of Dust and Powder, Plenum, New York, 1969.
[13] P. F. Knewstubb and T. M. Sugden, Nature, 181 (1958), 474; 1261.
[14] J. Deckers, A. Van Tiggelen, Nature, 181 (1958) 1460; 182 (1958) 863.
[15] R. R. Sowell, R. E. Cuthrell, D. M. Mattox and R. D. Bland, J. Vac. Sci. Technol. 11 (1974) 474.
[16] J. R. Vig and J. W. LeBus, IEEE Trans. PHP, PHP-12, No. 4, (1976) 365.
[17] H. W. Prengle, C. E. Mauk, R. W. Legan and C. G. Hewes, Hydrocarbon Processing, 82 (Oct. 1975).
[18] G. J. Jorgenson and G. K. Wehner, Trans. 10th AVS Symp., (1963) p. 388, Macmillan, New York, 1964.
[19] L. Holland, Vacuum Deposition of Thin Films, Chapman and Hall , London 1956.
[20] L. Holland, Brit. J. Appl. Phys., 9 (1958) 410.
[21] S. Schiller and U. Heisig, Bedampfungstechnik, VEB Verlag Technik, Berlin, 1975.
[22] H. K. Pulker, Habilitation Paper, University of Innsbruck, Austria, 1973, p. 80.
[23] A. R. Bayly, in: Ion Surface Interaction, Sputtering and Related Phenomena, R. Behrisch et al., (Eds.), Gordon and Breach, London, 1973, p. 255.
[24] P. Thevenard, Proc. SPIE, Vol. 400, S. Musikant (Ed.), New Optical Materials, (1983) paper No. 400-16.
[25] R. V. Pole, S. E. Miller, J. H. Harris and P. K. Tien, Appl. Opt., 11 (1972) 1675.
[26] P. D. Townsend, Proc. SPIE, Vol. 400, S. Musikant, (Ed.), New Optical Materials (1983) paper No. 400-17.
[27] H. K. Pulker, Glastechn. Ber., 38 (1965) 61.
[28] L. L. Hench, in: Characterization of Materials in Research Ceramic Polymers, V. Weis and J. Burke, (Eds.), Syracuse University Press, Syracuse, N. Y. 1975, p. 211.
[29] ASTM, F 21 – 62 T.
[30] ASTM, F 22 – 62 T.
[31] G. W. Longman and R. P. Palmer, J. Colloid Interface Sci., 24 (1967) 185.
[32] H. K. Pulker and K. Hayek, in: Basic Problems in Thin Film Physics, Proc. Internat. Symp. Clausthal-Göttingen, 1965, R. Niedermayer and H. Mayer, (Eds.) , Vandenhoeck & Ruprecht, Göttingen, (1966) p. 204
[33] H. K. Pulker, Thin film defects induced by glass substrates. Paper presented at the AVS 12. Natl. Vac. Symp., 1965.
[34] H. H. Karow, in Session V of Workshop on Optical Fabrication and Testing, Dec. 13-15, 1982 Palo Alto, CA; Techn. Digest, Copyright Opt. Soc. Am. 1982, WA 1-1.
[35] H. Bach, Glastechn. Ber., 56 (1983) 1;   56 (1983) 29;   56 (1983) 55.
[36] J. Koutsky, see Ref. [1], Vol. I, p. 351.
[37] D. M. Mattox, Thin Solid Films, 53 (1978) 81.

CHAPTER 5

5. GLASS AND THIN FILMS

5.1 CORRELATION BETWEEN GLASS AND THIN FILMS

For a long time, there has been a marked and a close interrelation between glass and thin films. Glass itself is used in the form of a film as glazing on porcelain and metal surfaces to protect and/or to decorate various articles.

Vitreous enamels bonded to glass surfaces are used to produce translucent colours for functional or decorative purposes. Such enamels have a low melting point and are applied by spraying, silk screening or decalcomania and are fired in a decorative lehr [1].

Soon after the rediscovery of gold ruby glass in the nineteenth century, coloured glass articles were made by casing white glass with stained glass layers [2]. This complex thermal bonding process could be performed by dipping the white glass articles into the staining glass melt, provided that the coefficients of thermal expansion were similar for both glasses, the induced thermal stress would otherwise bend the cased glass system or even destroy it by crack formation. One such application of cased glass was its use for ophthalmic purposes in the production of sun-glasses by the fusing of thin sheets of grey brown glass on to the outer surface of clear glass lenses [3].

Cased glass finds a further technical application in the production of rod or fibre-shaped light guides. Here, highly transparent fibres of high refractive index n_1 are surrounded with a transparent cladding of lower refractive index n_2. Light is conducted through the guide as a result of total reflection. No inorganic glass material combination is known which can be used in the ultraviolet, although certain glass-plastic combinations offer this possibility. Fused silica is used as the core material, with alkylpolysiloxanes and highly fluorinated glass-clear polymers as the cladding material [4].

Other examples of glass-plastic film combinations include laminated safety glass and special optical filters. In the case of laminated safety glass, these compound materials consist of several glass layers bonded by thin elastic polyvinylbutyrate intermediate layers [5]. Plastic foils are used for optical filters since dyed plastics have almost unlimited spectral possibilities. They are, however, less robust in use, more easily scratched and cannot be produced with the same high degree of planarity as polished glass. Glass-plastic composite filters thus combine the advantages of both materials and are suitable as ultra-violet suppression filters, as fluorescence suppression filters, as filters for the study of excitation conditions, and as conversion filters [6].

A further important effect of thin films on glass properties is to increase the strength. Although the theoretical strength of glass is already high, it becomes even higher under compressive stress, so that any treatment which places the surface under compression strengthens the glass. Chemical strengthening of certain glasses occurs on immersion in a bath of molten potassium nitrate. An ion exchange occurs between the potassium ions of the bath and the sodium ions of the glass surface, large K^+ ions are diffusing into the glass and at the same time the smaller Na^+ ions diffuse out of the glass into the salt melt. The incorporation of the larger ions causes the required compressive stress within a thin surface layer [7]. A similar result can be obtained by the incorporation of excess atoms by ion implantation or by special thermal treatments.

A process for making glass without passing through the molten phase* has been recently described by Dislich [8] of the Schott Labs. He found that multicomponent glasses can be produced by hydrolysis and condensation of the corresponding metal alkoxide complexes. This requires temperatures only up to the transformation range of the glass in question, usually 500 to 600° C. The method is particularly suitable for producing thin transparent multicomponent oxide layers of almost any composition on various substrates including glass. Some of these layers provide protection against atmospheric moisture attack and can be used as protective coatings on special glasses. Certain sensitive optical glasses, for instance, can be very effectively protected in this way against atmosperic attack by thin phosphate silicate glass protective layers.

Thin films possess characteristic physical and chemical properties, which depend on their composition and on their thickness. These intrinsic properties, together with contributions due to the substrates account for the effects described above as well as for many other observed characteristics. These few examples demonstrate how certain properties of glass can be modified and improved upon by special thick and thin solid films on or close to their surfaces.

Modern coating technologies allow carefully controlled and reliable production of ordinary single films as well as of simple and complex multilayer systems with special properties, such as glass optical components. Sophisticated new thin film products require both glasses and plastics to be used in many different ways. Examples include functional components of an optical system such as lenses, prisms or paraboloid mirrors whose surface properties such as light reflection may be changed by deposited films. In other cases, the glass or plastic may act simply as a mechanical support or substrate for special thin-film products, e. g. optical filters, polarizers or integrated electronic and optical circuits.

In practically all applications of coatings on glass, with the one important exception of the generation of surface replicas in replicated optics production, the films must have a reasonably good adherence to the glass or plastic surface. Deposited films with defined optical, electrical and mechanical properties should also be highly resistant to humidity as water-vapour adsorption often produces unwanted changes in film properties. Resistance to humidity is largely determined

---

*After that a multicomponent oxide glass can no longer be regarded as an «inorganic product of fusion» (ASTM definition) [9].

by the choice of materials used, but depends also on the film microstructure and can therefore be influenced by the chosen deposition technology and the applied parameters.

Hardness and abrasion resistance are also not exclusively determined by the material, and can be improved by a correct choice of the deposition method and by the control of parameters during film formation. The requirements in practice are quite varied. Some soft-film products are protected by cementing with a second glass plate, other films are allowed to be semi-hard because they are used on parts which are inserted into a closed housing, e. g. inner lenses of a photographic objective, and are therefore also protected. Hardness, abrasion and chemical resistance must be quite excellent for all layers on exposed surfaces such as the anti-reflection film on a front lens. Such parts will probably require cleaning from time to time, this cleaning usually consisting of some sort of rubbing action with a soft cloth or a lens tissue. Dust on such a surface may consist of both soft organic and hard fine-grained inorganic components. The result of such a cleaning treatment is abrasion so it is important to have the abrasion resistance of exposed films as high as possible. This important property will be discussed later in more detail. In many technical applications, the films are in contact with corrosive vapours and gases and should be chemically resistant. This requirement may involve questions of film structure and packing density, as well as a proper choice of materials.

5.2   ADHESION BETWEEN SUBSTRATE AND FILM

The question: «what is adhesion?» is answered by modern physics, chemistry and technical dictionaries with: «. . .the bond or the strength of the bond between two materials or two bodies; also the bond of individual molecules to the interface surfaces» [10]. The ASTM defines adhesion as the «condition in which two surfaces are held together by either valence forces or by mechanical anchoring or by both together» [11]. These bonding forces could be van der Waals'forces, electrostatic forces and/or chemical-bonding forces which are effective across the interface. In this monograph, the word adhesion will be used as a synonym for the adherence of a film to its substrate and, in a broader sense, for the adhesive strength. The degree of adhesion depends on the work necessary to separate atoms or molecules at the interface. A distinction is sometimes made between the maximum possible adhesion, the so-called basic adhesion of a system, which represents the maximum attainable value on the one hand, and the experimentally measured adhesion [12,13] on the other.

The macroscopic experimentally measured adhesion values are determined by the basic adhesion, the mechanical properties of the film and the fracture mechanism in the separation process [14,15]. The relation between the experimentally measured adhesion EA and the basic adhesion BA is given by

$$EA = BA - RS \pm MSE \tag{1}$$

where RS is the residual mechanical stress and MSE is the method-specific error of measurement. The basic adhesion cannot usually be determined exactly because the size of the measurement error often can only be estimated.

The experimentally measured adhesion is given in units of force or energy per unit surface area. Adhesion occurs due to a reaction between surfaces. The net adhesive energy W_{ad} is given by

$$W_{ad} = S_1 + S_2 - S_{1,2} \tag{2}$$

where S_1 and S_2 are the specific surface energies and $S_{1,2}$ is the interfacial energy of the materials 1 and 2. The sum of the specific interfacial energies can, in general, be positive or negative. This results in a specific adhesive force f_{ad} which either attracts or repels. A contact between identical materials leads to the formation of an interface which corresponds either to a grain boundary ($S_{1,1}$ or $S_{2,2}$) or to a phase boundary ($S_{1,2}$). The net energy depends on the specific energies of the included surfaces and the newly formed interfaces. In the case of a grain boundary, the energy ($S_{1,1}$) is usually lower than that found for an interface. Hence for a given interfacial energy ($S_{1,2}$) the adhesive force is small when the surface energies are low. Thus the following useful deductions about the expected specific adhesive force f_{ad} in various materials can be made.

f_{ad} is at its largest when identical materials with high surface energies are in close contact. This applies for metals with high melting points provided that the surfaces are not changed due to chemical effects from the surroundings.

f_{ad} is at its smallest when two materials with low surfaces energies come into contact. This occurs in particular when two different materials form an interface and is found mainly in high polymer materials, and to a lesser extent in materials with symmetrical molecular structures and consequently with compensating dipole moments such as polyethylene and polytetrafluorethylene.

The interfacial energy $S_{1,2}$ rises with increasing dissimiliarity of materials 1 and 2 with regard to the type of atom, atomic spacing and bonding character. Consequently, $S_{1,2}$ and therefore the specific adhesive force f_{ad} decreases in the following sequence: a) same materials, b) solid-solution formers, c) immiscible materials with different types of bonding. A plastic to metal contact provides an example of this last combination.

If the interfacial energy in an adhesive system is zero, then material 1 is identical with or closely similar to material 2. For example, when a white glass 1 is cased with a coloured glass 2 by fusing, then both of the surfaces disappear and the adhesive energy reaches a maximum $W_{ad} = 2 S_1$. In this case of an ideal fusion, adhesive forces are transformed into cohesive ones.

The total force of adhesion is obtained using the effective contact area A:

$$F_{ad} = A \ f_{ad} \tag{3}$$

A relation between the energy W_{ad} of adhesion and the total force of adhesion can only be derived when a reasonable and conclusive estimate of the path followed by the adhesive force $F_{ad}(x)$ over the dividing distance x between the film and the substrate surface can be made [15]:

$$W_{ad} = \int F_{ad}(x) \ dx \qquad (4)$$

The distance x is generally of molecular dimensions. The strength of the adhesion across the interface can be distributed very unevenly because the structure of the substrate surface and of the film are often heterogeneous. Contaminants covering very small areas and monomolecular contaminants on the substrate surface can also cause local changes in the adhesive strength. Thus, the experimentally determined adhesion values should be regarded as average values across the interface surfaces investigated. Any measuring method which gives a magnitude of adhesive forces significantly lower than that produced by the universal Van-der-Waals forces (about 10^9 dyn cm^{-2}) is suspect and an explanation should be sought in the form of imperfect contact, internal mechanical stresses, or stress concentrations which can produce progressive separation or peeling.

5.2.1 METHODS OF ADHESION MEASUREMENT

The individual methods of measurement used to determine the degree of adhesion can be classified and subdivided according to various criteria, e. g. mechanical and non-mechanical methods. In any practical application of a measuring method it is important to know, among other things, whether it can be carried out without destroying the test objects, whether it can provide reproducible results and the size of the measurement error. The simplicity of the measurement apparatus and the time necessary to make the measurement should also be taken into account. Another practical consideration is the choice of a measurement method that simulates as closely as possible the type of stress to which the coating/substrate system will be subjected in service. Tables 1 and 2 list a number of methods of measuring the adhesion of thin films which have appeared in the literature.

TABLE 1

MECHANICAL METHODS TO DETERMINE ADHESION

Qualitative	Quantitative
Scotch-tape test [13, 16-20]	Direct pull-off method [15, 25-39]
Abrasion test [18,21]	Moment or topple test [40-43]
Bend and stretch test [15,22]	Electromagnetic tensile test [44]
Shearing stress test [22-24]	Laser spalation test [45]
	Ultracentrifuge test [13,22, 46-50]
	Ultrasonic test [13,76]
	Peeling test [13, 51-54]
	Tangential-shear test [55,56]
	Scratch test [50,52, 57-73]

TABLE 2

NON-MECHANICAL METHODS TO DETERMINE ADHESION

Qualitative	Quantitative
X-ray diffraction test [71]	Thermal method [74,75]
	Capacity test [13,77]
	Nucleation test [52]

5.2.1.1 MECHANICAL METHODS

Mechanical methods for determining adhesion depend on applying a force to the coating/substrate system under examination. This force causes a mechanical stress at the interface which should remove the film from the substrate once the stress has been increased to an appropriate level. The definitions given here for «mechanical stress» are those commonly used in the field of physics and are thus different from those definitions which have become established in thin-film literature. The stress can be either tensile perpendicular to the interface or shearing parallel to the interface. In practical applications, a combination of these two types of stress usually occurs. That force or energy at which the separation of film and substrate first occurs is taken as an index of the experimentally measured adhesion [13]. As already mentioned, in addition to the stress produced by external forces, very strong internal stresses inherent in the film (intrinsic stress) also affect the interface and influence the experimental adhesion measurement results in an undetermined way. In extreme cases, very strong internal stresses alone can lead to detachment of the film [78,79] or to cohesive failures in the substrate [80].

In general, failure and thus separation can appear in any of the five regions shown in Fig. 1. Region 1 is the inside of the film and region 5 is the bulk material of the substrate distant from the interface. If the separation occurs in one of these two regions then failure is cohesive. On an atomic scale, region 3 can be either a very sharply defined interface or a very diffuse interface layer. If the adhesion is to be measured, separation must take place in this region. Even with a material transition of the monolayer-on-monolayer type, the changes in some physical characteristics (e. g. elasticity and electrical interface phenomena) take place continuously. When there is an interface layer (e. g. through diffusion) the transitional area increases in size. For this reason, a distinction is made between the mechanical behaviour of the interface layers in regions 2 or 3 and that of the inner regions of film 1 or substrate 5.

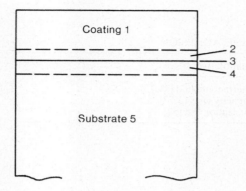

Fig. 1
The five regions [81] in which separation can take place.

Separation exclusively at the interface is improbable according to the «weak boundary theory». Rather, the separation should always occur in the bulk regions of coating or substrate (cohesive failure) or in a finitely thick «weak layer» (interface layer) between the two substances. A failure within the weak interface layer is regarded as an adhesive failure. The weak layer can be a brittle oxide layer or an absorbed and occluded layer of gas and/or some other kind of contamination.

The basis of another theory is the triggering of a fracture at pre-existing cracks in the substrate or within the film according to the so-called Griffith-Irwin criterion [25,86]:

$$\sigma^2 = \text{const. E G/l} \tag{5}$$

where E is the modulus of elasticity, G is the work per unit surface area required for crack propagation, l is the length of the longest cracks in existence and σ is the force per unit surface area required for crack propagation. The crack thus formed grows in a direction normal to that of the applied mechanical stress. The actual fracture is assumed to be a transformation of energy into dissipative energy (e. g. heat) stored as elastic deformation or supplied externally. It must also be assumed that an energy transfer takes place from the area around the fracture to the fracture zone. A fracture can also extend across several of the given regions. If the separation of the coating from the substrate takes place inside or around the interface layer, the force or energy necessary for the separation is a measure of the experimentally measured adhesion. If, however, the separation takes place deep inside the film or substrate, it is a cohesion fracture, i. e. the adhesion forces are stronger than the cohesion forces. From a qualitative point of view, the adhesion is considered to be poor if separation can be caused in the interface layer. However, if separation occurs in the bulk material, the adhesion is good. Quantitative values can only be obtained when the force used to lift the coating can be measured. Individual measurement methods are distinguished from one another by the way the load is coupled to the sample system and then by the way in which it is applied. These criteria allow a division into methods which can remove the coating through (a) a force applied perpendicularly to the interface layer or (b) a force applied at various angles to the interface layer. An upper limit of the transferable force, and thus a maximum measurable adhesion value, occurs because almost all the measuring methods (with the exception of, for example, ultracentrifuge or heat-expansion methods) require a mechanically attached traction piece. This limit depends on the load capacity of this traction piece and on the strength of the particular bonding method (e. g. gluing or soldering) used to attach it to the sample. Certain attachement techniques and individual transfer devices (Scotch tape, needles for scratching etc.) may cause further uncontrolled stresses to be generated in the interface layer, in addition to the controlled mechanical stress produced externally. Such additional stresses can be caused, for example, by drying, polymerizing or stiffening of the glue layer or the solder layer used for the traction piece. It can also occur because of plastic-elastic deformation of the traction piece itself during a test. The experimentally determined adhesion values thus depend on the measurement method used and unfortunately it is not always possible to compare measurements obtained by different methods.

5.2.1.2 NON-MECHANICAL METHODS

With the exception of the nucleation method, non-mechanical methods to measure adhesion are not very well developed, and their field of application is severely limited. These methods can, in any case, only be used for basic investigations.

The adhesion of epitactic films on monocrystalline substrates can be estimated from X-ray diffraction patterns. This idea is based on the well known observation that mechanical stress and deformation in the boundary films can be recognized in the diffraction pattern, and the influence on the adhesion can be estimated from the stress values calculated from it [71]. This more qualitative method of investigation has only a limited field of application.

In the heat of solubility method for adhesion determination, the film material is chemically dissolved from the substrate, e. g. by an exothermic reaction. The total energy thus converted consists of the energy released by the dissolving of the pure film material less the energy required to break the bonds between film and substrate. To enhance the contribution from the adhesion the ratio of the number of molecules on the film/substrate interface should be as high as possible in relation to the total number of molecules in the film. The thermal yield of the reaction is determined with a microcalorimeter, although tests with coated metal films on NaCl substrates have shown the sensitivity of such equipment to be, as yet, insufficient for exact measurements [74,75].

The adhesion of insulator films can be determined with the capacity method [87] by fixing an elastic electrode on the coated metal substrate and measuring the capacity of the condenser thus formed at high and low frequencies. The ratio of the difference between the two capacities ($C_{HF} - C_{LF}$) to the capacity of the arrangement at high frequency (C_{HF}) should be inversely proportional to the adhesion of the film. There are no details available regarding the accuracy and reliability of this method [13,77].

Nucleation methods for measuring adhesion are principally interesting for fundamental investigations. On an atomic scale, the adhesion is proportional to the force required to break the bonds between the film and the substrate atoms (or molecules) that are effective over the interface. The sum of these microscopic interaction forces over the boundary gives the basic adhesion [11]. Conclusions may be drawn regarding the nucleation mechanism and the adsorption energies of the film material atoms by investigating the kinetics of the film formation on the substrate. However, it should be remembered that the nucleation method allows the adsorption energy for only a single adsorbed atom to be determined. In order to obtain the basic adhesion, the adsorption energy of the adsorbed atoms adhering to the substrate surface must be added together, and the number of binding surface sites on the substrate has also to be known. This is generally in the order of 10^{12} – 10^{14} adsorption sites cm^{-2}. It is known that with evaporation in high vacuum the vapour of a film material condenses onto a solid substrate surface by a heterogeneous condensation. There are two different kinds of heterogeneous nucleation process. Purely statistical nucleation occurs spontaneously on even, homogeneous, defect-free substrate surfaces. However, selective nucleation takes place on defect sites in the crystal lattice and generally on sites with increased binding energy on the substrate surface. Furthermore, at low substrate temperatures, the surface mobility of the adsorbed atoms is greatly limited and statistical nucleation preferentially occurs, whilst at higher substrate temperatures and in conjunction with an increased ad-atom mobility only selective nucleation is significant. Two theories exist regarding nucleation, the one based on thermodynamic ideas (capillarity theory), the other employing atomic concepts. In accordance with both theories, methods were developed to determine the adsorption energies of ad-atoms, as follows:

ascertaining the adsorption energy by determining the supersaturation of the film material vapour required for nucleation (capillarity theory) [88,90,93,94],

ascertaining the adsorption energy by determining the rate of nucleation (atomic theory) [52,89,95–97,99,100],

ascertaining the adsorption energy by determining the saturation nucleus density (atomic theory) [52,90,95,98],

ascertaining the adsorption energy by determining the critical condensation (atomic theory) [52,93,101-103].

It may be concluded that in the adhesion investigations summarized under nucleation methods, either the existence of nuclei must be proved (e. g. critical condensation) or the population density of the nuclei on the substrate surface must be determined. Nucleus density determination is, however, only of use if the clusters (heaps of particles) do not wander (surface diffusion) and do not fuse with one another (coalesence). In practice, oscillation has been observed, however [104,105]. Determination of the density of (critical) nuclei is usually carried out using an electron microscope. Generally, the specimen must be fixed in an appropriate manner beforehand. Even with the high resolving power possible with electron optics, nuclei smaller than 10 atoms often cannot be recognized. Other artefacts can be caused by the electron beam or by unsuitable fixation. For example, after fixing certain samples with carbon films, wandering and coalescence of silver nuclei can be observed. On the other hand, fixation with tantalum films prevents artefact formation but this preparation technique impairs the resolving power [90]. Nucleation methods can be applied only to those film/substrate material combinations, where the nucleus density increases with time during the evaporation process until a saturation value has been achieved. The binding forces of the atoms at the substrate surface must be of the order of the van-der-Waals' forces, since otherwise cluster formation is no longer possible by surface diffusion but only by accumulation from the vapour phase and the calculated models can no longer be applied [52,88,90]. The technical and experimental expenditure for these methods is very large, and the reproducibility is rather bad. However, this is the only technique with which the adhesion values for very thin films (0.1–1 nm) can be obtained and the basic adhesion directly experimentally determined [36]. The nucleation methods cannot, however, be applied to systems in which strong interaction forces arise (e. g. oxidation systems). Nevertheless, in these cases it is possible to estimate the density of the bonds over the boundary layer from the experimentally determined adhesion force and the reaction process in the boundary layer which is usually known. The density of chemical bonds thus determined and the maximum nucleus density determined from the nucleation method are of approximately the same order of magnitude (10^{12} nuclei cm^{-2}). Presumably, the most energetically favourable sites for nucleus formation and the preferred sites for chemical bonding are identical [36,106].

5.2.2 CAUSES OF ADHESION

5.2.2.1 INTERFACIAL LAYERS

The quality of adhesion between solids, e. g. between a film and its substrate, depends to a large extent on the condition of the interfacial layer that is formed as can be seen in Fig. 2. The following types of interfacial layers can be distinguished [15,75,107].

Mechanical interfacial layer

This type of interfacial layer forms on rough porous substrates. The film material fills the pores and other morphologically advantageous places when there is sufficient surface mobility and wetting, and a mechanical anchor is formed. The adhesion depends on the physical characteristics (particularly the shear strength and the plasticity) of the combination of materials.

Monolayer on monolayer

This interface is characterized by an abrupt transition from the film material to the substrate material. The transition region has a thickness of 0.2 – 0.5 nm. Interfaces of this type form when no diffusion occurs; there is little or no chemical reaction and the substrate surface is supposed to be dense and smooth.

Chemical-bonding interfacial layer

This type of interfacial layer is characterized by a constant chemical composition across several lattice distances. The formation of the interface layer results from the chemical reactions of film atoms with substrate atoms which may also be influenced by the residual gas. A distinction has to be made between intermetallic bonds and alloys and chemical bonds such as oxides, nitrides etc.

Diffusion interfacial layer

This interfacial layer is characterized by a gradual change of the lattice and the composition within the film-substrate transition area. At least partial solubility is required for diffusion between the film and the substrate material to take place. The necessary energy of 1 to 5 eV must be supplied from elsewhere, e. g. when copper is evaporated onto an unheated gold substrate the heat of condensation is sufficient for diffusion to take place. Diffusion layers may have advantageous characteristics as transitional layer between very different materials, e. g. for reducing mechanical stress resulting from different thermal expansions.

Pseudodiffusion interfacial layer

This type of layer can be formed by implantation processes at high particle energies or in the case of sputtering and ion plating by gas-phase backward scattering of the sputtered substrate material atoms, in this way mixing with the vapour atoms of the coating material which leads finally to simultaneous condensation and recondensation on the substrate. Pseudodiffusion interfacial layers may have the same advantageous characteristics as diffusion interfacial layers, but unlike the latter, they can be formed from materials that do not mutually diffuse. Ion bombardment before coating can increase the «solubility» in the interface layer, thus increasing the diffusion by producing a higher concentration of point defects [108] and stress gradients [109].

A single isolated type of interfacial layer seldom occurs; in normal practice, combinations of the various types of interfacial layers often are found simultaneously.

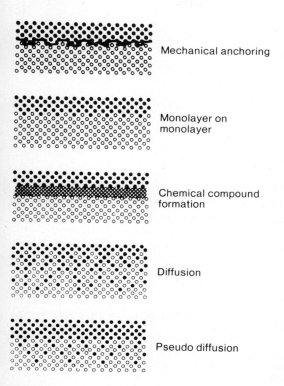

Mechanical anchoring

Monolayer on monolayer

Chemical compound formation

Diffusion

Pseudo diffusion

Fig. 2
Different types of interface layers formed between film and substrate.

5.2.2.2 TYPES OF BONDING

Adhesion forces lying between 0.1 and 10 eV can be classified as follows: physisorption, chemisorption and chemical bonding. In physisorption a film atom approaching a substrate atom is first attracted, then repulsed, and finally remains in balanced state. The attractive forces between electron shells which have remained intact (Keesom, Debye, London forces) are grouped together under the heading «van der Waals» forces.

The most frequent and important effect of the intermolecular forces is the London dispersion effect [123-125] between neutral atoms of the film and the atoms of the substrate surface. These quantum mechanical dispersion forces, which are based on a common influencing of the electron movement, produce the attraction. The adsorption energy E'_a of a film atom onto a surface atom may be expressed:

$$E'_a = -\frac{3}{2} \frac{\alpha_1 \alpha_2}{r^6} \frac{h\nu_1 \nu_2}{\nu_1 + \nu_2}$$

(6)

r = equilibrium distance
α = polarizability
ν = characteristic vibration frequency
h = Planck's constant
The interaction between the ad-atom and the substrate atom can be obtained by integration:

$$E_a = -\frac{N \pi}{4} \frac{\alpha_1 \alpha_2}{r^3} \frac{h\nu_1 \nu_2}{\nu_1 + \nu_2}$$

(7)

N = number of substrate atoms in a unit of volume.

Since the characteristic energies $h\nu_1$ and $h\nu_2$ are unfortunately unknown, the ionization potential v can be used as an approximation

$$E_a = -\frac{N \pi}{4} \frac{\alpha_1 \alpha_2}{r^3} \frac{v_1 v_2}{v_1 + v_2}$$

(8)

TABLE 3

CONTRIBUTION OF THE ELECTROSTATIC ADHESION COMPONENTS TO THE TOTAL PEELING ENERGY OF METAL FILMS ON GLASS [43].

Film	Work function of electrons (eV)	Charge density (nC cm^{-2})	Theoretical electrostatic adhesion energy (J m^{-2})	Theoretical van der Waals adhesion energy (J m^{-2})	Experimentally measured peeling energy (J m^{-2})
Au	4.7 5.3	$+ 2.6 \pm 1$	5×10^{-3}	0.95	1.4 ± 0.3
Cu	4.5	-10 ± 2	80×10^{-3}	0.4	0.8 ± 0.2
Ag	4.3	-13 ± 3	115×10^{-3}	0.8	1.0 ± 0.2

Substrate: Borosilicate glass (Corning 7059)

The charging effect [43] can also be classified as physisorption. When two materials with very different electron affinities are combined, an electrical double layer forms which also contributes to the adhesion. Physisorption contributes up to approximately 0.5 eV to the adhesion. The force lies between 10^4 and 10^8 dyn cm^{-2}.

The interaction between film and substrate atoms which is described as chemisorption can lead to strong bonds when electrons are shifted or exchanged. In true chemical bonding, such as covalent and ionic bonding as well as in metal bonding the bonding forces are very strong, depending on the degree of electron transfer. In covalent and ionic bonding the resulting bonds tend to be brittle, whereas in metal bonding ductile alloys are often produced. The energy contributed to the adhesion from chemical bonds ranges from 0.5 to about 10 eV. The forces are 10^{11} dyn cm^{-2} or greater. This apparently clear picture of differing interfacial layers and types of interaction is complicated by the fact that surfaces do not behave uniformly because they are under the influence of the so-called active and passive centres. Active centres include grain boundaries, dislocations, vacancies or crystallite faces with various free energies and activation energies of chemisorption. Passive centres are surface areas which have already been covered with foreign material so that practically no chemisorption can take place.

It should be noted that the quality of adhesion improves with time in some cases (e. g. silver on glass). This phenomenon can be explained in terms of slow diffusion leading to the formation of an oxygen-bonded interfacial layer. In other experiments of the changes in adhesion with time the adhesion of metal films on polymer substrates was investigated using the scratch test [108,110]. The results showed that there is sometimes a marked improvement in the adhesion of, for example, gold films over a period of time. In this case, the major part of the improvement was attributed to the slow formation of an electrostatic double layer. This was proved by bombarding the film with ions from a glow discharge which broke down the double layer, whereupon the adhesion returned to its original lower value [108]. These series of experiments showed that the electrostatic component contribute significantly to adhesion. This observation confirmed that adhesion seldom comes about only through reciprocal action but rather that it is determined by the combined effects of various intermolecular atomic interactions.

5.2.3 PARAMETERS INFLUENCING ADHESION

The adhesion of thin films is influenced by a large number of parameters. Some of these are defined by the choice of materials for the coating and the substrate. The others are influenced by the preparation of the substrate, the coating process and the handling of the film-substrate combination after the coating process is completed.

5.2.3.1 COATING AND SUBSTRATE MATERIALS

In most cases the choice of substrate is predetermined and a coating is applied to change certain of its characteristics, e.g. antireflection coating of lenses, corrosion protection of a metal or improvement of surface hardness. The choice of material combinations in each case is often quite limited by the application for which the system is intended. The particular combination of substances used (if the film is to be coated by evaporation) determines whether the interfacial layer is of the diffusion type or the chemical bond type, or whether weak reciprocal action forces will be effective across the interface. If it has to be expected that the adhesion between the chosen materials will be weak, then the adhesion can be improved by the addition of an appropriate intermediate layer (compound system). The most common application of intermediate layers is to improve the adhesion of gold films on oxide substrates, e.g. glass. Metals such as chromium which oxidize and alloy easily are usually used as the intermediate layer. The chromium adheres very well to the substrate because of oxidation, and the metallic chromium and gold interdiffuse to form an interfacial layer which also has very good adhesion [111]. The adhesion of evaporated aluminium films on glass can be greatly improved by using nickel or chromium in a similar way [62]. The evaporation of coatings of metal alloys with low internal stresses (e.g. 75 % Ti and 25 % Cr) is a special case particularly if the components of the alloy alone would have a very high internal stress [112].

5.2.3.2 SUBSTRATE PREPARATION

The formation of an interfacial layer and thus the adhesion are greatly influenced by the physical and chemical structure of the substrate surface and the neighbouring areas as well as by the morphology of the surface (flatness, waviness and roughness). The chemical compositon of a surface itself is almost always different from that of the bulk material. Prior treatment of the surface, e. g. cutting and polishing, changes not only the mechanical structure but also the chemical structure of the surface. This can have either advantageous or detrimental effects on adhesion. For example, a layer comprising the polishing agent (usually an oxide), reaction products, water and glass particles (Beilby film) forms on the substrate surface when glass is polished [113]. Therefore the substrate should be prepared in such a way as to provide a defined and reproducible surface. There are many physical and chemical cleaning and preparation methods that are capable of doing this. The desired cleaning is often achieved by using a cleaning process which is a combination of various individual steps. When working out a cleaning process it must be clear which contaminants are to be removed from which surface materials (metals, ceramics, etc.).

5.2.3.3 INFLUENCE OF THE COATING METHOD

The formation of the interfacial layer is very strongly influenced by the coating process. In general, for good adhesion between film and substrate, an energy input is necessary to produce the required activation energies to allow important physical and chemical processes to occur.

Depending on the chosen film formation method, which will be discussed in Chapter 6, this energy may be provided by incident atoms, molecules or ions which either have a high energy as a result of a special transformation process e. g. transfer to the gase phase, or which are accelerated by various means on their way to the substrate surface. The substrate temperature has also a very strong effect on adhesion. Thus the re-evaporation rate, the surface mobility, the diffusion and the chemical reactivity of the atoms are strongly affected by the substrate temperature. This indicates that the varied adhesion values obtained are caused by different physical and/or chemical bondings, depending on the energy of the incident atoms and on the substrate temperature [33].

5.2.3.4 AGING

In many cases, the film-substrate system is not totally stable once the coating process has ended, and it continues to change physically and chemically until it reaches a stable condition [114]. The adhesion of the film to its substrate often undergoes marked changes during this time. Three processes, which generally progress slowly, are responsible for this aging: chemical reactions in the interfacial layer area, solid body diffusion across the interfacial layer and changes in the crystal structure (recrystallization through self-diffusion). These processes are strongly dependent on temperature, their speed usually increasing with increasing temperature [61,63,115].

5.2.4 PRACTICAL ASPECTS OF ADHESION MEASUREMENT

The magnitude of the adhesive strength of a film to its substrate can vary, as already mentioned, between that of the van-der-Waals' forces at one extreme and the cohesive force of either component at the other. These limits are between about 10 kp mm^{-2} and 30.000 kp mm^{-2}. In measuring these relatively high forces acting between a thin film and a thick substrate, two particular problems must be overcome.

An exactly determined tensile or shear force must be applied to the coating.

It must be ensured that failure has really occurred in the interfacial layer and is not a cohesive failure of the coating or the substrate.

One simple qualitative test method is the scotch-tape test. In order to carry out quantitative measurements, however, more precise methods have to be used and existing commercially available measuring units are of use. Some of these methods are discussed in detail in the following sections.

5.2.4.1 SCOTCH-TAPE TEST

Possibly the most widely used and, certainly the simplest and quickest method, is the scotch-tape test, shown schematically in Fig. 3. An adhesive tape is

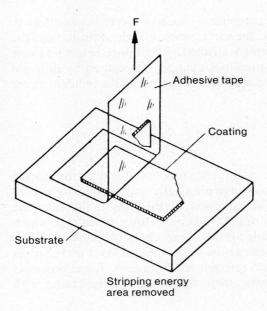

F

Adhesive tape

Coating

Substrate

Stripping energy
area removed

Fig.3
Schematic representation of the scotch-tape test.

stuck to the coating and then removed, a check being made to see if the coating adheres to the substrate or is torn off together with the tape. The test is essentially a qualitative «success or failure» method. If the tape is applied under a set load and is removed in a reproducible manner, then a semiquantitative result is possible based on the the energy required to remove that tape or the percentage of the coating removed. In practice, the test can only be applied to rather weakly adhering coatings, the upper limit being about 200 kp mm^{-2}.

5.2.4.2 DIRECT PULL–OFF METHOD

A method previously described in Section 5.2.1 may also be used where a mechanical linkage to the film is made by gluing or soldering a traction piece or so-called «stud» to it. This can then be pulled off in tension perpendicularly to the surface. The method is difficult to apply because of the problem of obtaining a uniform stress at the interface. Statistical evaluation of several measurements may help to minimize the possible error. The Quad coating adherence tester Sebastian 1 [116] is a commercially available tensile pull test unit which can be used for such measurements. The unit has been designed as a precision laboratory tool which can accommodate test surfaces of virtually any size or shape. Samples such as thin films on small glasses of 0.4 cm in diameter or large lens shapes can be accommodated provided that the test area can be placed flat on the test table mesa.

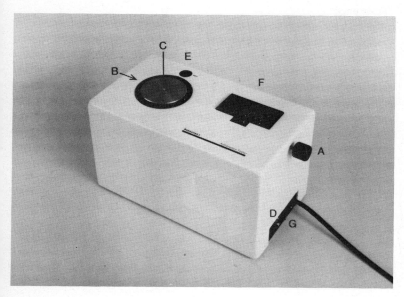

Fig.4
Sebastian 1 [116] direct pull-off adherence tester.
A = engagement knob, B = test table, C = stud insertion port, D = test button,
E = stud eject mechanism, F = LED display, G = connection panel

The adherence tester shown in Fig. 4 is supplied with an integral LED display which exhibits the load level and retains the breaking point load in a memory until reset. Precoated bonding studs for effective attachment to most coating surfaces are available. Bonding to the coating surface is accomplished by holding a stud, precoated with bonding agent, against the test surface during a thermal cure period as is shown in Fig. 5. The bonding stud contact end consists of a 0.28 cm diameter cone-shaped boss on a 0.16 cm diameter pull shank. It is imperative to employ a bonding agent which neither gives an unacceptable reaction nor introduces undue stress on bonding to the film. After thermal setting of the bond, the sample is ready for pull testing. This is done by inserting the stud as shown in Fig. 6 into the instrument. The shank of the bonding stud is clamped by use of special knob. After this load is applied at a predetermined rate until failure of the bond ensues. Under optimum conditions, an accuracy of ± 10 % can be achieved. The limit of the method, as with the scotch-tape test, is how well the stud can be made to adhere to the coating itself. With present day glues the maximum adhesion is probably 650 kp mm^{-2}.

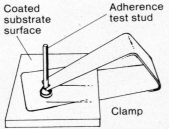

Coated substrate surface

Adherence test stud

Clamp

Step 1: Pre-coated pull stud is clipped to test surface.

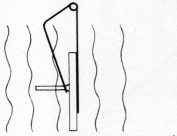

Step 2: Test stud is bonded to surface (epoxy cured or solder).

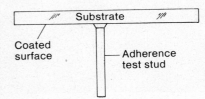

Substrate

Coated surface

Adherence test stud

Step 3: Stud is now bonded.

Fig.5
Bonding the stud to the film surface.

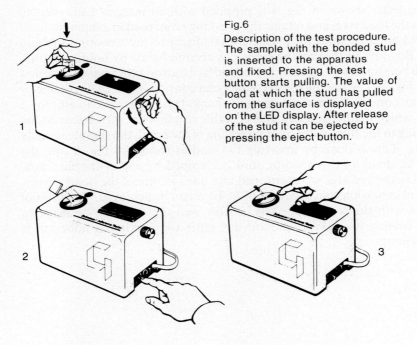

Fig.6
Description of the test procedure. The sample with the bonded stud is inserted to the apparatus and fixed. Pressing the test button starts pulling. The value of load at which the stud has pulled from the surface is displayed on the LED display. After release of the stud it can be ejected by pressing the eject button.

5.2.4.3   SCRATCH METHOD

A quite different approach is provided by the scratch test where a stylus is drawn repeatedly across a sample to produce a series of parallel scratches as is shown in Fig. 7. The load is increased with each traverse until the film is stripped

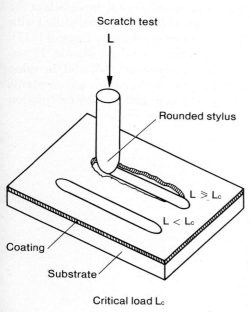

Critical load L$_c$

Fig.7
Production of scratches by a loaded stylus.

cleanly from the substrate. The scratches are then carefully inspected by optical or scanning electron microscopy to determine the critical load. This critical load has been related to the adhesion between film and substrate by an equation given by Benjamin and Weaver [66,110] whose model assumed that the interfacial area was subjected to a shear force by the action of the stylus. Fig. 8 shows a schematic representation of the model. According to these assumptions, it should be possible to calculate the adhesion F from equation (11) as follows, if the required data have been measured.

$$F = H \tan \theta_c \tag{9}$$

$$H = \frac{L}{\pi A^2} = \frac{L}{2\pi Rh} \tag{10}$$

$$F = \frac{L_c}{\pi A R} = K \frac{H \, L_c}{\pi \, R^2} \qquad \text{if } R \gg A \qquad (11)$$

A denotes the half-width of the scratch channel and is expressed by $A^2 = L/\pi H$, L is the stylus load and L_c is the critical load, R = is the radius of the stylus, H is the mean indentation pressure on the substrate, and the value of the numerical coefficient K lies between 0.2 and 1.0 depending upon the details of the model [66,110]. In the paper of Benjamin and Weaver [66], H was identified with the Vickers hardness H_v of the substrate. In a later investigation by Laugier [117], the frictional effects between the stylus and the sample were considered. H was again defined as the mean indentation pressure as with Weaver [110], and the value of K was assumed to be typically 5.5 depending again on the numerical constants in the model. Higher adhesive strength values were then obtained by calculation from the measured critical loads.

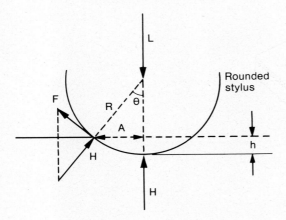

Fig.8
Schematic representation of Benjamin and Weavers model [66, 110].

The scratch test is one of the few test methods which can be applied to strongly adherent coatings. It is therefore of considerable importance. It has proved to be remarkably successful but has also been subjected to criticisms. Its advantages and limitations were recently reviewed by Perry, Laeng and Hintermann [118]. This excellent discussion is partially reproduced here, with the kind permission of the authors. The scratch test was first introduced about 30 years ago by Heavens and Collins [59,119] to study the adhesion of evaporated chromium films on glass. The present widespread use of the test can be attributed to Weaver et al. [66,110, 60-67] who adopted it a few years later to make a very extensive study of evaporated metal films on glass. They confirmed Heavens' result [59,119] that a minimum layer thickness (some 30 nm in their case) was necessary for maximum adhesion as measured by the scratch test, and also found that aging was necessary to achieve maximum adhesion by allowing the development, for example, of an oxide layer at the coating-substrate interface. The effects of an intermediate metal layer were also studied and the existence of a minimum interlayer thickness for good adhesion and further, the formation of an interaction zone between the two metal layers which could enhance or decrease the adhesion was demonstrated. Karnovsky and Estill [72] showed later that aging effects could be avoided by depositing the coating at elevated substrate temperatures.

As discussed above, Weaver et al. assumed that the coating-substrate interface was subjected to a shear force by the sliding stylus. Based on this assumption, they derived eqn. (11) which relates the adhesion to the critical loads. It should be noted that this equation has never been completely verified experimentally. The test

a)
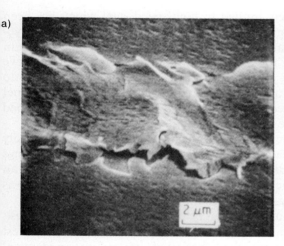

b)
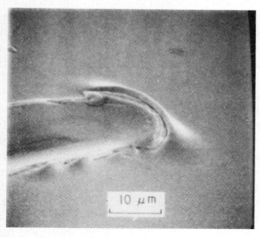

Fig.9 a and b

Examples of loss of adhesion during the scratch test (according to Butler et al [68, 41]).
a) a brittle coating, chromium on glass
b) a ductile coating, copper on glass

has come under closer scrutiny in recent years. It has been found that the loss of the coating can indeed be more complicated than the simple shear-stress model suggests. Butler et al. [41,68], again by studying metal coatings on glass substrates, found that two distinct types of coating removal can occur. Brittle coatings tend to crack and partially to detach due to the failure of the coating and of the coating-substrate interface to support the degree of deformation (Fig. 9a). The cracking is then followed by spalling of the coating, part of which can be pressed into the substrate again so that loss of adhesion is not necessarily accompanied by a complete physical removal of the coating from the channel by the stylus. In contrast, soft coatings tend to form a hillock ahead of the stylus and leave folds at the edge of the channel (Fig. 9b) without necessarily becoming detached. The coating can also be extruded sideways under the stylus and thinned to transparency so that it appears as though the coating has been removed although loss of adhesion has not occurred.

The scratch test has also been applied to high-speed steel which has been sputter coated with TiC. Greene et al. [69] encountered most of the above-mentioned effects: a minimum coating thickness for good adhesion (about 1-3 μm depending on conditions) aging for 30 hrs at room temperature to achieve maximum adhesion (which they attributed to short-range ordering leading to relaxation of stress at the interface) and tearing of the coating adjacent to the channel when the critical load was exceeded. In addition, they found that the critical load depended upon the argon pressure during sputtering.

The problem of defining the point at which the coating was completely stripped was also encountered because the absence or presence of the TiC coatings on the steel in the channel was difficult to ascertain by optical microscopy; an electron microprobe analyzer attachement to a scanning electron microscope was employed to confirm the stripping off of the coating.

In spite of the varied behaviour which can be encountered in the scratch test, the conclusion of both Butler et al. [41,68] and Greene et al. [69] was that the test could be used quantitatively to compare the adhesion of similar films on identical substrates. Ahn et al. [120] studied the scratch testing of single or multilayer coatings deposited by vacuum evaporation, sputtering or gas-phase polymerization onto an array of metallic substrates. They demonstrated yet again that the mechanisms of coating removal could be classified according to whether to coating was brittle or ductile. Brittle films could be removed completely but ductile films showed gradual thinning. The difficulty of working with ductile films was studied by Oroshnik and Croll [121] who used samples of quartz which had been coated with aluminium by vacuum evaporation. They found, as before, that the coating thinned locally in the centre of the scratch channel, and introduced the concept of threshold adhesion failure as the point at which thinning to transparency occurred.

One practical application of the scratch test was its use by Chopra [71] to make a quantitative comparison of the adhesive strengths of gold deposited onto glass and other substrates by sputtering and by evaporation. He was able to demonstrate the superiority of sputtering and also the marked effect of the bias voltage on the adhesive strength. In addition, he found that the critical loads increased quite rapidly with coating thickness and with substrate hardness.

Finally, to permit a quantitative comparison between the result of various tests to be made, Hamersky [70] studied the effect of stylus radius on the apparent adhesion. He employed stylus tips which varied in both size and material. Hamersky showed that the measured adhesion was only independent of stylus radius above about 0.2 mm.

It is clear that the scratch test is the only viable method of determining the adhesion of thin hard coatings which adhere well to hard glass substrates. It is equally clear that the test cannot be applied without great care being used in the interpretation of the results. Irrespective of the particular coating-substrate combination under examination, a number of points need to be kept in mind.

1.) The mode of failure depends primarily on the mechanical properties of the coating – whether it is brittle or ductile. Loss of adhesion appears to be more simple to detect in the case of brittle coatings although these do not all necessarily deform in the same manner under the stylus. Ductile coatings can be extruded sideways by the action of the stylus and can be thinned to zero without loss of adhesion. This complexity in the mode of failure has led many previous investigations to the erroneous general conclusion that the test cannot be applied to obtain quantitative data. If the mode of failure remains the same for a given system, then a quantitative comparison is feasible. However, this does confirm that the mode of failure needs to be studied with care.

2.) The detection of the removal of the coating by adhesive failure – either the onset of it or the critical load for total stripping – is crucial. The hard coatings fail in a brittle manner; these tend to spall off and can be pressed back into the substrate by the action of the moving stylus.

Fig.10
The LSRH scratch test equipment [122].

3.) Experimentally, the size, material and roughness of the stylus needs to be chosen [70] with due regard to the system being studied.

4.) In the context of the interpretation of the critical loads, the wide variation in the mode of failure mentioned above and also the dependence of the critical load on the coating thickness [71] indicate that eqns. (9-11) need to be validated and updated experimentally.

5.) Account needs also to be taken of the surface roughness of the substrate (that is the roughness of the interface) and it must be established also that failure does indeed occur within the interface region as defined here.

Thus we may conclude that the scratch test can be used to characterize the strength of the interface quantitatively. The determination of the onset of adhesion loss and of complete stripping of the channel represent major problems. In addition the relationship between the critical load and the energy required to propagate a crack at the interface, i. e. the strength of adhesion, needs to be determined.

The experimental evidence available to date appears to indicate that the results of different experiments can only be compared where the same coating-substrate combination is studied using a stylus of the same composition above a minimum radius.

Commercially available equipment can be used to perform the scratch test. This scratch test unit was developed at LSRH* in Switzerland [122], and provides the means to make scratches at a constant speed with a known load on the stylus. Usually the stylus consists of a cone-shaped diamond with a tip radius of 0.2 mm and a cone angle of 120°, that is the same diamond as is used for Rockwell hardness measurements. The Rockwell diamond can be replaced by another type of stylus if necessary. The scratching speed is 10 mm min^{-1} and the applied load to the stylus can be varied between 0.1 and 20 kp in steps of 0.1 kp. The LSRH scratch test unit is also equipped with an acoustic detector in the form of a piezoelectric accelerometer mounted directly above the stylus. So the acoustic emission accompanying the adhesive failure of some film-substrate combinations can be recorded as the scratches are made. The detector output signal is indicated on a voltmeter. The acoustic emission is due to the sudden release of elastic energy when cracks propagate in the film-substrate interface. However, no acoustic emission is detected when the film or the film system forms cracks perpendicularly to the interface as happens sometimes, especially at low loads. The possibility of using the acoustic signal to determine the critical load is particularly interesting in routine testing when the occurrence of adhesive failures is difficult to determine optically. The LSRH scratch test unit is shown in Fig. 10.

5.2.5 FINAL COMMENTS ON ADHESION

There are two important aspects of the adhesion of thin films. From the academic view point adhesion is an interesting phenomenon in itself. The nature

---

* LSRH = Laboratoire Suisse de Recherches Horlogères

and strength of the forces acting across the interface which actually effect this adhesion (basic adhesion) are of special interest.

In practice, the total adhesive strength of the entire film to the substrate has to be considered. Mechanical measuring methods tend to be developed with this in mind. However, the results are influenced by a multitude of unknown method-specific measurement errors so that the basic adhesion can only be calculated approximately. This does not, however, make these measurements less valuable for practical applications. Only through the introduction and expansion of modern investigatory methods will it be possible to carry out research on more complex surface and interface processes. At present, a great deal of information on the chemical composition and geometry of the interface layer, the bonding energy between film atoms and substrate atoms, the dipole moments of absorbed complexes, the distribution of electrical conditions, oscillations, surface area diffusion and the kinetics of sorption processes and surface reactions is required. A complete analysis of these effects is not possible at present. This means that only partial solutions to the problems of adhesion can be found while certain limiting conditions are imposed.

This gives the scientist a wide field to work on before he can explain the processes at the interface and deduce from them ways of improving adhesion. Those working in practical applications should develop more sophisticated testing methods which allow the energy applied to be measured exactly so that it can be compared with the values deduced theoretically.

REFERENCES

[1] H. Jebsen-Marwedel, Glas in Kultur und Technik, Druckhaus Bayreuth, 1981.
[2] W. A. Weyl, Coloured Glasses, Dawson's, London, 1959, p. 381.
[3] E. Daniels, Zeiss Informationen, No. 48, 1963, p. 50, and W. Grimm and M. Witzany, Deutsche Optiker Zeitung, 9 (1981).
[4] H. Dislich, Angew. Chem., Int. Ed. Engl., 12 (1979) 49.
[5] W. Vogel, Glaschemie, VEB Deutscher Verlag für Grundstoffindustrie, Leipzig, 1979.
[6] Deutsche Spiegelglas AG, Grünenplan, Company Publication on Conversion Filters, Daylight Composite Filters and Skylight-Filters.
[7] G. A. Chase, T. R. Kozlowski and R. P. Krause, Chemical strengthening of ophthalmic lenses, Am. J. Optom., 50 (1973) 470.
[8] H. Dislich, Angew. Chem., Int. Ed. Engl., 10 (1971) 363.
[9] ASTM definitions appear in ASTM designation: C 162-66.
[10] J. Kunsemüller, Meyers Lexikon der Technik und exakten Naturwissenschaften, B. I. Mannheim, 1971.
H. Franke, Lexikon der Physik, Franckh'sche Verlagshandlung, Stuttgart, 1969.
O. A. Neumüller, Römpps Chemie Lexikon, Franckh'sche Verlagshandlung, Stuttgart, 1972.
[11] R. J. Good, J. Adhes., 8 (1976) 1; ASTM Definition of term relating to adhesion, D 907-70 ASTM, Philadelphia, PA, 1970.
[12] K. L. Mittal, J. Adhes., 7 (1974) 377.
[13] K. L. Mittal, Electrocomp. Sci. Technol., 3 (1976) 21.
[14] D. M. Mattox, Thin Solid Films, 18 (1973) 173.
[15] D. M. Mattox, Interface formation and the adhesion of deposited thin films, SC Monogr. R-65 852, Sandia Corp., 1965.
[16] J. Strong, ASP Publ., 46 (1934) 18.
[17] K. E. Haq, K. H. Behrndt and I. Kobin, J. Vac. Sci. Technol., 6 (1969) 148.
[18] R. A. Mc. Kyton and J. J. Wallis, Thin film materials and deposition techniques for infra-red coatings, FA Techn. Rep. 74028, Frankford Arsenal, Philadelphia, PA, Oct. 1974.
[19] E. M. Ruggiero, Proc. 14th Ann. Microelectronics Symp., St. Louis Section, IEEE, May 24-26, 1965, IEEE New York, 6B-1.

90

[20] S. Bateson, Vacuum, 2 (1952) 365.
[21] F. Schlossberger and K. D. Frankson, Vacuum, 9 (1959) 28.
[22] D. Davies and B. Whittaker, Metall. Rev., 12 (1967) 15.
[23] W. G. Dorfeld, Thin Solid Films, 47 (1977) 241.
[24] H. Stuke and K. Wolitz, Beitrag zur Haftung von PVD-Schichten, Seminar Deutsche Gesellschaft für Metallkunde, Bad Nauheim, 1977.
[25] A. A. Griffith, Philos. Trans. Roy. Soc. London, 221 (1920) 163.
[26] H. G. Schneider and S. Meyer, Krist. Techn., 2 (1967) p K 15.
[27] P. Morzek, Wiss. Z. Techn. Hochsch. Karl-Marx-Stadt, 10 (1968) 175.
[28] R. Jacobsson and B. Kruse, Thin Solid Films, 15 (1973) 71.
[29] G. Katz, Thin Solid Films, 33 (1976) 99.
[30] R. B. Belser and W. S. Hicklin, Rev. Sci. Instrum., 27 (1956) 293.
[31] H. Kim and G. Tomandl, Conf. Proc. «Verbundstoffe», Deutsche Gesellschaft für Metallkunde, Konstanz, 1976, p. 264.
[32] I. Blech, H. Sello and L. V. Gregor, in L. I. Maissel and R. Glang (Eds.), Handbook of Thin Film Technology, McGraw-Hill, New York, 1970, p. 23-1.
[33] L. D. Yurk, Adhesion of Evaporated Metal Films to Fused Silica, Bell Telephone Laboratories Inc., Naperville, IL, 1976, pp. 60, 540. Abstract in J. Vac. Sci. Technol., 13 (1976) 102.
[34] J. R. Frederick and K. C. Ludema, J. Appl. Phys., 35 (1964) 256.
[35] S. P. Sharma, J. Appl. Phys., 47 (1976) 3573.
[36] R. Jacobsson, Thin Solid Films, 34 (1976) 191.
[37] K. Bhasin, D. B. Jones, S. Shinaroy and W. J. James, Thin Solid Films, 45 (1977) 195.
[38] K. Kendall, J. Adhes., 5 (1973) 77.
[39] K. Kendall, J. Phys. D, 4 (1971) 1186.
[40] D. W. Butler, J. Sci. Instrum., 3 (1970) 979.
[41] D. W. Butler, C. T. H. Stoddart and P. R. Stuart, Aspects Adhes., 6 (1971) 55.
[42] C. T. H. Stoddart, D. R. Clarke and C. J. Robbie, J. Adhes., 5 (1970) 270.
[43] H. Harrach and B. Chapman, Thin Solid Films, 22 (1974) 305.
[44] S. Krongelb, in K. L. Mittal, (Ed.), Adhesion measurements of thin films, thick films and bulk coatings, ASTM Spec. Tech. Publ. 640, 1978, p. 107.
[45] J. L. Vossen, in K. L. Mittal, (Ed.), Adhesion measurement of thin films, thick films and bulk coatings, ASTM Spec. Tech. Publ. 640, 1978, p. 122.
[46] J. W. Beams, W. E. Walker and H. S. Morton, Phys. Rev., 87 (1952) 524.
[47] J. W. Beams, Science, 120 (1954) 619.
[48] J. W. Beams, J. B. Breazeale and W. L. Bart, Phys. Rev., 100 (1955) 1657.
[49] I. R. Huntsberger, Off. Dig., Fed. Soc. Paint Technol., 33 (1961) 635.
[50] C. Weaver, in G. Kienel, (Ed.), Physik und Technik von Sorptions und Desorptionsvorgängen bei niederen Drücken, DAGV, Esch Taunus, 1963, p. 193.
[51] B. N. Chapman, Aspects Adhes., 6 (1971) 43.
[52] D. S. Campbell, in L. I. Maissel and R. Glang, (Eds.), Handbook of Thin Film Technology, McGraw-Hill, New York, 1970, p. 12-1.
[53] M. M. Poley and H. L. Whittaker, J. Vac. Sci. Technol, 11 (1974) 114.
[54] J. J. Garrido, D. Gerstenberg and R. W. Berry, Thin Solid Films, 41 (1977) 87.
[55] D. S. Lin, J. Phys. D, 4 (1971) 1977.
[56] E. Tsunasawa, K. Inagagi and K. Yamanaka, J. Vac. Sci. Technol., 14 (1977) 651.
[57] D. M. Mattox, J. Appl. Phys., 37 (1966) 3613.
[58] J. Oroshnik and J. W. Croll, Thin film adhesion testing by the scratch method, Surface Science Symp., New Mexico Section, Am. Vac.Soc., Albuquerque, NM, April 22, 1970.
[59] O. S. Heavens, J. Phys. Radium, 11 (1950) 355.
[60] C. Weaver and R. M. Hill, Philos. Mag., 3 (1958) 1402.
[61] C. Weaver and R. M. Hill, Philos. Mag., 4 (1959) 1107.
[62] C. Weaver and R. M. Hill, Philos. Mag., 4 (1959) 253.
[63] P. Benjamin and C. Weaver, Proc. R. Soc. London, Ser. A, 254 (1960) 177.
[64] P. Benjamin and C. Weaver, Proc. R. Soc. London, Ser. A, 261 (1961) 516.
[65] C. Weaver, Chem. Ind. (London), (1965) 370.
[66] P. Benjamin and C. Weaver, Proc. R. Soc. London, Ser. A, 254 (1960) 163.
[67] C. Weaver and D. T. Parkinson, Philos. Mag., 22 (1970) 377.
[68] D. W. Butler, C. T. H. Stoddart and P. R. Stuart, J. Phys. D, 3 (1970) 877.
[69] J. E. Greene, J. Woodhouse and M. Pestes, Rev. Sci. Instrum., 45 (1974) 747.
[70] J. Hamersky, Thin Solid Films, 3 (1969) 263.
[71] K. L. Chopra, Thin Film Phenomena, McGraw-Hill, New York, 1969, p. 313.
[72] M. M. Karnowsky and W. B. Estill, Rev. Sci. Instrum., 35 (1964) 1324.
[73] A. J. Perry, Thin Solid Films, 78 (1981) 77; 81 (1981) 357.
[74] B. N. Chapman, The adhesion of thin films, Ph. D. Thesis, Department of Electrical Engineering, Imperial College, London, 1969.
[75] B. N. Chapman, J. Vac. Sci. Technol., 11 (1974) 106.
[76] S. Moses and R. K. Witt, Ind. Eng. Chem., 41 (1949) 2338.
[77] T. R. Bullet and J. L. Prosser, Prog. Org. Coat., 1 (1972) 45.

[78] K. Kendall, J. Adhes., 5 (1973) 179.
[79] A. A. Abramov, V. N. Sachko, V. A. Solvov'ev and T. D. Shermergov, Sov. Phys.-Solid State, 19 (1976) 1023.
[80] S. C. Keeton, Delayed failure in glass, SCL Res. Rep. 7 100 10, Sandia Laboratories, 1971.
[81] R. J. Good in K. L. Mittal (Ed.), Adhesion measurement of thin films, thick films and bulk coatings, ASTM Spec. Tech. Publ. 640, 1978, p. 18.
[82] K. L. Mittal, in K. L. Mittal, (Ed.), Adhesion measurement of thin films, thick films and bulk coatings, ASTM Spec. Tech. Publ. 640, 1978, p. 5.
[83] J. J. Bikerman, The Science of Adhesive Joints, 2nd edn., Academic Press, New York, 1968.
[84] J. J. Bikerman, in L. H. Lee, (Ed.), Recent Advances in Adhesion, Gordon and Breach, New York, 1973, p. 351.
[85] R. J. Good, J. Adhes., 2 (1972) 133.
[86] G. Irwin, Fracturing of Metals, American Society for Metals, Metals, Park, OH, 1947; Trans. Am. Soc. Met., 40 (1948) 147.
[87] USSR Pat. No. 127065 (1960), to M. M. Nekrasov.
[88] J. P. Hirth, S. J. Hruska and G. M. Pound, in M . Francombe and Sato, (Eds.), Single Crystal Films, Pergamon Press, New York, 1964, p. 9.
[89] T. N. Rhodin and D. Walton, in M. Francombe and Sato (Eds.), Single Crystal Films, Pergamon Press, New York, 1964, p. 31.
[90] U. Schwabe, Keimbildung, Oberflächendiffusion und Epitaxie von Goldaufdampfschichten auf Natriumchloridspaltflächen, Ph. D. Thesis, University of Innsbruck, 1970.
[91] J. W. Gibbs, in Collected Works, Vol. I, Yale University Press, New Heaven, 1948.
[92] M. Vollmer, in Kinetik der Phasenbildung, Steinkopff, Dresden und Leipzig, 1939.
[93] J. D. Cockcroft, Proc. R. Soc. London, Ser. A, 119 (1928) 293.
[94] J. P. Hirth, in M. Francombe and Sato (Eds.), Single Crystal Films, Pergamon Press, New York, 1964 p. 173.
[95] D. Walton, T. N. Rhodin and R. W. Rollins, J. Chem. Phys., 38 (1963) 2698.
[96] D. S. Campbell, J. Vac. Sci.Technol., 6 (1969) 442.
[97] J. L. Robins and T. N. Rhodin, Surface Science, 2 (1964) 346.
[98] B. Lewis and D. S. Campbell, J. Vac. Sci. Technol., 4 (1967) 209.
[99] C. Neugebauer, in L. I. Maissel and R. Glang (Eds.), Handbook of Thin Film Technology, McGraw-Hill, 1970, 8-1.
[100] B. Lewis, Thin Solid Films, 1 (1967) 85.
[101] J. Frenkel, Z. Physik, 26 (1926) 117.
[102] P. Benjamin and C. Weaver, Proc. R. Soc. London, Ser. A, 252 (1959) 418.
[103] K. L. Chopra, J. Appl. Phys., 37 (1966) 2249.
[104] H. Reiss, J. Appl. Phys., 39 (1968) 5045.
[105] A. Masson, J. J. Metois and R. Kern, in Schneider and Ruth, (Eds.), Advances in Epitaxy and Endotaxy, VEB Deutscher Verlag für Grundstoffindustrie, Leipzig, 1971.
[106] D. M. Mattox and J. E. McDonald, J. Appl. Phys., 34 (1963) 2493.
[107] R. W. Berry, P. M. Hall and M. T. Harris, Thin Film Technology, Van Nostrand, Princeton, NJ, 1968, p. 584.
[108] C. Weaver, Faraday Spec. Discuss. Chem. Soc., 2 (1972) 18.
[109] G. Dearnley, Appl. Phys. Lett., 28 (1976) 244.
[110] C. Weaver, J. Vac. Sci. Technol., 12 (1975) 18.
[111] G. J. Zydzig, T. G. Uitert, S. Singh and T. R. Kyle, Appl. Phys. Lett., 31 (1977) 697.
[112] H. C. Tong, C. M. Lo and W. F. Traber, J. Vac. Sci. Technol., 13 (1977) 697.
[113] L. Holland, The Properties of Glass Surfaces, Chapman and Hall, London, 1964.
[114] H. Hieber, Thin Solid Films, 37 (1976) 335.
[115] P. M. Hall, N. T. Panovsis and P. R. Menzel, IEEE Trans. Parts, Hybrids, Packag., 11 (1975) 202.
[116] Sebastian I, Coating adherence tester, The Quad Group, The Quad Building, 2030 Alameda Padre Serra, Santa Barbara, CA 93103.
[117] M. Laugier, Thin Solid Films, 76 (1981) 289.
[118] A. J. Perry, P. Laeng and H. E. Hintermann in J. M. Blocher, G. E. Vuillard and G. Wahl, (Eds.), Proc. 8th Int. Conf. on CVD, The Electrochem. Soc., Inc., Pennington, NJ 08534, 1981, p. 475.
[119] O. S. Heavens and L. E. Collins, J. Phys. Rad., 13 (1952) 658.
[120] J. Ahn, K. L. Mittal and R. H. McQueen, in K. L. Mittal, (Ed.), Adhesion measurements on thin films, ASTM STP 640, 1978, p. 134.
[121] J. Oroshnik and J. W. Croll, [ibid.] p. 158.
[122] LSRH scratch test equipment described in e. g. A. J. Perry, P. Laeng and H. E. Hintermann, in J. M. Blocher, G. E. Vuillard and G. Wahl, (Eds.), Proc. 8th Int. Conf. on CVD, The Electrochem. Soc., Inc., Pennington, NJ 08534, 1981, p. 723.
[123] F. London, Z. Phys., 63 (1930) 245.
[124] C. Kittel, Einführung in die Festkörperphysik, R. Oldenbourg Verlag, München, Wien, 1973, p. 133.
[125] R. Eisenschitz and F. London, Z. Phys., 60 (1930) 491.

CHAPTER 6

6 FILM FORMATION METHODS

The formation of films on a glass surface can generally occur in a subtractive or additive way by chemical or physical processes. In the first class of film formation methods, the multicomponent glass loses some of its constituents on the surface and in regions near the surface through chemical or physical action. Films thus obtained are in optimal contact with the substrate, but generally have a somewhat altered composition compared with the bulk glass, and are always rather porous. In the case of additive film formation methods, which represent a very large class, the glass substrates are coated with films by deposition. Such films have in nearly all cases a composition which is different from that of the substrate. Their adhesion is good and in most cases they can be made rather compact. The various methods vary considerably in the medium used to produce the film, in the required technical equipment, in the rate of film formation, in the required substrate temperature and in all the other operations performed to reproduce and stabilize the properties of the final coating. In the following, a brief description is given of the various chemical and physical methods which can be used to produce films on glass surfaces.

6.1 SUBTRACTIVE METHODS

6.1.1 CHEMICAL PROCESSES

The attack by a humid atmosphere and by acid of glass resulting in the formation of a surface film with altered properties was observed by Fraunhofer in 1817 and later by Lord Rayleigh and Taylor [8].

6.1.1.1 SURFACE LEACHING

In 1904 a British patent was granted [1] for a method in which glass is treated with acidic solutions to form an antireflective layer on its surface. It was thought that the optical effect was produced through a low refracting inhomogeneous layer formed by leaching the glass surface. Investigations performed later in the laboratories of Schott und Genossen [2] showed, however, that the film formed by the acid attack is homogeneous. The film consists of solid rather compact silica gel with a refractive index between n = 1.45 and 1.46. It was surprising to discover that the transition zone between the film and the intact glass is only about 0.5 nm thick. In the leaching operation with acid solutions, not only are the alkali and the alkaline earth components in the surface region of the solid glass removed, but also to some extent little defined polishing residues which are smeared into the remaining polishing cracks and fissures. Therefore, polishing cracks invisible before leaching may appear after such treatment. Through the use of buffered solutions, however, film formation without development of polishing cracks is possible [3]. Extended investigations have improved the reliability of the chemical subtractive film formation methods [4,5]. However, the optical property of the compact silica gel film (n = 1.45) is insufficient to reduce the reflection of low refracting glass. For this reason skeleton-type silica gel films were created with lower film refractive indices which are most suitable as antireflection films even on low refracting glass. Such films can be made by leaching the glass with basic solutions of p_H- values between 7 and 8 [4b]. Chemically leached films on glass have been produced on an industrial scale with good optical quality. For some years, they were the only antireflection films of reasonable mechanical and chemical resistivity. However, the large number of different optical glasses which required different leaching solutions and leaching times, the danger of occasionally developing polishing cracks and the loss in optical quality by absorption of humidity and grease, especially in the case of the porous skeleton type films, may explain why production soon ceased. But this process might still be used for special applications. For instance recently it was discovered that chemically leached antireflection layers on high power laser optical components exhibit a very high laser damage resistance e. g. [6].

6.1.2 PHYSICAL PROCESSES

6.1.2.1 HIGH-ENERGY PARTICLE BOMBARDMENT

High-energy particle bombardment at low residual pressure produces surface modifications on multicomponent glass and fused silica. According to a British patent [7], treatment of glass in a noble gas glow discharge at an accelerating voltage of 50 kV produces an antireflection film. It was also shown that a glow discharge treatment of glass at voltages between 3 and 5 kV in streaming dry air at reduced pressure, after more than 1 hour produces an antireflection film [7]. Unfortunately, no systematic investigation on the formation mechanism was performed. It is concluded, however, that the films consisted of glass constituents. It could be assumed that the alkali and other easy removable components will be sputtered leaving behind a silica-enriched less dense glass network on the glass surface. Although such layers were found to be very stable and resistant to concentrated acids, such as chromic sulfuric acid or nitric acid, the method has never been used in technical production of antireflection films.

Recently systematic studies of the surface properties of glass and other insulators have been made after bombardment with ions in the 40 to 100 keV range. Refractive index modifications were observed which are caused by phase changes, microporosity, colour centre introduction, compaction or compositional changes. The optically modified layer thickness was found to be dependent on the incident ion range. With treatment by medium weight ions such as e. g. Ne, Ar, Kr, modifications within the layers occur quite uniformly. Noble gas ion bombardment of fused silica produces a glass network compaction and an increase in the refractive index. But crystalline quartz shows a rather stable disordered structure after such treatment. In the case of soda lime glass, bombardment with argon ions can cause a decrease of the refractive index from 1.52 to 1.34 due to a sodium depleted surface layer. Preferred sodium sputtering and enhanced diffusion over the implanted ion range could be observed cementing the early assumption on the mechanism of the process. A survey of this topic can be found in a paper by Mazzoldi [8].

6.2 ADDITIVE METHODS

The subtractive methods practically can only be used to produce single films with lower refractive indices than those of the multicomponent glass substrates. Consequently in optics their use is limited to reflection-reducing films. This limitation does not exist with the additive film formation methods. With additive methods, it is possible to deposit in any sequence low- or non-absorbing compound films with various refractive indices as well as absorbing an highly reflecting metal or alloy films and mixtures of both types of films. It is obvious therefore that the subtractive methods are almost completely superseded by additive ones. The additive film formation methods can be classified into chemical and physical methods and further into wet or dry processes the latter may run at atmosphere or under vacuum.

6.2.1 CHEMICAL FILM FORMATION PROCESSES

6.2.1.1 DEPOSITION OF METAL FILMS FROM SOLUTIONS

The oldest wet film deposition process is chemical silvering of glass and plastics. It is still in use today for the fabrication of relative cheap glass mirrors metallized on the reverse side. There are innumerable formulas reported which were summarized some years ago [9]. The various recipes have only minor variations, and all involve a solution of a silver salt (usually $AgNO_3$) as one reagent solution and a reducing compound (usually sucrose or formaldehyde) dissolved in a second solution. The two reagent solutions are either mixed just before using or are sprayed from a two-nozzle gun on the glass surface to be silvered. A reduction reaction occurs on the surface and an extremely fine-grained metallic silver film is precipitated, which shows a high reflectance. The meticulously cleaned surface should be sensitized before coating. This can be done by polishing with tin oxalate or tin oxide or by dipping in stannous chloride solution acidified with hydrochloric acid.

For the widely used Brashear process [10-12], the two following solutions are required. The silver nitrate solution contains 20 g $AgNO_3$ and 10 g KOH in 400 ml water. To prevent precipitation, about 50 ml NH_4OH is added. The solution should not be stored for long since explosive fulminates can form. The reducing solution is formed by dissolving 2 g sucrose and 4 ml HNO₃ in 1000 ml boiling water. The solution is cooled to room temperature before use. Immediately before silvering, four parts of silver-salt solution are mixed with one part reducer. After formation of the silver film, the deposit is stabilized by an electroplated copper film and/or a lacquer coating. Sometimes the silver film is cemented with a second glass plate. These treatments prevent the silver film from degrading chemical reactions with traces of sulphur compounds from the atmosphere. In a similar way, copper [13] and gold [14] films can also be deposited by reduction reactions in solutions.

Another wet technique to form metal films of Ni, Co, Pd, Pt, Cu, Au, Ag and alloys of these elements also containing P and B is the autocatalytic chemical reduction [15,16]. In this process, a solution of a chemical reducing agent (usually NaH_2PO_2) is required to provide the electrons for the reduction reaction: $Me^{n+} + ne^- \rightarrow Me$, e. g. [17]:

$$2H_2PO_2^- + 2H_2O + Ni^{2+} \rightarrow Ni + H_2 + 4H^+ + 2HPO_3^{2-} \tag{1}$$

However, in contrast with the homogeneous chemical reduction reactions described above, in autocatalytic plating the reaction takes place only on a catalytic surface. Therefore the metal deposited must itself be catalytic to guarantee that film growth will continue after the process has started. For bath compositions and operating conditions, the literature on electroplating should be consulted [15]. The tanks containing the baths must be made of a non-catalytic material. Not all substrate surfaces, nor all metal surfaces are suited to autocatalytic film deposition, nevertheless proper treatment can be performed with practically all materials to initiate film deposition. Once the process has started there is no theoretical limit for the thickness of the deposit. Thickness growth of the films takes place linearly with time similar to electroplating at constant current density. The characteristics of the process are a very constant deposition rate and little build up on edges of the substrate. The resulting films are relatively dense, smooth and shiny. As mentioned above, autocatalytic deposition takes place only on catalytic surfaces. With insulating substrates like glass and plastics, the problem could be solved by special treatment. Chemical conditioning processes have been developed not only to roughen the surface but also to change its chemical nature. This is also important for film adhesion, which shows useful results. It seems that the bonds are mechanical and chemical in nature. For glass, ABS plastics (acrylonitrile-butadiene-styrene) and polypropylene, the process starts with cleaning in mixtures of chromic and sulphuric acid. After thorough rinsing with water, sensitizing and nucleation are performed. Sensitizing is the adsorption of a reducing agent, frequently a stannous compound (e. g. $SnCl_2$). The subsequent nucleation is the adsorption of a catalytic material often a Pd compound (e. g. $PdCl_2$), but also Au compounds are used in which case sometimes the catalyst must be brought into its final state by an additional process. Recently it has been suggested that the expensive Pd be replaced by a cheaper Cu-Sn complex [18]. Sensitizing and nucleation is generally a two-step process. However, the procedure can also be performed in a one-step process using one aqueous solution containing $SnCl_2$, $PdCl_2$ and HCl. The mechanism is not very well understood. According to [19], it can be assumed that $PdCl_2$ is reduced to colloidal Pd, which is stabilized by the Sn^{4+} hydrolysis products obtained by the oxidation of Sn^{2+}. The adsorbed Pd aggregates are surrounded by Sn^{4+} ions and/or a colloidal matrix of SnO_2 xH_2O. All stannous compounds must be removed by dissolving, leaving the surface covered with Pd aggregates as in the two-step process. The surface thus prepared is suitable for coating by the autocatalytic reduction. In practice today, mainly Ni films are produced by this technique.

6.2.1.2 DEPOSITION OF OXIDE FILMS FROM SOLUTIONS

The formation of thin oxide films from solutions by immersion or dipping and by spraying is also a relatively old method [20-23]. The development of the processes was started in Germany and in the United States. Both processes are used in industrial thin-film production mainly for optical and opto-electrical applications.

6.2.1.2.1 IMMERSION- OR DIP-COATING

The fundamental step in this process is the simple hydrolysis of suitable metal organic compounds whose hydrous reaction products are then converted into the final oxide coatings. The industrial development of this kind of film formation was started more than 40 years ago in the laboratories of the glass company Schott und Genossen in Germany [20,23]. From theses investigations, production processes resulted in the fabrication of single and multilayer coatings of pure and mixed oxides [24-36]. Films produced in this way are mainly used to change the optical properties of the glass surface.

6.2.1.2.1.1 FORMATION, STRUCTURE, OPTICAL AND MECHANICAL PROPERTIES

The basic chemical reactions can be demonstrated in the case of Al_2O_3 film deposition from an aluminium alkoxide. In order to achieve completely transparent oxide films, hydrolysis and condensation have to take place slowly and simultaneously, thus initiating the formation of a gel-like microstructure. The single stages may be rather complex and are represented by the following sequence of reactions [37]:

$$Al(OR)_3 \ + \ H_2O \ \longrightarrow \ Al(OR)_2OH + ROH\uparrow \tag{2}$$

$$Al(OR)_2OH \ + \ H_2O \ \longrightarrow \ Al(OR)(OH)_2 + ROH\uparrow \tag{3}$$

$$Al(OR)(OH)_2 \ + \ H_2O \ \longrightarrow \ Al(OH)_3 + ROH\uparrow \tag{4}$$

$$2\,Al(OH)_3 \ + \ \Delta T \ \longrightarrow \ Al_2O_3 + 3H_2O\uparrow \tag{5}$$

It can be assumed that during hydrolysis (eqns. 2 to 4), polycondensation will also occur to a certain degree according to:

$$X \; Al(OR)(OH)_2 \xrightarrow[-H_2O]{} -Al(OR)-O-Al(OR)-O-Al(OR)-O- \qquad (6)$$

$$+H_2O \; \Big\downarrow \; -ROH$$

$$\begin{array}{cc} OH & OR \\ | & | \\ -Al(OR)-O-Al-O-Al-O- \end{array} \qquad (7)$$

$$+ \Delta T \; \Big\downarrow \; -H_2O$$

$$\begin{array}{c} -Al(OR)-O-Al-O-Al(OR)-O- \\ | \\ O \qquad OH \\ | \qquad | \\ HO-Al-O-Al-O- \end{array} \qquad (8)$$

At the end of the deposition process, hydrolysis decreases and polycondensation increases supported by substrate heating ($+ \Delta T$):

$$\begin{array}{c} | \\ O \\ | \\ Al(OH)-O-Al-O-Al- \\ | \qquad\quad | \\ OH \qquad\quad O \\ | \\ HO-Al-O-Al-O-Al-O- \\ | \qquad\qquad | \\ O \qquad\qquad OH \\ | \end{array} \qquad (9)$$

$$+ \Delta T \; \Big\downarrow \; -H_2O$$

$$\begin{array}{c} | \qquad\qquad | \\ O \qquad\qquad O \\ | \qquad\qquad | \\ Al-O-Al-O-Al-O- \\ | \qquad | \\ O \qquad O \\ | \qquad | \\ Al-O-Al-O-Al-O- \\ | \qquad\qquad | \\ O \qquad\qquad O \\ | \qquad\qquad | \end{array} \qquad (10)$$

$$X \; (Al_2O_3)$$

The oxide films obtained contain, depending on the applied substrate temperature, no or only a little incorporated water. Basically it is possible to start from solutions of any inorganic or organic metal compounds if these tend to form polymolecules or polysolvated aggregates in solution which yield gels of poor crystallization tendency during drying. It is obvious therefore that generally the hydrolysis method is not the only one for fabrication of films from solutions.

TABLE 1

CHARACTERISTICS OF NON ABSORBING AND ABSORBING METAL OXIDE FILMS PRODUCED BY DIP-COATING [28, 36].

Starting material	Film material	Colour in transmission	Structure	Remarks
Al-sec-butylate, Al$(NO_3)_3$ 9H_2O	Al_2O_3	colourless	amorphous crystalline	forms mixtures with other oxides
Y$(NO_3)_3$	Y_2O_3	colourless	–	–
La$(NO_3)_3$	La_2O_3	colourless	–	–
Ce$(NO_3)_3$ 6H_2O	CeO_2	colourless	crystalline	forms mixtures with other oxides
Nd$(NO_3)_3$	Nd_2O_3	colourless	–	attenuation of transmittance by absorption bands of Nd^{3+} between 500 and 600 nm
In$(NO_3)_3$	In_2O_3	colourless	crystalline	semiconductor
Si$(OR)_4$	SiO_2	colourless	amorphous	forms mixtures with other oxides
Ti$(OR)_4$	TiO_2	colourless	crystalline	forms mixtures
$TiCl_4$	TiO_2	colourless	crystalline	with other oxides
$ZrOCl_2$	ZrO_2	colourless	crystalline	–
$HfOCl_2$ 8H_2O	HfO_2	colourless	crystalline	traces of Cl in the layers
$ThCl_4$	ThO_2	colourless	crystalline	–
Th$(NO_3)_4$	ThO_2	–	–	–
$SnCl_4$	SnO_2	colourless	crystalline	semiconductor
Pb$(OOCCH_3)_2$	PbO	colourless	amorphous	diffuses into glass at 500°C
$TaCl_5$	Ta_2O_5	colourless	–	–
$SbCl_5$	Sb_2O_5	colourless	–	–
Cu$(NO_3)_2$ 3H_2O	CuO	brown	–	–
$VOCl_2$	VO_X	greenish-yellow	–	optical properties strongly dependent on preparation conditions
Cr$(NO_3)_3$ 9H_2O CrOCl	CrO_X	yellow-orange	–	–
Fe$(NO_3)_3$ 9H_2O	Fe_2O_3	yellow-red	crystalline	–
Co$(NO_3)_2$ 6H_2O	CoO_X	brown	–	–
Ni$(NO_3)_2$ 6H_2O	NiO_X	grey	–	–
$RuCl_3$ H_2O	RuO_X	grey	–	semiconductor
$RhCl_3$	RhO_X	grey-brown	–	–
$UO_2(OOCCH_3)_2$	UO_X	yellow	–	–

However, practical experience has shown that hydrolysis of metal acid esters and/or alcoholates which are obtained by treating metal compounds with the corresponding reactive solvents are particularly suitable. Oxides which can be produced as thin films of good optical and mechanical quality are listed in Table 1 [28,36]. The oxide films of the first group are practically free from absorption in the visible range and are assumed to be stoichiometric compounds. On the other hand, the oxide films of the second group show pronounced absorption in the visible range and may also be partially non-stoichiometric, because their extinction can be influenced, for example, by the baking process. Of all high-index materials, TiO_2 has proved to be of special practical interest. For this reason, such coatings have been investigated with particular thoroughness [27]. With TiO_2 films deposited from alcoholic ethyl and butyl titanate solutions, the crystalline structure formed during the thermal treatment is strongly influenced by the surface composition of the glass substrate. On alkali-free glass, the formation of anatase and (at higher temperatures) of rutile is predominant. In the case of substrates containing sodium compounds, Na^+ ions diffusing into the films induce the brookite structure mixed with various portions of another structure type which is attributed to the Na_2O $xTiO_2$ type described in the literature [38,39]. As discovered by electron diffraction experiments, the relative structural composition, that is, the percentage of anatase, brookite, Na_2O $xTiO_2$ and rutile in the TiO_2 films is strongly determined by the type of solution applied, the heat treatment and the release of alkali ions from the substrate. A rutile component in the films is obtained on alkali free amorphous SiO_2 substrates of temperatures exceeding 500°C from ethyl titanate solutions doped with $BiCl_3$ [27]. This transformation into rutile goes to completion at temperatures between 600 and 700°C.

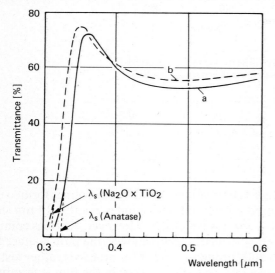

Fig. 1
Spectral transmittance of TiO_2-films deposited from ethyl-dichloro-titanate solution on (a) fused silica and (b) window glass (coated on both sides)

TiO_2 films with different crystal structure also show marked differences in their optical properties. Figure 1 shows the spectral transmission of slowly tempered TiO_2 films deposited from an ethyltitanate solution on both sides of (a) fused silica and (b) window glass. The absorption of the latter has been eliminated by calculation. For the various types of film, differences in refractive index n and even more pronounced differences in the position of the ultraviolet absorption edge, shown in Fig. 2, have been found. [27].

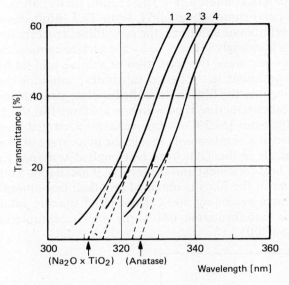

Fig. 2
Ultraviolet absorption edge of TiO₂-films deposited from ethyl-dichloro-titanate solution on window glass precoated with SiO₂ layers of different thickness:

SiO₂ layer thickness	(nm)	0	6	12	18
Observed curve, when	rapidly	3	3–4	4	4
heated up	slowly	1	2	2–3	3–4

Compared with pure anatase-type coatings, Na₂O-type coatings formed under slow tempering conditions exhibit a lower refractive index n and a 14 nm shift of the uv absorption edge towards shorter wavelengths. However, coatings composed substantially of brookite are optically nearly identical with the pure anatase films. The diffusion of Na⁺ ions into and through a TiO₂ film proceeds rather fast even after the formation of crystals of a certain structure type in the film is terminated. The concentration gradient of the alkali which has penetrated thicker layers can be detected optically by a decrease of the refractive index within such layers from the film/air boundary towards the substrate. Silica layers, produced in a similar way from solutions of silic acid esters on glass, behave completely differently. At a minimum thickness of about 20 nm they make an excellent diffusion barrier. In Table 2 are listed the structural and optical characteristics of TiO₂ films which were prepared under different conditions on mainly precoated plate glass [28].

TABLE 2

STRUCTURAL AND OPTICAL CHARACTERISTICS OF 60 NM TiO_2 FILMS WHICH WERE TEMPERED AT 500°C DEPOSITED ON PLATE GLASS PRECOATED WITH SiO_2 LAYERS [27, 28].

SiO_2 base coat thickness (nm)	TiO_2 film deposited from diethyl-dichloro-titanate heating rate:		TiO_2 film deposited from tetrabutyl-titanate heating rate:	
	50 deg min^{-1}	10 deg min^{-1}	50 deg min^{-1}	10 deg min^{-1}
0	brookite (+Na_2OxTiO_2) λ_s=323nm, n_{550}=2.25	Na_2OxTiO_2 λ_s=311nm, n_{550}=2.24	Na_2OxTiO_2((+anatase)) λ_s=315, n_{550}=2.05	Na_2OxTiO_2(+anatase) λ_s=315nm, n_{550}=2.03
6	brookite + anatase λ_s = 324 nm, –	Na_2OxTiO_2 (+ brookite)((+anatase)) λ_s = 315 nm, –	anatase (+Na_2OxTiO_2) λ_s = 323 nm, –	anatase + brookite λ_s = 323 nm, –
12	anatase λ_s=325nm, n_{550}=2.28	brookite (+Na_2OxTiO_2) λ_s=317 nm, –	anatase λ_s = 325 nm, –	anatase (+Na_2OxTiO_2) λ_s = 325 nm, –
24	anatase λ_s=325nm, n_{550}=2.28	anatase λ_s=325nm, n_{550}=2.28	anatase λ_s=325nm, n_{550}=2.20	anatase λ_s=325nm, n_{550}=2.20

λ_s = absorption edge, n_{550} = refractive index, structure type in parentheses means minor quantity, in double parentheses means trace.

In the case of starting compounds other than ethyl titanate, the optical properties of the TiO_2 films may be different. The use of butyl titanate or of other Ti compounds, for example, yields after tempering TiO_2 films with slightly changed refractive indices, since n depends on the packing density in each case. The ultraviolet absorption limit λ_s, however, is independent of the starting compound but depends on the specific crystal phase formed during baking. The changes in refractive index n_{550} and geometrical thickness d during solidification at different temperatures are shown in Figs. 3 and 4. As can be seen from the curves

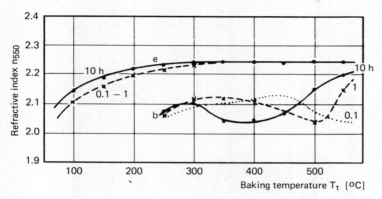

Fig. 3
Refractive index n_{550} of TiO_2-films deposited from ethyl-dichloro-titanate (e) and butyl-titanate (b) as function of baking temperature T_t

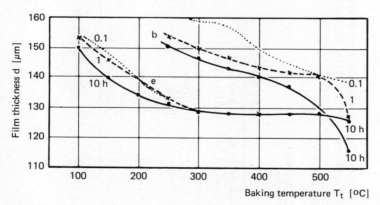

Fig. 4
Film thickness d of TiO_2-films deposited from ethyl-dichloro-titanate (e) and butyl-titanate (b) as function of baking temperature T_t

in both figures for films made of ethyl titanate, the formation of the oxide layer almost ceases at a rather low temperature of about 250°C. The reason is obviously that the required hydrolytic reaction of the liquid film with water vapour from the atmosphere starts even at room temperature and is practically complete at moderate tempering. Butyl titanate, however, is much more resistant to hydrolysis and the transformation phenomena produced here essentially by pyrolysis extend over the whole temperature range applied up to the softening region of the glass.

Since alkali diffusion from a substrate into a crystallizing film requires time, the influence on the resulting structure type decreases from substrate/film boundary to film/air boundary depending on film thickness d. Thus, thin TiO_2 films deposited on alkali glasses exhibit lower mean refractive indices $\bar{n}$ than thicker films. This effect is shown in Fig. 5 for TiO_2 films made from butyl titanate solution and baked 1h at 450°C and 500°C respectively [28]. As discussed above, the refractive index of TiO_2 layers fabricated in this way is dependent on various

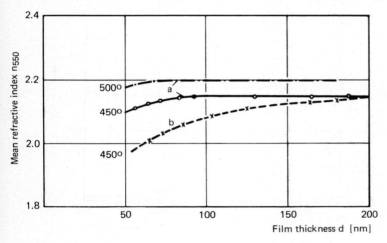

Fig. 5
Mean refractive index n_{550} as function of layer thickness d of TiO_2-films deposited from butyl-titanate solution on (a) fused silica and (b) plate glass

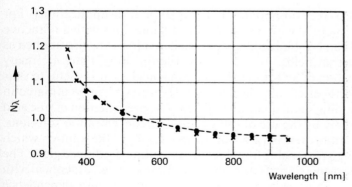

Fig. 6
Reduced dispersion N_λ of various TiO_2 materials in the visible region:
(x) films with anatase structure, (●) films with $Na_2O \times TiO_2$-structure,
(- - -) bulk material (anatase, mean values over the crystallographic axes)

104

production parameters; it is clear therefore that the dispersion n_λ must exhibit corresponding variations. Introduction of a reduced dispersion $N_\lambda = n_\lambda/n_{550}$, however, yields a nearly identical run for the various individual curves. This fact is demonstrated in Fig. 6.

Finally for certain TiO_2 coatings with anatase structure deposited on fused silica, the optical constants n and k were determined [28] from intensity measurements. The values are shown in Fig. 7. As can be seen in the figure, there is again a slight dependence of n on thickness d.

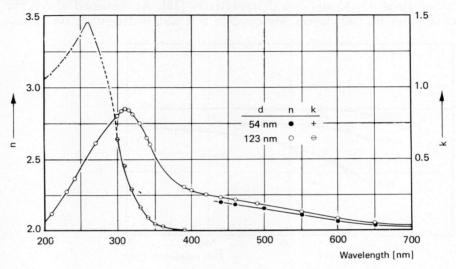

Fig. 7
Optical constants n, k of TiO_2-films of thicknesses of d = 54 nm and 123 nm deposited on fused silica, calculated from intensity measurements (T_λ and R_λ, T_t = 430 °C). The films were made from a butyl-titanate solution.

SiO_2 films are also very important in many thin film applications. For example, it is the only material with which hard and stable films with a refractive index smaller than 1.5 can be obtained. Since silica derivatives are also applied as coatings in other industrial fields, for instance textile finishing, there are many demands for the fabrication of SiO_2 films. For coating on glass, it is useful to start from methyl or ethyl orthoesters of silic acid [40]. Hydrolysis can be initiated in alcoholic solution by adding small quantities of acid. The precipitation of silic acid gel is delayed considerably by the use of sufficiently low concentration solutions. Owing to polycondensation, there is a gradual rise in viscosity of the solution which is important in obtaining a reasonable film thickness by lifting the dipped glass. The refractive index and the dispersion of tempered SiO_2 films are nearly identical with those of fused silica. As can be seen in Fig. 8, there is only a minute dependence of n on tempering temperature. It can be concluded that the amorphous SiO_2 films attain practically the density of fused silica at temperatures between 500 and 550°C. As can be seen in Fig. 9, after tempering the layer thickness shows

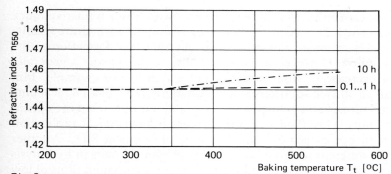

Fig. 8
Refractive index n_{550} of SiO_2-films as function of baking temperature T_t

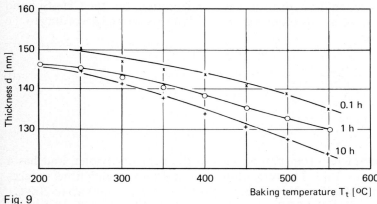

Fig. 9
Film thickness d of SiO_2-films as function of baking temperature T_t

substantial decrease which is higher than that for TiO_2 films fabricated from titanium butylester. Volatile reaction products, as physically adsorbed water and water molecules formed during transition of Si-OH groups into -Si-O-Si-combinations, are released here only with considerable supply of energy. After baking, a slight increase in refractive index ($\Delta n \approx +0.005$) is observed, clearly caused by water vapour absorption of the cold film.

As with fused silica, the ultraviolet transmission of SiO_2 films is dependent on purity. However, the excellent uv transmittance value of purest silica glass cannot even be attained with carefully prepared SiO_2 films because of incorporated traces of hydrolytic and pyrolytic residues which produce measurable absorption below 205 nm. In the infrared region there are strong absorption bands in the films between 7.8 and 11.5 μm. Those are the same frequencies at which absorption is also observed in pure fused silica. Si-O vibrations are responsible for this absorption. However, there are also weaker bands in the SiO_2 films between 2.7 and 3.6 μm and at 6 μm suggesting incorporated –OH groups and adsorbed H_2O molecules.

There are some possibilities of fabricating films with lower refractive index than that of a compact SiO_2 layer. One method is to add materials which are volatile at higher temperatures or which can be leached out after film formation to the silicon alkoxide solution. However, a reduction of n by this means is generally associated with a decrease in film hardness and stability. For example, film deposition from alkali silicate solutions and elimination of the alkali by leaching yields SiO_2-skeleton films with a refractive index of about n $\simeq$ 1.3, but the optical and mechanical stability is insufficient. For these reasons, in technical application the reduction should not exceed 3 % of the value of compact SiO_2. Another method is to use the different reactivity of differing chemical bonds in hydrolysis to obtain by polycondensation special substituted silicon-compound films of low refractive index. An example is the formation of methylpolysiloxane films [33] from solutions of methyl-tri-alkoxy silane. In the compound CH_3-Si $(OR)_3$, the Si-OR bonds are reactive for hydrolysis and condensation, but the CH_3–Si bond remains intact under the usual conditions. The polymer is formed according to the following formalism:

$$X CH_3 - Si (OR)_3 \xrightarrow[- ROH, - H_2O]{+ H_2O} \begin{array}{cc} CH_3 & CH_3 \\ | & | \\ -O-Si-O-Si-O- \\ | & | \\ O & O \\ | & | \\ & -O-Si-O... \\ & | \\ & CH_3 \end{array} \qquad (11)$$

The material is completely transparent, also in the near ultraviolet, and has a refractive index of n $\simeq$ 1.42. The remaining organic group in such films reduces hardness and thermal stability. Nevertheless, the films are resistant up to 200 °C.

Replacing the number of Si-O bonds in the structural unit of the inorganic polymer (12)

$$\begin{array}{c} O \\ | \\ O-Si-O \\ | \\ O \end{array} \qquad (12)$$

$$\begin{array}{c} O \\ | \\ O-Si-R \\ | \\ O \end{array} \qquad (13)$$

$$\begin{array}{c} R \\ | \\ O-Si-R \\ | \\ O \end{array} \qquad (14)$$

by organic residues R leads to a decrease in silicate character of the corresponding polymers formed by polycondensation of the monomers of type (13) and (14). Such ilms should be interesting because of special properties. However, attempts to extend the structural principles of silicate chemistry further into organo-silicon chemistry by transition to heteroorganopolysiloxanes, for example of type (15), shown with bivalent Me for simplicity, have so far been technically less successful:

$$
\begin{array}{ccc}
R & R & R \\
| & | & | \\
-Si-O-Me-O-Si-O-Me-O-Si- \\
| & | & | \\
R & R & R
\end{array}
\qquad (15)
$$

where Me = B, Al, Ti, Sn, Pb, P, As [41].

Polymers of this type often have more or less irregular compositions and this makes the starting material/end product correlation very diffucult [42].

For optical film application mainly in interference systems, many different and sometimes exotic refractive indices are required in the design, which are not known in naturally existing compounds (see e. g. Table 1). It is therefore an important advantage of the dip-coating process that solutions of different compounds can be mixed. In this way, the refractive index of SiO_2 films can be increased by mixing with almost all metal oxides suitable for fabricating stable films of higher index. In particular, the addition of TiO_2 produces mixed oxides of particularly high resistance and stability. Crystalline structures are obtained with more than 30 mol % TiO_2 in the mixed film.

The refractive indices of SiO_2 films as a function of the molar proportion are shown in Table 3. The measurements reveal this function to be practically linear [28]. It is interesting to note here that oxides whose deposition as one-component films causes difficulties can very often be deposited as homogeneous mixed oxides by embedding them in a matrix forming high quality films, as for example those of SiO_2, TiO_2 and others.

ABLE 3

EFRACTIVE INDEX OF MIXED FILMS OF SiO_2 AND TiO_2, PREPARED BY DIP-OATING, AS A FUNCTION OF MOLAR PROPORTION [28, 36].

ilm material iO$_2$ / SiO$_2$ (mol%)		Refractive index n_{550}
0	100	1.455
10	90	1.53
20	80	1.61
30	70	1.69
40	60	1.77
50	50	1.85
60	40	1.92
70	30	2.01
80	20	2.08
90	10	2.18
00	0	2.25

With hydrolysis and polycondensation of mixtures, it is also possible to produce dip-coatings of magnesium aluminium spinel [30]:

$$Mg(OR)_2 + 2\,Al(OR^1)_3 \longrightarrow$$

```
                    OR
                    |
                    Mg
                   /    \
              RO          OR¹
        R¹O    ¦            |      OR¹
          \    ¦            |     /
            Al               Al
          /     \          /     \
      R¹O         O ----        OR¹
                  |
                  R¹
```

(16)

$$Mg[Al(OR)(OR^1)_3]_2 + 8\,H_2O \longrightarrow Mg[Al(OH)_4]_2 + 2\,ROH + 6\,R^1OH \tag{17}$$

$$Mg[Al(OH)_4]_2 \xrightarrow[+\,\Delta T]{} \underset{spinel}{Mg[AlO_2]_2} + 4\,H_2O \tag{18}$$

$$R = -CH_3, \qquad R^1 = -CH_2\text{-}CH(CH_3)_2$$

Magnesium methoxide is allowed to react with aluminium sec-butoxide in a molar ratio of 1:2 in alcohol. The magnesium aluminium alkoxide (16) is formed in solution. The choice of the organic residues is a matter of purely practical importance, since they are removed later in any case. On evaporation of the solvent in the presence of atmospheric water vapour, hydrolysis occurs (17) and a gel is formed. X-ray diffraction experiments show the principal reflections, although very broad, even on heating at 250°C. They become more pronounced on tempering at 400°C and are very sharp after treatment at 620°C. The layers obtained are glass-clear and very hard. The spinel is thus formed at temperatures well below those normally used for such syntheses. Magnesium and aluminium alkoxides react with one another even at room temperature, whereas high temperatures of about 850°C are required for the reaction of salts or oxides. The alkoxide method also seems to be more direct, since no crystalline phase other than spinel is formed, even at low temperatures. It can be assumed that something like a co-condensation occurs after hydrolysis instead of homo-condensation. If the latter reaction type occurs, which is common according to the literature [43], the reaction product would be a mixture of single oxides instead of a homogeneous multicomponent material. Small amounts of metal alkoxides may undergo condensation reactions with alcohol elimination even at room temperature, but this is not important to the final result, since the reaction continues in the formation of the polyorgano-oxymetal-oxane intermediate product (16), as was discovered by [43] and [30]. The temperature treatment is evidently required for the removal of the last traces of water from hydroxyl groups in the compound. In the case where metal alkoxides react with one another, as they do very often, the process is also applicable to other elements. Thus was the formation of eucryptite [LiAl (SiO$_4$)] through such a process starting from lithium-ethoxide, aluminium sec-butoxide and Si(OR)$_4$ performed by [30], and in a similar way the preparation of BaTiO$_3$, SrTiO$_3$ and SrZrO$_3$ was possible [44]. These examples underline the scope of the method.

If, finally, more metal compounds participate in the reaction and if also Si(OR)$_4$ is present, then a polyorgano-oxymultimetallo-oxanesiloxane is formed which is then converted by polycondensation into a polymultimetallo-oxanesiloxane which is none other but a multicomponent glass containing no organoxy groups.

In this way, it is possible to produce borosilicate glass films at temperatures below 500°C by a dip-coating process [30,32,36]. The following sequence of reactions shows the synthesis:

$$m\ Si(OR)_4 + n\ B(OH)_3 + o\ Al(OR)_3 + p\ NaOR + q\ KOR \longrightarrow$$

$$\xrightarrow{complexation} \{(Si_mB_nAl_oNa_pK_q)\ (OR)_{4m+3o+p+q}\ (OH)_{3n}\} \tag{19}$$

$$\xrightarrow{hydrolysis} \{(Si_mB_nAl_oNa_pK_q)\ (OH)_{4m+3o+p+q+3n}\}$$

$$\xrightarrow[-H_2O,\ condensation]{heat} \{(Si_mB_nAl_oNa_pK_q)O_{(4m+3o+p+q+3n)/2}\}$$

According to analysis, the residual carbon content of the borosilicate glass is only 0.0002 % which is surprisingly low considering the organic starting materials. This, however, indicates that hydrolysis is practically quantitative despite the strong cross linking. The residual water content can be determined by infrared spectroscopy and was found to be 0.04 %. At higher tempering temperatures ($\sim$ 700°C), the amount is reduced to about 0.03 %.

Using the purest starting materials, the glass films obtained show extremely low contamination levels.

Other examples for multicomponent glass layers prepared by dip-coating are [30,44]:
phosphate silicate glass layers   (SiO$_2$, Al$_2$O$_3$, P$_2$O$_5$, BaO, B$_2$O$_3$, CaO, MgO)
lead silicate glass layers   (SiO$_2$, PbO, Na$_2$O) and
alkali aluminosilicate glass layers (SiO$_2$, Na$_2$O, Al$_2$O$_3$).

6.2.1.2.1.2 COATING PROCEDURE

A thin-film production method is of interest only if the various layers that can be produced with this method cover a possibly large variety of physical properties and technical requirements and/or if also substrates of special type and of very different size can be coated. For the production of homogeneous dielectric coatings with desirable optical properties from liquid films, the solutions used must possess special physical and chemical characteristics. To obtain such proper characteristics, according to Schröder [28], the four following requirements must be fulfilled.

1.) The initial compounds must have an adequate solubility and the solution formed may only have a minor tendency towards crystallization during solvent evaporation. These requirements are generally met by those compounds which either already dissolve in a colloidal or polymeric state or reach such a state by reaction with the solvent, or which remain after solvent evaporation as a gel-like non crystalline residue.

2.) To obtain good wetting of the substrate, the contact angles formed with the solution have to be sufficiently small. Wettability can be improved in some cases by the addition of wetting agents to the solution. The wettability of the substrate is decreased by insufficient cleanliness. Scratches and surface roughness (grinding structures) with a large depth of more than 10 μm may also cause considerable surface irregularities in the coatings when using solutions with a low degree of polymerization.

3.) The solution must have an adequate durability and the processing conditions must be maintained constant. Durability is sometimes difficult to realize with solutions of colloidal or polymeric character, but using stabilizers it is often possible.

1. DRAWING   2. CHEMICAL REACTION   3. BAKING

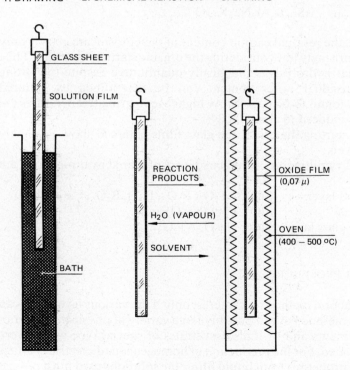

Fig. 10
Graphs showing the oxide film formation in the dip coating process

4.) To obtain reproducible solid and homogeneous oxide films, drying and heating must be carried out carefully. During this process the solidification of the film structure should occur without the appearance of cracks or haze and at the same time high bonding strength to the substrate must develop.

Dip coatings can be made by two different process technologies:

1.) lifting processes, in which the parts to be coated are withdrawn from the solution and simultaneously covered with a liquid film, and

2.) lowering processes, in which the parts to be coated remain at rest and the liquid level is lowered.

The lifting process is the more economical for large substrates and it is also universally applicable. The scheme of this coating procedure is shown in Fig. 10. Large glass panes of dimensions of 3 to 4 meters first optimally cleaned are inserted into a dipping container with the coating solution and are then carefully pulled out at a constant speed. The lifting operation must be completely smooth and shockless and has to be so slow that the liquid film adhering to the substrate surface is left in a flow zone which is very short, as can be seen in Fig. 11. The adherent liquid film

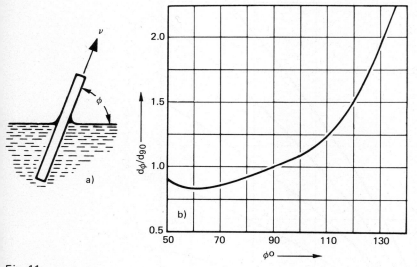

Fig. 11
a) Dip coating at inclined lifting
   ϕ = angle of inclination
   v = lifting speed
b) Relative film thickness d_ϕ/d_{90} as a function of the substrate inclination ϕ during lifting

partly flows down the plate and solidifies after solvent evaporation, hydrolysis and condensation. Its thickness distribution in the vertical direction is dependent only on evaporation of the solvent and becomes stationary during the lifting, so that the solid film finally takes up a constant thickness along the lifting direction. With regular operation at quiet liquid surface, film formation is also very uniformly along the horizontal dipping line. The uniformity of the deposited film can be controlled by watching the interference fringes of the system which are formed during evaporation of solvent and volatile reaction products. During lifting, the fringes should run as horizontally as possible and, at constant lifting speed, remain always at about the same distance above the surface of the coating solution. As well as having dependence on lifting speed, the thickness d of the coating is mainly a function of the concentration c of the solution and of the angle of inclination Φ of the coated surface with the surface of the solution. At constant temperature and humidity, the approximate relation between film thickness and lifting speed for vertical movement is:

$$d \simeq k \, v^{2/3} \tag{20}$$

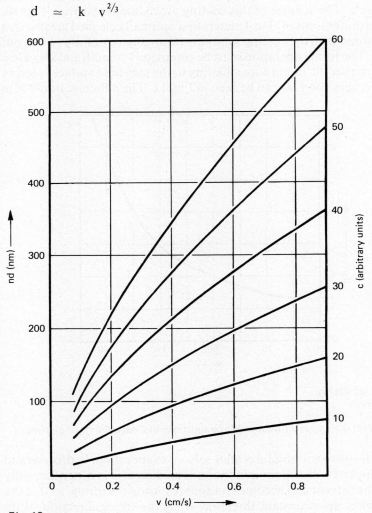

Fig. 12
Optical thickness of dip coating films as a function of lifting speed for solutions of various concentrations

Here the constant of proportionality k contains all the other properties affecting the layer thickness such as viscosity, surface tension and vapour pressure of the solution. The simple relation was found to be valid because from extensive experimental investigations [28] of the influence of v, Φ and c on film thickness d, it followed that all suitable solutions have the same fundamental characteristics which can be seen in the Figs. 11 – 13.

The investigations of Schröder [28] also showed that a non-linear relation may exist between the film thickness d and the concentration c of the coating solution, as can be seen in Fig. 13 for a TiO_2 forming solution. This non-linearity comes from

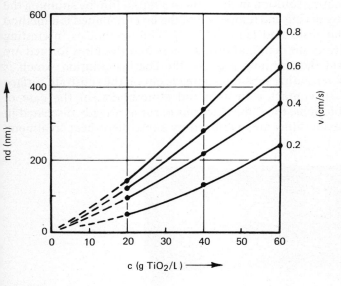

Fig. 13
Optical thickness of TiO_2-films as a function of coating solution concentration for different lifting speeds

an increase in the viscosity η of the solution with increasing concentration c. In particular, more concentrated and viscose solutions can show during lifting of the pane a remarkable enrichment of solute compounds in this region, because of solvent evaporation and imperfect mixing in the dipping zone. This cannot be avoided because stirring of the solution would result in surface disturbances of the liquid and therefore in irregularities of the coating obtained. The thickness increase can be eliminated, however, by a continuous change in the lifting speed. Supporting parts, required to hold the substrates, should not be immersed because of danger that solution residues can rinse in strips over the uniformly coated surface.

On a narrow-brimmed area at the lower end of a glass pane withdrawn from the coating solution, a strong increase in film thickness may occur caused by adhering liquid residues. The width of this useless zone depends on the shape of the edge, lifting speed and viscosity of the solution and is normally between 2 and 8 mm. With large glass panes this small useless zone is simply cut off after the coating process is finished. Finally, the film is hardened in an oven at higher temperatures between 250 and 650°C until a transparent oxide film has been formed. A limitation seems to exist in the film thickness. According to Ref. [30], it is difficult to produce films thicker than 0.5 μm although some increase is possible through the use of additives in the solutions or by multiple coating.

6.2.1.2.2 SPIN COATING

Spreading out the coating solution in the form of a liquid film by spinning the wetted surface followed by drying and tempering of the layer obtained is a method for coating small circular disks and lenses [45,46]. This method is interesting because there are practically no marginal disturbances and the films formed are very uniform.To perform the spinning process, the coating solution which is poured onto the horizontal rotating substrates spreads out on the substrate surface absolutely evenly [47]. In contact with a humid atmosphere, in the case of oxide-forming coating solutions, the same reactions occur as already discussed in Section 6.2.1.2.1. The oxide films are stabilized by subsequent heat treatment.

Fig. 14
Spin Coater

Although this process is able to produce optical films of good quality, it is less economical and therefore no longer used in the production of inorganic films. However, in the deposition of organic films, such as photoresists, the spinning method is widely employed [48,49].

For this purpose, various spin coaters are commercially available. Fig. 14 shows a spin coater made by Convac GmbH (Germany).

6.2.1.3 DEPOSITION OF ORGANIC FILMS FROM SOLUTIONS

The first attempts to deposit organic dielectrics as surface films onto glass substrates were stimulated by Langmuir's observation [50] that some substances of high molecular weight insoluble in water and also having polar groups spread over a water surface as a monomolecular layer could be transferred to the surface of a solid substrate. The transfer method was developed mainly by Blodgett [51,52] and became known as Langmuir-Blodgett method. To deposit such films, a small quantity of a solution of suitable material is placed on the surface of carefully purified water in a wide open tank. After evaporation of the volatile solvent, the remaining monolayer of the organic compound is compressed by means of a slider until it forms a quasi-solid with the thickness of one molecule. A glass substrate is then repeatedly dipped through the water surface with constant but slow speed. In this operation, one monolayer is transferred to the substrate on each pass through the water surface. The procedure is shown schematically in Fig. 15.

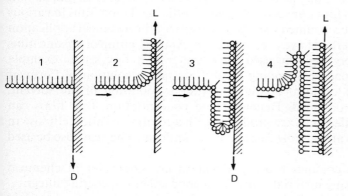

Fig. 15
Illustration showing the Langmuir-Blodgett method [54] of depositing organic films
D = dipping, L = lifting
⚬ — polar group ⎱
❘ — non polar part ⎰ of an organic molecule

The clean glass remains uncoated when dipped the first time. During lifting, however, the organic molecules adhere with their hydrophilic polar end to the glass surface, they are turned to 90° and are pulled away from the water surface to the glass surface. In this way the substrate surface is coated with an oriented monomolecular layer. During the next dipping, a second layer is formed in a similar way but with the hydrophobic end of the molecule sticking on the hydrophobic layer surface. Films of great perfection can thus be built up monolayer by monolayer. The films are generally free from pinholes and can be deposited to thicknesses of about 1 μm or even more. It is also possible to dope the film either by interposing one or more layers of a different material or by mixing various other molecules in the coating solution. With some materials, it is also possible to form multilayers. In early experiments, efforts were made to produce antireflection coatings with this technique [53] but such coatings did not seem to offer any practical applications. Langmuir-Blodgett films, however, have been used in many scientific studies. Films of fatty acids as well as those of condensed compounds have been investigated. Shearing force measurements, determinations of molecule dimensions and simulations of higher-order molecular structures have been performed [54,55]. Technological applications of such films have up so far been little explored. However, the ability to produce films whose constitution may be carefully controlled offers exciting possibilities. Possible technological applications are enormous. Organic coatings could be used principally for various practical devices ranging from electronic devices to high-temperature superconductors [56,57]. There is evidence [58,59] that, if anthracene could be formed into high quality films of about 0.2 to 1 μm in thickness, it would probably be an efficient large area electroluminiscent device. Organic films have also been proposed for rectifiers and memory devices [60]. Smooth, dense and homogeneous films are an important requirement in integrated optics [61] and for superlattices [62] as well as for active films in thin film transistors [63] and for the insulation films in various MIS devices [64]. It can be assumed that the prospects for the technical application of this film formation method have been improved. Another group of organic films are the replica and supporting films used in electron microscopy, see for example [65]. These films are made by dipping a substrate into a solution of Formvar (Polyvinylformal) in chloroform or dioxan. After evaporation of the solvent, polymerized films are obtained. Depending on concentration, thin films can usually be formed. Similar films are produced with a solution of nitrocellulose in amylacetate. Such films are pinhole free and fairly uniform. They can also be used as protective coatings.

Protective organic coatings bearing a pattern are important in chemical etching [66]. Such masking materials must have good adhesion, coating integrity, adequate resolution and chemical resistance to the etchant. Very often, applied masking materials for high quality thin-film patterning are photoresists [67]. These are organic polymers whose solubilities in certain special solvents change drastically as a consequence of exposure to ultraviolet radiation. Irradiated areas become less soluble in the developing solution in the case of negative photoresists and conversely positive photoresists become more soluble in exposed areas. The films are produced by the spinning method. The wet films obtained are first dried in air and then heated in an oven to about 80°C. The uniformity of films produced in this way is quite excellent.

6.2.1.4 CHEMICAL VAPOUR DEPOSITION AT LOW TEMPERATURES

Chemical vapour deposition (CVD) processes have become a very important group of film-formation methods. Basically CVD is a material synthesis in which constituents of the vapour phase react chemically to form thin solid films as a solid-phase reaction product which condenses on the substrate. The reaction should take place very near to or on the substrate surface (heterogeneous reactions) and not in the gas phase to avoid powdery deposits. Activation of the reaction can be performed by various means such as the application of heat, high-frequency electrical fields, light or X-ray radiation, electric arc, electron bombardment or catalytic action of the substrate surface. A marked influence of the process parameters such as substrate temperature, gas pressure, concentration of the reactants, flow rate, and so on, on the properties of the deposited films has been observed. Practically any controllable reaction with one or more vapours leading to a solid reaction product may in principle be used to produce thin films. The search for suitable reactions is still in progress. Many of the reactions which are used can be put into one of the five following categories:

1. decomposition reactions:

$$AB_{(g)} \longrightarrow A_{(s)} + B_{(g)} \tag{21}$$

2. reduction and oxidation reactions:

$$AB_{(g)} + C_{(g)} \longrightarrow A_{(s)} + BC_{(g)} \tag{22}$$
$$C = \text{hydrogen or metal}$$

$$AB_{(g)} + 2D_{(g)} \longrightarrow AD_{(s)} + BD_{(g)}$$
$$D = \text{oxygen or nitrogen}$$

3. hydrolysis reactions:

$$AB_{2\,(g)} + 2\,HOH_{(g)} \longrightarrow AO_{(s)} + 2BH_{(g)} + HOH_{(g)} \tag{23}$$

4. polymerisation reactions:

$$x\,A_{(g)} \longrightarrow A_{X(s)} \tag{24}$$

5. transport reactions:

$$A_{(s)} + B_{(g)} \underset{T_2}{\overset{T_1}{\rightleftarrows}} AB_{(g)} \qquad \text{with } T_2 > T_1 \tag{25}$$

Oxidation reactions of a metallic substrate are excluded. Generally the substrate plays only a passive role in CVD processes. Many examples of such reactions are listed in the review papers [68], [69] and in the proceedings of CVD conferences such as [70]. In addition, there are different fabrication techniques in use which can be classified as:

1 atmospheric-pressure CVD
2 low-pressure CVD

3   high-temperature CVD
4   low-temperature CVD
5   plasma-assisted CVD
6   photon-enhanced CVD

and some combinations of these techniques. Special reactors have been developed to perform these fabrication techniques, see for example [69].

The fundamental principles of CVD encompass an interdisciplinary range involving reaction chemistry, thermodynamics, kinetics, transport mechanisms, film growth phenomena and reactor engineering. The question of whether or not a process will run in the desired direction can be answered first by thermodynamic calculation of the free energy of the chemical reactions ΔG_r^o using the standard free energy of formation of the compounds G_f^o [69, 146]:

$$\Delta G_r^o = \Sigma \, \Delta G_f^o \text{ products} \; - \; \Sigma \, \Delta G_f^o \text{ reactants} \tag{26}$$

The free energy of the chemical reaction ΔG_r^o is related to the equilibrium constant k_p which itself is related to all the partial pressures in the reaction systems:

$$- \, \Delta G_r^o = 2.3 \, R \, T \, \log k_p \tag{27}$$

$$k_p = \prod_{i=1}^{n} \, p_i \text{ products} \, / \, \prod_{i=1}^{n} \, p_i \text{ reactants} \tag{28}$$

To calculate thermodynamic equilibrium in multicomponent systems, the so-called optimization method and the nonlinear equation method are used, both discussed in [69]. In practice, however, kinetic problems have also to be considered. A heterogeneous process consists of various occurrences such as diffusion of the starting materials to the surface, adsorption of these materials there, chemical reactions at the surface, desorption of the by-products from the surface and their diffusion away. These single occurrences are sequential and the slowest one determines the rate of the whole process. Temperature has to be considered. At lower substrate temperatures surface processes are often rate-controlling. According to the Arrhenius equation, the rate is exponentially dependent on temperature:

$$r = A \, e^{-\Delta E / R T} \tag{29}$$

with A = frequency factor, ΔE = activation energy (usually 25 to 100 kcal mol^{-1} for surface processes). At higher temperatures, the diffusion of the reactants and of the by-products may become the rate-determining factor. Their temperature dependence varies between $T^{1.5}$ and $T^{2.0}$ [69]. Further influences may come from different crystallographic orientations and the topography of the substrate surface.

In the case of glass, however, no great variations in behaviour between different types are expected because of their very similar structure and surface composition. Chemical vapour deposition reactions had already been tried by the last century, for instance in the refinement and deposition of silicon by reduction of SiF_4 and $SiCl_4$ with alkali metals [71] and in the refining of Ni using Ni-carbonyl in the Mond process [72,73]. The major impact of chemical vapour deposition on thin-film technology took place, starting some 50 years ago, when refractory compounds such as metal carbides, nitrides, silicides, borides and oxides as well as mixed phases of some of these compounds, came into extensive technological application, and with semiconductors, starting more than 20 years ago when it was demonstrated that epitaxial layers of silicon and germanium could be grown by this method. Today the use has been extended to the fabrication of metal and alloy films, of elemental and compound semiconductors, superconductors, refractory and hard materials, transparent conductors and of other special films.

Considering the three following decomposition reactions [68]:

$$Ni\,(CO)_4 \xrightarrow{\;200\,-\,300°C\;} Ni\,+\,4\,CO \tag{30}$$

$$SiH_4 \xrightarrow{\;800\,-\,1300°C\;} Si\,+\,2\,H_2 \quad \text{and} \tag{31}$$

$$CH_4 \xrightarrow{\;1000\,-\,2000°C\;} C\,+\,2\,H_2 \tag{32}$$

it is clear that rather high temperatures are required for film deposition. This requirement is also valid for many other reactions used in CVD. The high-temperature heating of the substrates may, depending on the kind of material, induce a lot of undesirable thermal damage. The temperature also influences the film structure. It is known that deposits formed at high temperatures are generally polycrystalline and very high temperatures are necessary for growing single crystal films. On the other hand, materials deposited at relatively low temperatures, e. g. below 600°C, tend to be amorphous. A distinction is often made in the literature between high-temperature and low-temperature processes.

For coating glass, only low-temperature processes can be used. However, many low-temperature reactions need substrate temperatures between 300°C and 600°C, so that coating of inorganic glass in fact is possible but coating of organic substrates is still impossible. For most organic glass surfaces, temperatures of only 100°C maximum can be tolerated. Today there are rather few coating processes known which run below 100°C substrate temperature.

In general the substrate is kept somewhat hotter than the surroundings so that the heterogeneous nucleation reaction at the substrate surface occurs preferentially and the unwanted homogeneous nucleation in the gas phase is suppressed. However, conventional CVD is also frequently accompanied by a homogeneous gas-phase reaction because film formation takes place throughout the whole of the higher-temperature zone of the reactor. The products of the gas-phase reaction may be incorporated into the growing film structure and can degrade the film quality. These two problems are generally recognized as the fundamental disadvantages of conventional CVD. To overcome these difficulties, the required activation energy for initiating the chemical reaction and the energy necessary for good film adhesion and growth may be supplied to the substrate in a more accurate and at the same time more protective manner from a glow discharge (plasma-assisted CVD) or from radiation (light- or laser-enhanced CVD).

Fortunately a general trend to lower processing temperatures can be observed. A broader applicability of CVD for coating of glass depends on further development of low-temperature processes which enable film deposition at substrate temperatures below 500°C. Various applications of CVD processes for coating of glass have been reviewed recently [74].

6.2.1.4.1 ATMOSPHERIC-PRESSURE AND LOW–PRESSURE CVD

6.2.1.4.1.1 SPRAY COATING

Traditionally, spray coating is not considered to be a CVD process. In spray coating, highly dispersed liquids or vapours are allowed to react on mixing in a humid atmosphere or are allowed to undergo pyrolysis. The difference is one of degree: in CVD the reactants are in the gas phase whilst in spray coating the reactants can range from droplets right down to atom clusters. They are sprayed or blown onto the face to be coated.

Such reactions are of great practical interest since in some cases large surfaces can be coated without protection against air and at relatively low temperatures below 500°C where a great variety of different glasses may be used.

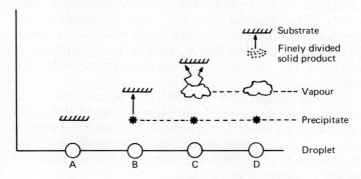

Fig. 16

Schematic representation of the deposition processes initiated with increasing substrate temperature.

According to Viguié and Spitz [75], both a chemical and a physical factor determine whether or not a spray process can be classified as chemical vapour deposition. Processes that may occur with increasing temperature are shown schematically [75] in Fig. 16. In process A the droplets reach the substrate, the solvent vapourizes and leaves a dry deposit which will react and form a film.

In process B the solvent evaporates before the droplets arrive at the surface and the precipitate impinges upon the surface where it reacts. In process C the solvent evaporates as the droplets approach the substrate and the solid precipitate melts then vapourizes or sublimes and the vapour diffuses to the substrate where the heterogeneous reactions occur; and finally in process D at the highest temperatures the compound to be deposited vapourizes before it reaches the substrate surface and a homogeneous chemical reaction takes place in the vapour phase.

Only process C is true chemical vapour deposition and in the case of solvent-free vapours blown to a surface only a heterogeneous surface reaction counts as CVD. In practical applications of such deposition technologies, however, a detailed knowledge of the nature of the basic processes is often difficult to obtain. One of the oldest technical vapour process for producing films of TiO_2, SiO_2 and their mixtures is the gas reaction process (Gaszersetzungs-Verfahren) which was invented in the laboratories of Schott und Genossen more than fourty years ago by Geffcken and Berger [20]. A rather elaborate machine is required to perform this process. The glass plates and lenses are supported on a heated horizontal rotating disk having a maximum speed of 60 r-p-m. Burners or gas-mixer nozzles are supported and geared to sweep back and forth horizontally across the work pieces, one depositing TiO_2 and the other SiO_2. Mixed oxides can be formed by the simultaneous operation of two nozzles so that TiO_2 and SiO_2 are deposited alternately in very thin films. The titanium dioxide nozzle has two inlets for dry air, one for a mixture of dry air and titanium tetrachloride vapour and one for moist air. To prevent chemical reaction of $TiCl_4$ with humid air inside the nozzle a concentric curtain of dry air between these two reaction partners is maintained. The glasses to be coated are exposed to a temperature of about 250°C and the mixture from the burner reacts on the heated surface to form a TiO_2 film.

The burner for SiO_2 has an inlet for air containing methyl silicate vapour and another inlet for a mixture of 25 percent hydrogen in air. The mouth of the burner is supplied with a grid flame arrester and the gas mixture is burned at the grid surface. A special gas mixer assures complete mixing before combustion. The heat and water resulting from the combustion causes the deposition of the SiO_2 films. A threaded collar at the top of each nozzle provides a fine adjustment of the distance to the work piece. By the use of interchangeable fixtures on the machine it is possible to coat concave, convex or flat surfaces. The gas flow to the burners is controlled by automatic pressure regulators and rotameter-type flow gauges. Titanium tetrachloride and methyl silicate vapours are introduced by bubbling air through the liquids at constant temperature. Metal halogenides can be replaced by metal organic compounds to prevent the formation of unwanted aggressive by-products. Many single oxide and mixed oxide films can be deposited by this method. Various types of single and multilayer antireflection coatings were made by the gas reaction process [76] during World War II. In optical film production the method was later replaced by physical vapour deposition. In the last decade, however, the technical application of spray coatings have markedly increased.

In particular the deposition of transparent conducting films based on In_2O_3, SnO_2 or on mixtures of both oxides is often done by spray coating. The method is of great practical interest because although the areal thickness distribution of the deposited films is not outstanding, in certain cases both small pieces and also large glass panes can be coated even continuously in air and at low temperatures [77,78,79,80]. A schematic diagram of modern equipment for spray deposition is shown in Fig. 17. The solution of typically chlorides and acetylacetonates in

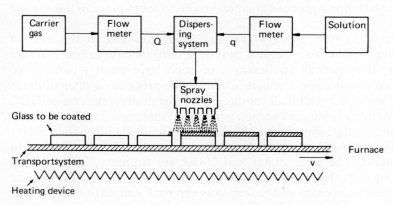

Fig. 17

Schematic diagram of equipment for chemical spray deposition

water, butylacetate, butanol or toluene is dispersed and transported by means of a carrier gas. Very often air, nitrogen or argon is used as a carrier gas depending on the type of reaction. The conversion of the spray solution into an aerosol consisting of very fine droplets is usually performed pneumatically in the dispersion system. As regards small droplet size, better results are obtained using intense ultrasonic agitation to form the aerosol, see for example [81]. The properties of this aerosol are dependent on the nature of the liquid and on the frequency and intensity of the ultrasonic device. With increasing ultrasonic frequency the mean diameter of the droplets decreases. Compared with droplets produced by a pneumatic transducer, ultrasonically produced droplets may be an order of magnitude smaller. Generally a very narrow droplet spectrum is obtained and the relative uniformity of the coating is improved accordingly. A further advantage of this method is that the gas flow rate is independent of the aerosol flow rate, which ist not the case with pneumatic spraying. A modern spray deposition process can be described by relatively few parameters such as flow of carrier gas Q, concentration of the solution C, solution flow q, droplet radius r, distance between nozzle and substrate d, temperature of the gaseous environment T_e, temperature of the substrates T_s and their speed through the furnace v. Typical values of such parameters, according to [75], are listed in Table 4.

TABLE 4

PARAMETERS OF SPRAY DEPOSITION

Parameter	Piezoelectric transducer	Pneumatic transducer
flow of carrier gas	3 - 6 1 min^{-1}	3 - 6 1 min^{-1}
solution concentration	0.1 - 0.4 mol l^{-1}	0.1 - 0.4 mol l^{-1}
solution flow	30 - 60 cm^3 h^{-1}	500 - 800 cm^3 h^{-1}
droplet radius	1 - 4 μm	5 - 50 μm
distance nozzle to substrate	3 - 15 mm	40 mm
temperature of the gaseous environment	400 - 550°C	500 - 550°C
speed of substrate through furnace	10 - 40 mm min^{-1}	10 - 40 mm min^{-1}

The spraying method involves a chemical reduction or a hydrolysis reaction between the organic or the inorganic highly dispersed starting material and the mixture of carrier gas and reactive gas at the heated substrate surface where the expected metal or oxide film is formed.

$$
\begin{array}{c}
\text{CH}_3 \qquad \text{CH}_3 \\
| \qquad\quad | \\
\text{C-O}\quad \text{O=C} \\
// \quad \backslash \quad / \quad \backslash \\
\text{CH} \quad \text{Pd} \quad \text{CH} + \text{H}_2 \longrightarrow \text{Pd} + 2\ \text{CH}_3-\text{C}-\text{CH}_2-\text{C}-\text{CH}_3 \\
\backslash \quad / \quad \backslash \quad // \qquad\qquad\quad \| \qquad\quad \| \\
\text{C=O} \quad \text{O-C} \qquad\qquad\qquad\quad \text{O} \qquad\quad \text{O} \\
| \qquad\quad | \\
\text{CH}_3 \qquad \text{CH}_3 \qquad\qquad\qquad\qquad \text{(acetylacetone)}
\end{array}
\tag{33}
$$

(acetylacetonate)

$$
\text{SnCl}_4 + 2\ \text{H}_2\text{O} \longrightarrow \text{SnO}_2 + 4\ \text{HCl}
\tag{34}
$$

Investigations to make films of other compounds are rare, see for instance [81].

Typical conditions for the deposition of spray coatings of some metals and oxides [75,81] are listed in Table 5.

The substrate temperature is a very important parameter. If it is high, the deposition rate can usually be increased whereas lower temperature values favour the incorporation of reaction by-products into the growing layers. In any case, however, it must clearly be lower than the softening point of the glass. In all applications the influence of the substrate temperature on the film properties is precisely discussed but the influence of the nature of the glass surface on film properties is only rarely mentioned, although it may be important as diffusion source for alkali ions. For instance, in the case of undoped SnO_2 and In_2O_3 films these positive ions may act as unwanted p-type doping agents thus compensating the native donors [82,83]. It is especially critical when using the cheap soda lime silicate glass with 70 – 75 wt. % SiO_2 and a high sodium content of 10 – 15 wt. % Na_2O. At elevated temperatures ($T_s > 450°C$), an increase in the diffusion rate of alkali atoms may occur producing the above-mentioned effects. This can be

TABLE 5

TYPICAL CONDITIONS FOR THE DEPOSITION OF METAL AND OXIDE FILMS ON GLASS, FUSED SILICA AND ALUMINIUM OXIDE BY SPRAY COATING [75, 81] (AA) = ACETYLACETONATE.

METALS

Final film	Starting material	MP/BP (°C)	Solvent	Carrier gas	Reaction temperature (°C)
Pt	$(AA)_2Pt$	–	butanol	$N_2 + H_2$	340 - 380
Pd	$(AA)_2Pd$	–	butanol	$N_2 + H_2$	300 - 350
Ru	$(AA)_3Ru$	–	butanol	$N_2 + H_2$	380 - 400

OXIDES

Final film	Starting material	MP/BP (°C)	Solvent	Carrier gas	Reaction temperature (°C)
Fe_2O_3	$(AA)_3Fe$	181/200	butanol	Air	300 - 500
Cr_2O_3	$(AA)_3Cr$	–	–	Air	520 - 560
Al_2O_3	$(AA)_3Al$	–	butanol	Air	480
Al_2O_3	Al-isopropoxide	248/-	butanol	Air	420 - 650
In_2O_3	$(AA)_3In$	260/280	acetylacetone	Air	480
Y_2O_3	$(AA)_3Y$	190/-	butanol	Air	600
V_2O_3	$(AA)_3V$	–	–	Air	300 - 360
VO_2	$(AA)_3OV$	180/200	butanol	Air	360
SnO_2	$SnCl_4$	–	methanol	$N_2 + Air$	300 - 500
TiO_2	butyl-orthotitanate	–	acetylacetone butanol	$N_2 + Air$	400
ZrO_2	butyl-orthozirconate	–	butanol acetylacetone	Air	450

critical even with borosilicate glasses. Fused silica substrates can be used to overcome this problem, but it is also possible to use special cleaning procedures for the glass as for instance treatment with nitric acid which leaches out the alkali from surface and the regions near the surface. Another approach to avoid this problem is the depletion of glass surfaces of ions by the application of an electric field at high temperature and the cooling of the sample using the applied field [84]. For highly doped films such as SnO_2 with F or SnO_2 with Sb and in the case of mixed films such as ITO (indium tin oxide), alkali contamination is not important because the conductivity is usually orders of magnitude higher than for undoped layers [80].

6.2.1.4.1.2 ATMOSPHERIC-PRESSURE CVD

In true CVD processes, as compared with the spray coating discussed above, the reactants are transported to the place of reaction, in our case the glass surface, in the form of vapours and gases and not in the form of liquid droplets. Although there are also exceptions, the vapour of the reactive compound, an easily volatalized liquid, is generally prepared by injection of the liquid, into water or oil-bath heated evaporators. From there, the vapour is transported to the reaction zone by a carrier gas.

Using this method and a hydrolysis reaction of metal chlorides, an apparatus was built [85] which enables small glass surfaces of up to 10 cm^2 in size to be coated with SnO_2 and In_2O_3 layers. It is a general rule that the unwanted gas phase nucleation (homogeneous reaction) in atmospheric-pressure CVD can be suppressed by using high carrier-gas flow rates, minimum temperatures, and cold wall reactors.

In recent investigations, a CVD process was developed for the continuous deposition of tin-oxide layers under flat glass production conditions [86]. The substrate for all depositions was a float-glass ribbon moving with a speed between 8 and 15 m min^{-1}. Its temperature was about 600°C. The surface was typical of glass which has left the float bath but had not yet entered the annealing furnace and was not treated or cleaned before film deposition. The base reaction is again the hydrolysis of SnCl$_4$. The reactive vapours are transported by nitrogen gas in separate thermostatically controlled pipes to the deposition device formed by five identical slit nozzles. They expand in the first chamber of the nozzle and are then accelerated to the slit of 0.5 mm width. A schematic diagram is shown in Fig. 18.

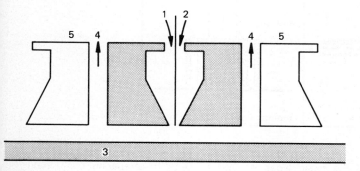

Fig. 18
A cross section through the chemical vapour deposition nozzle: 1, SnCl$_4$–N$_2$ inlet; 2, H$_2$O–N$_2$ inlet; 3, glass surface; 4, reaction gas evacuation; 5, identical nozzle.

The reactive vapours leave the nozzle in a laminar gas flow and mix only by diffusion. The optimum distance between the glass surface and the slit nozzle depends on the total gas flow rate and is for a flow rate of 1 m^3 h^{-1} per 1 m nozzle length in the range of 3 mm. The chemical reaction to form SnO$_2$ takes

126

place mainly on the hot glass surface. The gaseous reaction by-products and the carrier gas are evacuated at flow rates which have no negative influence on the laminar gas flow below the nozzle. On the moving glass ribbon a SnO_2 film is formed with deposition rates beween 0.5 and 1 $\mu m \ s^{-1}$. The investigations showed that the deposition of SnO_2 onto flat glass is less sensitive to variations in the chemical parameters than to the control of gas flow conditions and the design of the deposition plant [86].

Reactor systems for carrying out CVD processes must provide several basic functions that are common to all types of reactors. They must allow controlled transport of the reactant and diluent gases to the reaction zone, provide heat to the substrate to maintain a defined temperature and safely remove the gaseous by-products. These functions should be fulfilled with sufficient control and maximal effectiveness, which requires good engineering design and automation.

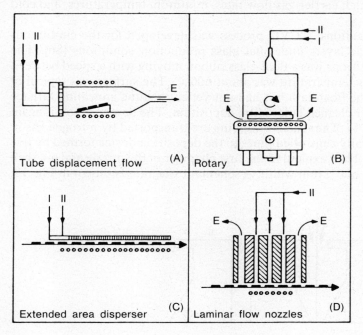

Fig. 19
Operating principles of the basic types of CVD reactors for preparing thin films.
A = horizontal tube displacement flow reactor
B = rotary vertical batch type reactor
C = continuous reactor employing premixed gas flow fed through an extended area slotted disperser plate
D = continuous reactor using separate reaction gas streams that are directed towards the substrates by laminar flow nozzles
I = reaction gas 1
II = reaction gas 2
E = exhaust gases
OOOO = resistance heater

The reactor in which the chemical coating processes actually take place is the essential part of the system and must be designed according to the specific physical and chemical process parameters. To coat glass by atmospheric-pressure CVD, generally low temperature reactors are used which can be classified according to their gas flow characteristics and operation principles into four basic types [69]:

1. horizontal tube displacement flow type;
2. rotary vertical batch type;
3. continuous-deposition type using premixed gas flow; and
4. continuous-deposition type employing separate gas streams.

A schematic diagram is shown in Fig. 19. The CVD plants can be divided further into hot wall reactors which are preferred when the deposition reaction is exothermic in nature and cold-wall reactors mainly employed when the deposition occurs by an endothermic reaction. By these means, the reaction is directed to the part to be coated and undesirable contamination of the walls is prevented.

6.2.1.4.1.2.1 COMPOUND FILMS

Reported in the CVD literature are many low temperature reactions which form oxides in particular but also nitrides and oxynitrides. Particularly suitable and widely used starting materials are metal alkoxides, metal alkyls, metal halides, metal acetylacetonates, metal carbonyls and metal hydrides. The most frequently used reactions are pyrolysis and pyrolytic oxidation of the metal alkoxides at relatively low temperatures, and hydrolysis or oxidation of the metal halides, generally at higher temperatures. Alkoxy compounds are often preferred to other organometallic compounds because they are highly volatile but also sensitive to hydrolysis. Using alkoxy pyrolysis in the presence of oxygen, the following oxides can be deposited at substrate temperatures between 250 and 750°C:

TiO_2	[87,88,89,90,91]
ZrO_2	[92]
HfO_2	[90,92]
SiO_2	[93]
Al_2O_3	[94,95,96,97,98,99]
Ta_2O_5	[100,101,102]
Nb_2O_5	[92]
ZnO	[99,103]

With oxidation and pyrolysis of metal alkyls, the following oxides have been obtained:

TiO_2	[104]

ZnO   [105,106]
PbO   [107]
SnO_2 [108]

Tin oxide films have been deposited onto soda-lime-silica glasses in a cold-wall reactor at substrate temperatures between 350°C and 490°C [108]. The mechanism of formation was carefully studied using four different tin alkyls as starting compounds namely dibutyl tin diacetate (DTD), tetrabutyl tin (TBT), tetraethyl tin (TET) and tetramethyl tin (TMT). The investigations showed for the mechanism a dependence on the molecular structure and reactivity of the initial tin compound to oxygen. The influence of substrate temperature on deposition rate (Arrhenius plots) showed three characteristic regions for which three processes were proposed [108]:

1. reaction between chemisorbed oxygen and chemisorbed tin compound;
2. reaction between chemisorbed oxygen and gaseous tin compound;
3. reaction of an intermediate complex consisting of oxygen and tin compound at the surface.

One of these processes becomes dominant as the substrate temperature changes, depending on the degree of the activation state of the tin compound. As was found [108], DTD and TBT follow process type 3. in the high-temperature region and type 2. in the low-temperature region, while TET and TMT follow process type 2. and type 1. in the high-temperature region an in the low-temperature region, respectively.

The pyrolytic decomposition of metal acetylacetonates at temperatures between 400 and 500°C resulted in the formation of:

Fe_2O_3 [109,110] and
NiO   [109].

Amorphous Fe_2O_3 has also been prepared from Fe-pentacarbonyl, $Fe(CO)_5$, in an argon or nitrogen atmosphere and in the presence of CO_2 with or without H_2O and O_2 at a substrate temperature of only 90 – 160°C [111,112,113,114].

The oxidation of hydrides at low temperatures between 400 and 450°C yielded:

B_2O_3 [115].

Although most oxide films prepared from metal halides by hydrolysis and oxidation require high temperatures, there are some exceptions. Metal-halide hydrolysis at temperatures lower than 300°C has been used to deposit TiO_2 films [102,116,117].

Various mixed oxides can be produced by these techniques. Also, for instance, luminiscent coatings of rare-earth doped YVO_4, Y_2O_3 and Y_2O_2S

phosphors have been prepared by pyrohydrolysis (H_2O) and pyrosulfurlysis (H_2S) of suitable organo-metallic compounds at temperatures of about 500°C [118].

Ta-doped Al_2O_3 films have been deposited by pyrolysis of tantalum pentaethoxide, Ta $(OC_2H_5)_5$ and triethyl aluminium, $Al(C_2H_5)_3$ [100] or aluminium chloride $AlCl_3$ [119].

From the various possibilities of forming mixed oxide films, for example to obtain special refractive indices, only a few are used. Polymeric structures like silicate glass films can also be made at fairly low temperatures.

Phosphosilicate glass films have been obtained with low-temperature oxidation of hydrides:

$$SiH_4 + PH_3 + O_2, \quad T_s = 325 - 450°C \qquad [120].$$

Borosilicate glass films have been obtained in a similar way:

$$SiH_4 + B_2H_6 + O_2, \quad T_s = 325 - 450°C.$$

With N_2 or Ar as the diluent, the system is able to form borosilicate glasses over the whole composition range of $(SiO_2)_{1-x}(B_2O_3)_x$. All the possible reactions in the system have been carefully investigated [105,106,115,120,121–129].

Aluminosilicate glasses have been obtained from tetraethoxysilane and triisobutyl aluminium vapours in an oxygen atmosphere [130,100,131].

$$Si (OC_2H_5)_4 + Al (C_4H_9)_3 + O_2, \quad T_s \simeq 250 - 500°C.$$

Arsenosilicate glass films have been prepared from the following starting compounds:

$$Si (OC_2H_5)_4 + AsCl_3 + N_2 \text{ (or Ar)}, \qquad [132 - 136]$$

$$Si (OC_2H_5)_4 + AsCl_3 + CO_2 + H_2, \qquad [133]$$

$$Si (OC_2H_5)_4 + AsCl_3 + N_2, \qquad [134]$$

$$SiCl_4 + AsCl_3 + CO_2 + H_2, \qquad [133]$$

$$SiH_4 + AsCl_3 + O_2 + Ar, \qquad [137,138]$$

$$SiH_4 + AsH_3 + O_2 + Ar \text{ (or } N_2), \qquad [136,139,140]$$

The last reaction in this sequence is generally preferred because it proceeds at a temperature of 500°C with rates of 50 to 150 nm min^{-1} for concentrations of up to 15 mol % As_2O_3 in the glass film [136].

Extensive infrared spectroscopic measurements have indicated clearly that borosilicate glass films deposited from the hydrides at temperatures of only 325 –

475°C are in fact chemically formed silicates rather than physical mixtures of B_2O_3 and SiO_2. The same also holds for other glass films prepared in a similar way, such as phosphosilicate and arsenosilicate [106,115,128,129,141,142].

Most nitride and oxynitride films are deposited at substrate temperatures much higher than 500°C. The formation reactions take place generally only between 900°C and 2500°C. Under such conditions, coating of glass is impossible. There are some exceptions, however, where it is possible to develop low temperature reactions. Thus, for instance, amorphous germanium nitride Ge_3N_4 has been deposited by ammonolysis of $GeCl_4$ at substrate temperatures between 400 and 600°C [143]. Such films can be used in electronics [144].

Also, the pyrolysis of the dialkylamides has been used to deposit TiN, ZrN and NbN films in the substrate temperature range between 250 and 800°C [145].

6.2.1.4.1.2.2 METAL FILMS

Nearly all metals and many alloys have been deposited by chemical vapour deposition. Exceptions are the alkali and alkaline earth metals. Detailed descriptions of the conditions used to prepare such deposits are given in several reviews and bibliogràphies of CVD work, e. g. [69,146,147]. The most important processes for depositing metals are:

1. thermal decomposition and pyrolysis of organometallic and metal carbonyl compounds;

2. hydrogen reduction of metal halides, oxyhalides, carbonylhalides and other oxygen containing compounds; and

3. chemical transport reactions based on temperature differences in heterogeneous reaction systems.

For low-temperature depositions, decomposition and pyrolysis reactions are usually used. Most coatings are made on metal substrates but it is very likely in most cases that deposition on glass is also possible.

Aluminium films can be prepared by pyrolysis of triisobutyl aluminium with Ar as the carrier gas at relatively low substrate temperatures:

$$Al\,(C_4H_9)_3 + Ar; \quad T_s \geqslant 260°C, \qquad [148]$$

or better diluted with hydrocarbons to make the reaction less explosive:

$$Al\,(C_4H_9)_3 + C_7H_{16} + Ar; \quad T_s \simeq 300°C \qquad [149-151].$$
$$25\,\% 75\,\%$$

Chromium films can be deposited by the pyrolysis of chromium hexacarbonyl:

$$Cr(CO)_6, \quad T_s = 200 - 500°C; \quad\quad [152 - 155] \text{ or}$$

by thermal decomposition of dicumene chromium:

$$Cr(C_6H_5-CH(CH_3)_2)_2, \quad T_s = 300 - 400°C, \quad\quad [156].$$

Cobalt films have been prepared by pyrolysis of cobalt acetylacetonate:

$$Co(C_5H_7O_2)_2 + H_2, \quad T_s = 325 - 340°C \quad\quad [157, 158]$$

The reaction is carried out at atmospheric pressure using a vapourization temperature of 150°C, a condenser temperature of 170°C, a carrier gas flow-rate of 1.2 to 2.8 l min^{-1} and a deposition or substrate temperature of 340°C. These conditions are considered optimum because they produced thin metallic cobalt films having the same magnetic properties as bulk cobalt. A deposition time of 8 – 10 min resulted in films of about 0.6 μm thickness on glass substrates. Cobalt films can also be prepared by pyrolysis of cobalt nitrosyl tricarbonyl:

$$CoNO(CO)_3 + H_2, \quad T_s = 150 - 400°C \quad\quad [159,160].$$

Nickel films are deposited by pyrolysis of Ni-tetracarbonyl. The optimum deposition temperature for practical coating ranges from 180–200°C. The atmosphere contains a mixture of Ar with various other gases:

$$Ni(CO)_4 + Ar + CO_2 + NH_3 + H_2, \quad T_s = 180 - 200°C \quad\quad [161,162].$$

Coating of plastic is also possible because of the low decomposition temperature of the carbonyl nickel of 100°C. The rate of decomposition can be increased by adding catalysts such as $TiCl_4$, $SiCl_4$ or H_2S [163,164].
The pyrolysis of nickel acetylacetonate also yields Ni films:

$$Ni(C_5H_7O_2)_2, \quad T_s = 350 - 450°C \quad\quad [165].$$

The reaction is mainly performed at reduced pressure.

Iron films are obtained by thermal decomposition of the pentacarbonyl at substrate temperatures lower than 200°C:

$$Fe(CO)_5, \quad T_s = 60 - 140°C \quad\quad [166,167].$$

The process is optimal at low pressure.

Vanadium films can be made in a similar way to iron films by pyrolysis of the hexacarbonyl:

$$V (CO)_6, \qquad T_s = 100°C; \qquad [168].$$

Lead films can be deposited by pyrolysis of lead tetraethyl or tetramethyl:

$$Pb (C_2H_5)_4 \quad or \quad Pb (CH_3)_4, \qquad T_s = 300 - 420°C \qquad [169].$$

The films obtained are not dense, however, and contain various amounts of codeposited carbon.

Copper films have been obtained by thermal decomposition of copper acetylacetonate at reduced pressure:

$$Cu (C_5H_7O_2)_2, \qquad T_s = 260 - 450°C \qquad [165,170].$$

Displacement coatings can also be obtained on alkali-oxide-containing glass surfaces, especially those with sodium oxide. Treatment of the glass surface with cuprous chloride vapour at about 500°C, or a temperature near the softening point of the glass, results in replacement of the alkali metal atoms with copper. Subsequent treatment of the glass with hydrogen at about 400°C reduces the copper compound in the glass, producing an attractive stain which may range from a light pink to a deep ruby red in intensity [171,172]. Borosolicate glass (e. g. Pyrex) does not stain as readily as soft glasses, and no colour is produced with borosilicate glass types containing no alkali oxide.

Silicon and *Germanium* films can be produced by pyrolysis of their hydrides at 620°C and 500°C respectively [173]:

$$SiH_4, \qquad T_s = 620°C, \qquad GeH_4, T_s = 500°C.$$

Germanium films can also be fabricated from GeI_2 with CO_2 as carrier gas using the following disproportionation reactions:

$$2 \, GeI_{2 \, (g)} \xrightarrow[T_s = 450°C]{} Ge_{(s)} + GeI_{4 \, (g)} \qquad [174].$$

Arsenic, Antimony and *Bismuth* films can be deposited by hydrogen reduction of their volatilized chlorides [175]. The hydrogen reduction process is best carried out with the trichlorides. These materials are commercially available in grades of sufficient purity for most coating work. The chloride vapourization temperatures and the substrate temperatures for deposition are as follows:

$AsCl_3$ . . .	0 – 30°C . . .	$T_s = 300 - 500°C$
$SbCl_3$ . . .	80 – 110°C . . .	$T_s = 500 - 600°C$
$BiCl_3$ . . .	240 – 250°C . . .	$T_s = 250°C$

A mixture of hydrogen and hydrogen chloride may have to be used in depositing Bi to avoid reduction of $BiCl_3$ in the liquid state. Such metal films can also be made by thermal decomposition of the hydrides with hydrogen as carrier gas:

$$SbH_3 + H_2 \quad \ldots \quad T_s = 150°C$$

$$\left.\begin{array}{l} AsH_3 + H_2 \\ BiH_3 + H_2 \end{array}\right\} \quad \ldots \quad T_s = 230 - 300°C$$

A considerable excess of H_2 as carrier gas is required in depositing antimony to prevent the tendency of SbH_3 to explode when heated to about 200°C. The decomposition of SbH_3 has been used to prepare small amounts of high-purity antimony with only less than 1 ppm metallic impurities [176].

Tin films can be deposited by pyrolysis of tin alkyls or alkyl hydrides such as:

$$Sn\, H_2\, (CH_3)_2, \quad T_s = 540°C \qquad [177].$$

The thermal decomposition of tetramethyl tin in a static system at 440 to 493°C resulted in the deposition of films which were carbon contaminated with longer deposition times [178,179].

Molybdenum films have been obtained by carbonyl pyrolysis:

$$Mo\,(CO)_6 + H_2, \quad T_s = 400 - 600°C \qquad [153,180].$$

Extensive investigation of this process showed that the substrate temperature and the pressure of the decomposition products have a pronounced influence on the composition of the resulting deposit. The pyrolysis can be performed only in a carbonyl atmosphere, but the mixture of H_2 was found to be virtually necessary in any practical deposition process, both as a carrier gas and as a carbon-removing agent [153,181]. Films with low carbon content of about 0.1 at % have been obtained by a minimum temperature of 400 – 450°C and a CO partial pressure of 0.1 mbar. The amount of carbon can be further reduced by using higher deposition temperatures of up to about 600°C.

Tungsten is very similar to molybdenum in its chemical reactions and can be vapour-deposited by the same processes. The typical low temperature process is pyrolysis of the hexacarbonyl $W(CO)_6$ [153, 181-184]:

$$W(CO)_6 + H_2, \quad T_s = 350 - 600°C$$

Tungsten deposited from carbonyl is less prone to form deposits containing carbon or to form powdery films at higher temperatures and pressures than is molybdenum. A wider range of plating temperatures and pressures and higher deposition rates can thus be used. A deposition temperature between 350 – 600°C, a hydrogen to carbonyl ratio of 100:1 and a total pressure lower than 3 mbar have been recommended [181,182].

Thermochemical investigations have shown that temperature levels of about 900°C are required for the deposition of pure metal films at atmospheric pressure unless the carbonyl is fed as an extremely dilute gas mixture. At these high temperatures, however, gas-phase nucleation is a serious problem in obtaining dense and adherent films [184].

Another method of depositing tungsten films is the reduction of the hexachloride or fluoride with hydrogen:

$$WF_6 + H_2, \qquad T_s = 400°C, \qquad [185].$$

Platinum and also some platinum group metals can be deposited by decomposition or, better, reduction with hydrogen of their carbonyl chlorides [186,187]:

$$Pt(CO)_2Cl_2 + CO + H_2, \qquad T_s \geqslant 125°C$$

In the case of very unstable carbonyl halides the pyrolysis of acetylacetonates was used to deposit films [188, 165]:

$$Pd(C_5H_7O_2)_2 + CO_2, \qquad T_s \simeq 350-450°C.$$

Osmium and *Rhenium* films have been obtained by reduction with hydrogen of the tetra- and trioxide respectively:

$$OsO_4 + H_2, \qquad T_s = 500°C \qquad [189] \quad \text{and}$$
$$ReO_3 + H_2, \qquad T_s = 350°C \qquad [190].$$

Gold and *Silver* films have been prepared by pyrolysis and metal halide reduction [100,191].

Finally, an example for the low-temperature deposition of alloy films is the production of Ni/Fe films of the permalloy type by pyrolysis of Ni and Fe carbonyl mixtures on glass at a substrate temperature of 300°C [192,193].

6.2.1.4.1.3 LOW-PRESSURE CVD

Chemical vapour deposition processes can be carried out over a pressure range from super-atmospheric down to pressures characteristic of physical vapour deposition. In many cases the substrates are heated to sufficiently high temperatures to cause the vapourized plating compounds to react leaving behind the solid reaction products as the desired coating, and the by-products as gases which have to be removed. With less stable starting compounds, these may tend to be reduced or decomposed at a larger distance from the heated surface producing non-adherent powdery deposits possibly contaminated with non-volatile partial decomposition products. This difficulty can often effectively be eliminated by reducing the pressure of the plating atmosphere, and thereby increasing the mean free path of the gas molecules as well as preventing the condensation of non-volatile partial-reduction or partial-decomposition products. In low-pressure

CVD, generally no carrier gas is required. Deposition rates are dependent on material and on the process used. In some cases 0.5 to 10 μm.min^{-1} are obtained, but the rate is often much lower.

A uniform distribution of the film thickness not only depends on uniformity of substrate temperature, but also on uniformity of the rate of supply of fresh starting compound to all surfaces of the parts to be coated. Non-turbulent gas flow at nearly atmospheric pressure forms a deposit in pronounced streamlines. Although this can be prevented by substrate rotation, deposition at reduced pressure is generally more useful in improving the uniformity of film thickness. The following facts may explain the observed advantages of low-pressure CVD. The diffusion of gas atoms is inversely proportional to the pressure. A reduction of the pressure from atmospheric to 1 mbar increases the diffusivity by a factor of about 10^3. The distance across which the reactants must diffuse, the boundary layer [194], increases at less than the square root of the pressure. Effectively an increase of gas-phase transfer of reactants to the substrate and the removal of by-products from it by more than one order of magnitude is achieved. Thus the reaction rate is determined by the surface reaction and less attention must be paid to mass-transfer problems which are limiting factors in atmospheric-pressure CVD [69,195]. At very low pressures of the order of 10^{-3} mbar or less, film deposition occurs only on line-of-sight surface. This is not necessarily disadvantageous, and may even be an advantage if only one side of a plate is to be coated.

Low-pressure CVD enables a uniform deposition of a large number of substrates in one run. The density of the arrangement of the substrates on their holder can be high and the geometrical shape of the substrates may even be complex [196,197].

6.2.1.4.2 PLASMA-ACTIVATED AND PHOTON-ACTIVATED CVD

6.2.1.4.2.1 PLASMA-ACTIVATED CVD

The main reason for the application of plasma-activated chemical vapour deposition lies in the fact that it enables the reaction to proceed at temperatures lower than the corresponding thermal reactions. This property is sometimes an advantage for processes that would otherwise suffer from high temperatures. Typical substrate temperatures in plasma-enhanced CVD depositions are at about 300°C and may sometimes be even lower. As well as a lower temperature and compared with other techniques, a more energy-efficient functioning, low pressures in the millibar region give the additional benefit of large diffusion constants.

Two kinds of plasmas have been used for chemical purposes: isothermal and non-isothermal plasmas. The first type is not interesting for low-temperature application because all charged and neutral particles are in thermal equilibrium. The advantage of a plasma, however, lies in generating chemically reactive species at low temperatures which is due to the non-equilibrium nature of non-iso-thermal plasmas. In such a glow discharge plasma, typically sustained at pressures between 0.1 and several millibars, the free electrons exhibit temperatures of more than 10^4 degrees Kelvin while the temperature of translational and rotational modes of atoms, ions, radicals and molelcules is only few hundreds of Kelvins, typically between 298 and 573 Kelvin. The interaction with these high-energy electrons leads to the formation of reactive species in the gas phase and on the surface that would normally occur only at high tempertures. The principles of generating low-temperature plasmas are found for example in [198] and the fundamentals of plasma chemistry are discussed in [199] for example. A short review on plasma CVD has recently been published [200]. The possible lower substrate temperature makes this relatively new process attractive for coating glass. The films obtained in the plasma reactions are usually amorphous in their structure with very little short-range ordering. Unfortunately at this time there is little systematic work available showing the influence of the various parameters on film formation and film properties. This is a consequence of the very complicated conditions in such processes. A plasma-assisted CVD process has many parameters and there are also different deposition reactor designs in use. It is therefore almost impossible to compare the individual results.

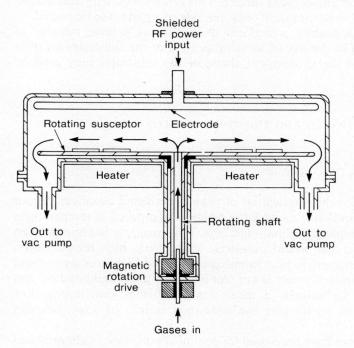

Fig. 20
Applied Materials plasma CVD deposition system cross section [203].

An exception is the systematic investigation of the formation of *silicon nitride* films [194, 201-203]. All the plasma-assisted depositions were carried out in the reactor shown in Fig. 20. It was found that the major variables affecting the film deposition rate and uniformity are rf power density and distribution, gas-phase composition and distribution and total pressure in the reaction zone. Most important is the distribution of power density, because the lifetimes of the reactive states are generally very short, so that the activated and excited species must be continuously generated and regenerated across the deposition zone. For deposition, a power input between 300 and 600 W was used. The reactive gas mixtures were composed of SiH_4 (100 – 150), NH_3 (150 – 300) and N_2 (300 – 1000). Typical physical and chemical properties of films grown at about 300°C substrate temperature with a rate of 20 – 40 nm min^{-1} are shown in Table 6. For

TABLE 6

PHYSICAL AND CHEMICAL PROPERTIES OF SILICON NITRIDE FILMS FROM $SiH_4 + NH_3 + N_2$ [195]

Property	Si_3N_4 Atmospheric-Pressure CVD 900°C	$Si_xN_yH_z$ Low-Pressure Plasma-Enhanced CVD 300°C
density	2.8-3.1 g cm^{-3}	2.5-2.8 g cm^{-3}
refractive index	2.0-2.1	2.0-2.1
dielectric constant	6-7	6-9
dielectric strength	$1 \cdot 10^7$ V cm^{-1}	$6 \cdot 10^6$ V cm^{-1}
bulk resistivity	10^{15}-10^{17} Ω cm	10^{15} Ω cm
surface resistivity	$> 10^{13}$ Ω sq.$^{-1}$	$1 \cdot 10^{13}$ Ω sq.$^{-1}$
stress at 23°C on Si	1.2-1.8$\cdot 10^{10}$dyn cm^{-2} tensile	1-8$\cdot 10^9$dyn cm^{-2} compressive
thermal expansion	$4 \cdot 10^{-6}$ °C^{-1}	> 4–$7 \cdot 10^{-6}$ °C^{-1}
colour transmitted	none	yellow
step coverage	fair	conformal
H_2O permeability	zero	low-none
thermal stability	excellent	variable > 400°C
composition Si/N ratio	Si_3N_4 0.75	$Si_xN_yH_z$ 0.8-1.0
ir absorption Si-N maximum Si-H minor	~ 870 cm^{-1} –	~ 830 cm^{-1} 2180 cm^{-1}

comparison values for Si_3N_4 films prepared by atmospheric-pressure CVD at 900°C from SiH_4, NH_3, N_2 gas mixtures are also listed. Microchemical analysis data have shown that the chemical composition of the plasma silicon nitride films is slightly Si-rich at a Si/N-ratio between 0.8 and 1.0 compared to 0.75 for stoichiometric deposits. As indicated by infrared spectrometry, there are also larger amounts of hydrogen and traces of oxyen in the films. Therefore the true film composition may be represented by $Si_xN_yH_z$ [195]. Films deposited at 300°C are thermally stable up to about 400°C. As a consequence of compressive film stress, blistering is observed during higher heating resulting in the formation of circular holes. The films contain practically no pinholes and sodium ions are very effectively masked by them. A potential nonelectronic applicaton would be a coating over photolithographic masks to reduce sticking and scratch damage [195].

Amorphous *silicon* and *germanium* films can be prepared in a plasma of their hydrides [204]. Large amounts of hydrogen are found in the films depending on substrate temperature during deposition, for example $T_s = 250°C$ and 0.1 mbar pressure yields 0.7×10^{22} atoms cm^{-3}, while $T_s = 25°C$ and a pressure of 1.0 mbar can even yield 1.7×10^{22} atoms cm^{-3} hydrogen in the films. Mass spectrometric investigation indicated that the plasma contains SiH_2^+ and other fragmentation products so that a film containing large amounts of hydrogen may be somehow similar to polysilane [204]. The hydrogen concentration in the amorphous films has been analysed by infrared spectroscopy, mass spectroscopy and other techniques [205]. Such hydrogenated, amorphous silicon films are used for solar cells. Amorphous silicon films have been doped by the addition of dopants as PH_3 and B_2H_6 to SiH_4 [206,207]. A p-n junction was fabricated with amorphous Si with a photovoltaic response qualitatively similar to that of crystalline silicon p-n junction.

Various oxide films have also been prepared:

Aluminium oxide in the form of amorphous films have been deposited at a substrate temperature of 480°C by vapourizing $AlCl_3$ into an oxygen plasma [208].

Silica films have been obtained from SiH_4/N_2O mixtures at substrate temperatures between 250 and 350°C with a rf power of 250 W and a pressure of 0.35 mbar [209]. Small amounts of nitrogen may be incorporated into the SiO_2 film depending on the N_2O/SiH_4 flow-rate ratio. Optimum properties of the films were achieved with a N_2O/SiH_4 ratio of at least 15 to 1. Density $\rho = 2.18 - 2.44$ and refractive index $n = 1.40 - 1.50$ of the deposits are very similar to bulk SiO_2 glass. The films show compressive stresses between 1 and 3×10^9 dyn cm^{-2} which inhibit cracking.

Optically transparent *crystalline* and *glassy oxide films* have been obtained from some metal alkyls and alkoxides in a microwave discharge plasma at a substrate temperature of 200°C [210]. Chemical compositions were controlled by infrared absorption and refractive index measurements and the data obtained were compared with the stoichiometric compounds. The optical data of such plasma-deposited oxides are listed in Table 7 [210].

$$Si(OC_2H_5)_4 + \frac{microwave}{plasma}, \quad T_s = 200°C \qquad [210]$$

Titanium Oxide, TiO_2, and *Silicon Oxide*, SiO_2, films have recently been produced from the volatile chlorides in the presence of O_2 in the plasma [211]. These processes could be performed even at room temperature. The resulting oxide films were deposited on glass with rates of 50 nm min^{-1}. The film refractive indices obtained were $n_{550} = 2.4$ for TiO_2 and $n_{550} = 1.45$ for SiO_2.

With the same process *SnO_2 films* were also fabricated [211]. *Indium Oxide* films were made using the organometallic trimethyl compound to give a conducting oxide film and it was shown that this method can be combined with a sputtering source (PVD process) to give mixed oxide films of relatively good conductance, especially in the knowledge that the process took place at room temperature [211].

$$In(CH_3)_3 + plasma, \quad T_s = 30°C \qquad [211]$$

Further inorganic films made by plama-assisted CVD are listed in [212] and results of recent developments in production technology are found in [213].

TABLE 7

COMPARISON OF THE OPTICAL PROPERTIES OF PLASMA-DEPOSITED OXIDE FILMS WITH THE CORRESPONDING GLASSES OR MINERALS [210]

Film	Production Techniques	Principal infrared frequency (cm^{-1})	index of refraction
GeO_2	plasma-formed film	850	1.582 ± 0.002
	fusion-formed glass	850	1.534 - 1.607
B_2O_3	plasma-formed film	1350	1.470 ± 0.002
	fusion-formed glass	1350	1.464 (3)
Ti_xO_y	plasma-formed film	950 - 700	1.7
TiO_2	anatase	1200 - 500	1.7
Sn_xO_y	plasma-formed film	1425	1.536 ± 0.002
SnO_2	cassiterite	850 - 500	1.7
SiO_2	plasma-formed film	1045; 800	1.458 ± 0.002
	fusion-formed glass	1080; 800	1.458 ± 0.002

6.2.1.4.2.2 PHOTON-ACTIVATED CVD

Photon-assisted chemical vapour deposition induces chemical processes by heating substrates or by single and multiphoton excitation of vibrational modes in the gas molecules of the starting compound. With an optical heat source the surface temperature of a substrate, T_s, and therefore the deposition rate is a function of the absorbed light intensity and of the irradiation time. Several different optical cases can therefore be defined, namely the deposition of a reflecting film on an absorbing substrate, and an absorbing deposit on a reflecting substrate. In the first case, during metal film condensation on the substrate, the absorbed intensity decreases and the resulting lower surface temperature decreases the deposition rate. Changes in absorptivity by a factor 5 to 10 are common, sometimes causing self-limiting of the film thickness. In the opposite case of an absorbing film on a reflecting substrate, the deposition rate is enhanced as the film is deposited.

Intensive conventional ultraviolet light sources as well as various types of lasers are used in photon-assisted CVD [214-219].

The photons of uv light beams with wavelength at about 250 nm have just the right energy to break up organometallic molecules such as metal alkyls, providing free metal atoms for condensation of metallic films. The longer infrared CO_2-laser wavelengths between 9 to 11 μm are particularly efficient for substrate heating. uv irradiation of insulating surfaces in contact with gas mixtures of Cd $(CH_3)_2$ or Al $(CH_3)_3$ with He having pressures of a few millibars results in metal film deposition caused by photodissociation essentially at room temperature [215]. It was also possible with a focused beam of the frequency-doubled output, $\lambda = 257.2$ nm, of a 415.5 nm argon ion laser to write metal lines one micrometer in width with a deposition rate of 100 nm s^{-1} [215]. Carefully performed investigations showed that not all the metal comes from the photolysis of molecules in the gas phase. Some organometallic molecules are initially adsorbed on the substrate surface and a photolytic break-up of these adsorbed molecules creates preferred nucleation sites for further metal atoms freed in the surrounding gas. Thus the deposition pattern appears to be sharpened by this prenucleation effect [215].

Spots and lines have been deposited of Ni, Fe, W, Al, Sn, TiO_2 and TiC on quartz substrates. For Ni films from Ni $(CO)_4$ rates of 1 to 17 μm s^{-1} for CO_2-laser powers of   0.5 to 5 W were measured [217].

SiO_2 films can be prepared from SiH_4 and N_2O with traces of mercury vapour in the gas mixtures as photosensitizer for the uv light radiation $\lambda = 253.7$ nm. The pressure is 0.5 mbar, the rate is about 17 nm min^{-1} and the substrate temperature ranges between 75° C and 175° C. Very little mercury is incorporated into the films. The index of refraction varies strongly with the flow rate of N_2O and reaches an asymptotic value of n = 1.48 at flow rates higher than 60 ccm/s. This is still a little higher than for a stoichiometric film. The reason is assumed to be a slight oxygen deficit in the films [216].

In further experiments with glass substrates, the low-temperature formation conditions for Si– (from SiH_4) [218] and for SnO_2–films (from $(CH_3)_2SnCl_2$ and O_2) [219] were investigated.

6.2.1.5 PHYSICAL VAPOUR DEPOSITION

In physical vapour deposition, PVD, coatings are produced on solid surfaces by condensation of elements and compounds from the vapour phase. The principles are based generally on purely physical effects, but PVD may also be associated occasionally by chemical reactions. Some of these chemical reactions are used intentionally in a special physicochemical film deposition technology, reactive deposition. Reduced to its essence, physical vapour deposition involves three steps:

1. Generation of the vapours either by simple evaporation and sublimation or by cathodic sputtering.
2. Transport of the vapourized material through the reduced atmosphere from the source to a substrate. During their flight, collisions with residual gas molecules can occur depending on vacuum conditions and source to substrate distance. The volatilized coating material species can be activated or ionized by various means and the ions can be accelerated by electric fields.
3. Condensation occurs on the substrate and finally a deposit is formed by heterogeneous nucleation and film growth possibly during higher energy particle bombardment or under the action of impinging reactive or non-reactive gas species or both together.

Let us distinguish between different basic types of film deposition techniques. These are:

> coating by evaporation
> sputtering, and
> ion plating.

All these techniques and their variants operate appropriate only under vacuum. Technical vacua are sufficient for the feasibility of such deposition processes. However, to eliminate or to control precisely the influence of the residual gas on film composition and on properties, clean vacuum conditions and highly defined residual atmospheres are required. Whereas in a simple evaporation process generally the highest vacuum is of advantage, in most sputtering processes and some related techniques a working gas is required generally in the pressure range between about 10^{-2} and 10^{-3} mbar to perform the coating process. Without regard to other differences, the various PVD processes are characterized by the very different energies of their volatilized film material atoms. These energies range from about 0.1 eV in evaporation over 1 to 10 eV in sputtering, to possibly much more than 10 eV and sometimes even up to 100 eV in the case of ion plating. Among other reasons, these specific differences in vapour species energies are responsible for characteristic differences in film properties, as will be shown later.

With PVD technologies, depositions can be made over a wide substrate temperature range between heated glass of several hundred degrees down to liquid nitrogen cooled or even colder samples. Thus there is no problem in coating glass and plastics with PVD technologies when the material and substrate specific conditions are properly chosen.

6.2.1.5.1 VACUUM TECHNOLOGY

The term vacuum means theoretically a space entirely devoid of matter. Natures closest approach to a perfect vacuum is found in free space but even there one finds a few hydrogen atoms per cubic centimeter. In technical usage, however, the term vacuum applies to any space at less than atmospheric pressure. As can be seen in Table 8, there are various degrees of vacuum: rough, medium, high and ultrahigh. These regions are characterized by the mean free path of the residual gas molecules and the type of gas streaming. The units for expressing the amount of pressure remaining in a vacuum system are based on the force of atmospheric air under standard conditions. This force amounts to 1.03 kg cm^{-2}. Gas pressure is stated in terms of the height of a column of mercury supported by that pressure in a barometer. Atmospheric pressure of 1.03 kg cm^{-2} will support a column of mercury 760 mm. One torr is the equivalent of one millimeter or of 1/760 of atmospheric pressure. In modern vacuum technology the common unit is the bar. 1 mbar corresponds to 0.75 torr and this is equal to 10^2 pascal.

The lowest pressures attainable are around 10^{-13} mbar. The residual gas in a clean ultrahigh vacuum system consists practically only of hydrogen. It appears that hydrogen is the most universal contaminant, permeating practically all materials. There is no way known to condense or trap it completely.

Pressure, type of gas flow and mean free path of gas molecules are phenomena which can be described with the aid of the well known kinetic theory of gases. The theory is based on the following idealized assumptions.:

Gases consist of particles (atoms and molecules) which are perfectly elastic spheres.

The particles are in a complete and continous state of agitation.

The system is in equilibrium.

External forces such as gravity and magnetism are ignored.

TABLE 8

DATA CHARACTERIZING THE VARIOUS TYPES OF VACUA

	Low vacuum	Medium vacuum	High vacuum	Ultrahigh vacuum
pressure (mbar)	$1013 \ldots 1$	$1 \ldots 10^{-3}$	$10^{-3} \ldots 10^{-7}$	$< 10^{-7}$
mean free path (cm)	$< 10^{-2}$	$10^{-2} \ldots 10$	$10 \ldots 10^5$	$> 10^5$
flow type	viscous flow	transition viscous flow to molecular flow	molecular flow	molecular flow
remarks	normal pressure $p = 1013.25$ mbar according to DIN 1343	strong variation of the thermal conductivity	Beginning with $p < 10^{-3}$ mbar, the surface absorbed gas atoms or molecules surpass the number of molecules in the gas phase	

A detailed description of the theory is beyond the scope of this monograph and can be found for example in [220-225]. Only some results important in vacuum technique are reported here.

The gaseous state (ideal gas) can be described by the universal gas equation (35):

$$p \ V = \frac{m}{M} \ RT \tag{35}$$

p = pressure, V = volume, m = mass, M = molecular weight, R = molar gas constant, T = temperature in Kelvin.
The molar gas constant can be expressed by eqn. (36):

$$R = N_A \ k \tag{36}$$

With Avogadros number $N_A = 6 \times 10^{23}$ mol^{-1} and Boltzmann's constant k = 13.8 mbar l K^{-1}, the gas constant is R = 83.14 mbar l mol^{-1} K^{-1}.

From eqn. (35), it follows that a given mass of gas at constant temperature can be described by the product p V = constant. Simply expressed, it states that by halving the volume, the pressure is doubled, or when the pressure is halved the volume is doubled. In vacuum systems generally, gas mixtures are found so that the total pressure of the system is given by the sum of the partial pressure:

$$p = p_1 + p_2 + p_3 + \ldots + p_i \tag{37}$$

The average velocity of a gas molecule is determined by the molecular weight and the absolute temperature of the gas. Air molecules, like many other molecules at room temperture, travel with velocities of about 500 m s^{-1} but there is a distribution of molecular velocities. This distribution of velocities is explained by assuming that the particles do not travel unimpeded but experience many collisions. The constant occurrence of such collisions produces the wide distribution of velocities. The quantitative treatment was carried out by Maxwell in 1859, and somewhat later by Boltzmann. The phenomenon of collisions leads to the concept of a free path, that is the distance traversed by a molecule between two successive collisions with other molecules of that gas. For a large number of molecules, this concept must be modified to a mean free path which is the average distance travelled by all molecules between collisions. For molecules of air at 25° C, the mean free path λ at 1 mbar is 0.00625 cm. It is convenient therefore to use the following relation as a scaling function:

$$\lambda = \frac{6.25 \times 10^{-3}}{P \ mbar} \ cm \tag{38}$$

An important consequence of this theory is the survival concept. If molecules are moved between two fixed points by shooting them through a gas of the same composition, there is a probability that they will get through. In the case of a gas pressure which corresponds to 0.01 λ, for that distance 99 out of 100 molecules will traverse without collision. If, however, the pressure is increased so that that distance is λ, then only 37 out of 100 molecules will travel unhindered.

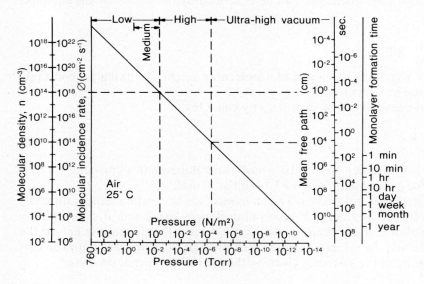

Fig. 21
Relationship of several concepts defining the degree of vacuum [225].

A relationship of several concepts defining the degree of vacuum is shown in Fig. 21 [225]. The rate at which a gas flows in a vacuum system at a given temperature depends on the pressure of the gas and on the size and geometry of the system. It is convenient to characterize the type of flow by a dimensionless parameter called the Knudsen number K*. This parameter is defined as the ratio of the mean free path of the gas particles to a characteristic dimension of the constraint through which the gas is flowing, for instance the diameter of an orifice or tube. During the beginning of evacuation in a vacuum system, the pressure may be high enough to keep the Knudsen number below 0.01. In this case, collisions between molecules occur more frequently than collisions between molecules and the surrounding walls. Consequently, intermolecular collisions become important in determining the transfer of momentum through the gas, and the flow depends on the coefficient of viscosity. Such flow is called viscous flow or, if no turbulence is present, streamline flow. In the later stages of evacuation, when the Knudsen number becomes greater than one and molecular flow exists, the interaction of gas molecules with the walls is, in general, very complex. Concerning the occurrence of specular and/or diffuse reflections, the surfaces of the walls of a vacuum system which are chiefly atomically rough favour the diffuse type. Depending on type of gas and on type of wall material, the gas molecules may also condense and re-evaporate. For large Knudsen numbers, almost all collisions by the molecules are with the walls and the distance between collisions is the mean characteristic dimension of the container, rather than the mean free path characteristic of the pressure. Nevertheless, the molecules maintain a Maxwellian velocity distribution.

6.2.1.5.1.1 VACUUM PUMPS

Six different types of pump are used to evacuate a vacuum system. The first and most common is the mechanical displacement pump [226,227]. The second is the vapour stream pump [228]. The third is the modern turbomolecular pump [229]. The fourth type is the chemical getter pump. The fifth is the ion pump; and the sixth is the cryogenic pump [230]. In technical coating systems, the first three types of pump are preferred. The development of these pumps is narrowly associated with the names: W. Gaede, I. Langmuir, C. Burch, K. Hickman and W. Becker [221,225]. Recently, cryogenic pumps have also been used to evacuate technical coating plants.

6.2.1.5.1.1.1 MECHANICAL DISPLACEMENT PUMPS

Vacua were first produced by mechanical pumps that separated a definite volume of the gas from the recipient to be exhausted, compressed it and released it to the atmosphere. Of the rotary oil-sealed pumps, the sliding vane pump is mainly used. It has a cylindrical rotor located off-center in a cylindrical housing, the stator. The rotor carries two vanes 180° apart, spring loaded so as to ride on the stationary housing. The gas inlet and exhaust valves are separated practically by a contact area where the stator is machined to the same radius as the rotor. Oil serves as a sealant and lubricant between closly fitted moving parts. The speed of rotation is between 350–1400 rpm. The action of the rotary vane pumps is shown in Fig. 22. Rotary vane pumps have an extremely high compression ratio which allows them to pump drawn-in gases against atmospheric pressure. This special feature makes

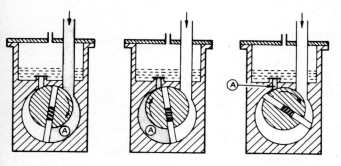

Fig. 22 a
Schematic representation of the operating principle of a rotary vane pump.

Fig. 22 b
BALZERS - PFEIFFER rotary vane pump UNO 1,5 A.

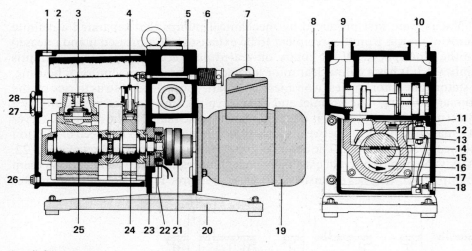

1 = oil filling screw	9 = vacuum connection	19 = motor
2 = cap	10 = exhaust connection	20 = base plate
3 = relief pressure valve	11 = pump valve (like item 4)	21 = coupling
4 = pump valve	12 = intake channel	22 = dynamo
5 = support	13 = magnetic valve	23 = radial shaft seal
6 = gas ballast valve	14 = vane	24 = pumping stage II
7 = ON-OFF switch with motor protection	15 = rotor	25 = pumping stage I
	16 = working chamber	26 = oil drain plug
8 = high vacuum safety valve	17 = pump cylinder	27 = oil level sight glass
	18 = control connection	28 = oil level

Fig. 22 c
Sectional drawing of a BALZERS - PFEIFFER rotary vane pump DUO 030 A.

these basic pumps suitable for a wide range of operations as vacuum pumps as well as backing pumps in vacuum systems. Single and compound forms are fabricated. Parallel connection of two equal pump mechanisms will provide twice the displacement but the same ultimate pressure. In contrast, series connection provides the same displacement but greater pumping speed at low pressure. The practically preferred compound form consists of two pump stages connected in series. The lowest pressure obtained by a single stage is about 10^{-2} mbar. However, a twin-stage (series) pump may reach an ultimate pressure of about 2×10^{-3} mbar. The pumping speed curves are shown in Fig. 23. The gas ballast was introduced

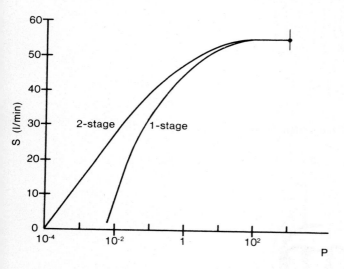

Fig. 23
Pressure versus pumping speed diagram of UNO and DUO rotary vane pumps.

by Gaede 1935 for pumps handling condensable or soluble vapours. A gas ballast valve admits a controlled and timed amount of air into the compression stage of the pump. By this means, the formed compressed gas-vapour mixture reaches the ejection pressure before condensation of the vapour takes place. The principle of the gas-ballast pump and the action of its one-way gas-ballast valve is shown in Fig. 24. The application of gas-ballast decreases the ultimate vacuum of the pumps. In practice, however, this disadvantage is unimportant because the gas-ballast valve is usually open only during the early stages of pumping.

Another type of mechanical pump is the Root's blower. This type of pump contains two counter rotating lobes, each with a figure of eight cross section. The lobes do not touch each other or the housing because of a clearance of about 0.25 mm between the lobes and the housing. The Root's blower is driven by a single shaft and the second rotor is synchronized and driven through a set of timing gears.

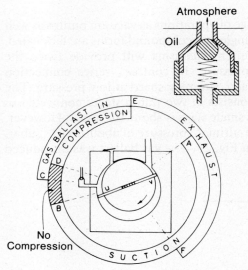

Fig. 24
Schematic representation of the action of the gas ballast.

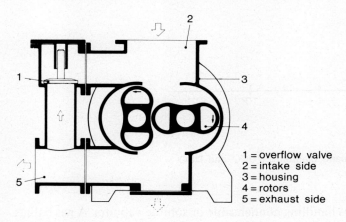

1 = overflow valve
2 = intake side
3 = housing
4 = rotors
5 = exhaust side

Fig. 25 a
Schematic representation of the operating principle of a Roots
blower.

Fig. 25 shows the action of a Root's pump. Gas inlet and outlet are separated
by only a narrow gap which enables a back flow of gas from the exhaust region to
the inlet region. Although the efficiency of compression (maximum compression
$K_m \sim 50 - 70$) is much lower than with oil-sealed pumps, the absence of rubbing
contacts allows higher speed of rotation between 1000 and 3000 rpm leading to
much higher pumping speeds. Common practice is to back the Root's blower with
an oil-sealed mechanical pump. The blower is turned on at atmospheric pressure
and the highest pumping speed is in the 10^{-2} to 10^{-1} mbar region.

6.2.1.5.1.1.2 DIFFUSION PUMPS

When the pressure in the receiving volume decreases to values where the mean free path is greater than the dimensions of this volume, the residual gas molecules experience more collisions with the walls than with each other. In this case it becomes difficult to pump in the conventional sense of the word. It becomes necessary instead to wait until the gas molecules travel into the inlet opening of the pump and then give them a preferred direction of motion by momentum transfer. This is actually done in the diffusion pump. The idea of evacuating a volume by momentum transfer from streaming to diffusing molecules was first described by Gaede [231] but development to improve their performance is still continuing. The basic elements of a diffusion pump are shown schematically in Fig. 26.

Modern diffusion pumps use special silicon oils or other stable compounds as the work fluid instead of a mineral oil. A heater causes the oil to boil and to vapourize in a flue. The oil vapour emerges from nozzles such that the vapour issue is directed away from the high vacuum side. During the exhaust from the nozzles, the vapour expands. This expansion changes the normal molecular velocity

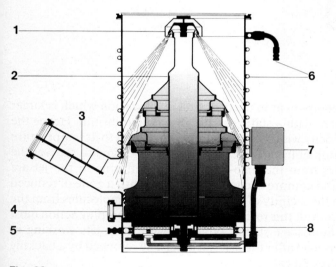

Fig. 26 a

Schematic representation of the operating principle of a diffusion pump.
1 = cold cap
2 = jet system with fractionating high vacuum stage
3 = fore vacuum baffle
4 = pumping fluid sight glass
5 = compressed air rapid cooling
6 = water cooling
7 = connection for thermal protection switch
8 = heater

Fig. 26 b
BALZERS oil diffusion pumps.

distribution by creating a component in the direction of expansion which is larger than the velocities associated with a static gas at thermal equilibrium. Hence the vapour jet moves with a velocity which is supersonic with respect to its temperature toward the walls of the pump housing. The walls are water-cooled to recover the oil. Residual gas molecules from the high-vacuum side diffusing to the jet are knocked down the pump and compressed to the exit side. Thus a zone of reduced gas pressure is generated in the vicinity of the jets and more gas molecules from the high-vacuum side diffuse towards this region. To enhance the pumping action most diffusion pumps employ multistage stacks, typically with three jets working in series. The accumulated gas load after the last stack must be removed by a backing pump, for example the oil sealed rotary pump.

The ultimate pressure attainable by a simple diffusion pump is not as low as the vapour pressure of the oil at the temperature of the upper parts of the pump. In the lower pressure region, these pumps emit as much contaminant gas as they remove. This happens because organic pump oils have the disadvantage that they dissolve all kinds of gases and vapours which may form metastable compounds as well as true solutions. Furthermore, thermal decompositon of the oil may result in the formation of additional volatile products. Release of these gases and vapours into the high-speed jet causes backstreaming into the vacuum system, thus limiting the ultimate pressure. To minimize these detrimental effects, self-purifying pumps are used. In these diffusion pumps, the oil is fractionated by a special device so that the most volatile fraction stays in the outermost boiler ring and the least volatile fraction remains until it reaches the center of the boiler. Thus the lowest fraction emerges from the center and top nozzle which is closest to the vacuum recipient. Nevertheless, even after this a slight oil backstreaming can be observed. Oil backstreaming is a serious problem in thin-film deposition because it may cause contamination of the substrate and the deposit and degrade adhesion and film quality. As discovered [231,232], the greatest contribution to backstreaming comes from oil condensation on the top jet cap and from molecule-wall and intramolecular collisions in the boundary layer of the oil jet. Backstreaming into the vacuum chamber can be reduced by using a water-cooled cold cap and various baffles and cold traps which, however, obviously lower the net pumping speed. Most common are chevron baffles as shown in Fig. 27. They may be merely air-cooled or have a cooling jacket for water, freon, or liquid nitrogen circulation.

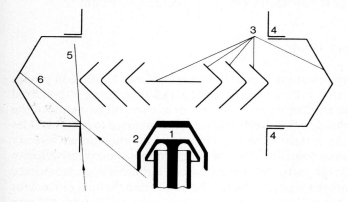

Fig. 27
Chevron baffle.
1 = upper jet of the diffusion pump
2 = cold cap
3 = Baffle surfaces can be refrigerated by different coolants,
e.g. water or liquid nitrogen
4 = water-cooled walls
5 and 6 = possible extreme directions of primary backstreaming

Refrigerated baffles are superior to water-cooled. With proper working fluid such as silicones, polyphenyl ethers or perfluoralkyl ether and in conjunction with a liquid-nitrogen-cooled baffle, it is possible with diffusion pumps to maintain an ultimate pressure of about 5×10^{-9} mbar. This pressure, however, is hardly obtained in coating plants because of outgassing of the seals. The ultimate pressure in a clean plant may range at about 5×10^{-8} mbar. The vacuum generated by a diffusion pump equipped with cold cap and baffle is, however, not completely free of pump fluid molecules. The requirements of maximum conductance of the baffle and minimum oil backstreaming conflict for all given models. Although most baffles present the

152

optimum compromise, in engineering practice there are two reasons for back-streaming of the pump fluid. The first, evaporation of the trapped oil vapour, can easily be minimized by better cooling of the baffle. The second reason, scattering in the baffle, is more serious. A portion of the oil molecules leaving the pump penetrate the optically dense baffle without touching its surface. This is caused by collision with one another or with other molecules. This scattering effect is independent of the temperature and can be suppressed only by sufficiently tight baffles, but at the expense of the conductance. The BALZERS combination [228] of pump, cold cap and baffle presents a compromise which meets the needs of practical application. At a pressure of 1×10^{-7} mbar, the backstreaming due to scattering amounts to 1×10^{-7} mg cm^{-2} min^{-1} at most. In addition, there is a pressure-independent backstreaming rate. With nitrogen as the scattered gas, this is 1×10^{-2} (mg cm^{-2} min^{-1}) mbar^{-1}. Hence, with a scattered gas pressure of 10^{-6} mbar, an additional backstreaming rate of 1×10^{-8} mg cm^{-2} min^{-1} will result in the maximum. These values refer to the upper edge of the baffle. Pumping port and plate valve can reduce the backstreaming to a considerable extent.

6.2.1.5.1.1.3 MOLECULAR PUMPS

The molecular pump functions in a fashion analogous to a diffusion pump. Instead of a molecule being diffused to the pump and then given a preferred direction by a momentum transfer from an oil vapour stream, the principle of the molecular drag pump is based on the directional velocity imparted to gas molecules which strike a fast moving surface. The overlapping of fhe directional velocity component with the thermal velocity component determines the overall velocity and the direction in which the particles are thrust away. If there is a second surface opposite the first, this process is repeated there. The undirected thermal movement of the gas molecules prior to collision with the moving surface is thus turned into a directed movement of the gas stream.

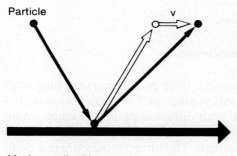

Particle

v

Moving wall with speed v

Fig 28
Principle of the molecular pump.

The pumping process is shown schematically in Fig. 28. In the range of molecular flow, the mean free path of the gas particles is larger than the distance between the two surfaces or walls; therefore the molecules collide with one another rather seldom while collisions with the walls are much more frequent. In this way, the molecules attain their wall velocity as an additional component. In the area of laminar flow, the effect of the moving surface is, on the contrary, limited by frequent collisions of the molecules with one another. The influence of the moving walls on the gas molecules is highest in the area of molecular flow. This principle is applied in modern turbomolecular pumps, invented by Becker [234–236] in 1957. These contain alternate axial stages of rotating and stationary disks and plates with inclined spiral channels. The preferred directions to a gas molecule is given by a fast-moving turbine. Fig. 29a shows sectional drawing of a BALZERS-PFEIFFER turbomolecular pump TURBO 270.

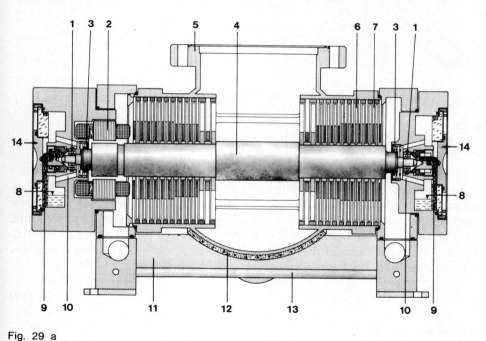

Fig. 29 a

Sectional drawing of the BALZERS - PFEIFFER turbo-molecular pump TUR-BO 270.

1 = bearings	6 = rotor disk	10 = oil return line
2 = motor	7 = stator disk	11 = fore-vacuum channel
3 = labyrinth chambers	8 = oil reservoir	12 = heater
4 = rotor	9 = oil supply to	13 = water-cooling
5 = UHV connection	bearings	14 = closure

Fig. 29b shows the rotor of a turbomolecular pump. The channels of the rotor disks are arranged laterally reversed to the stator disks. Each channel of a disk is an elementary molecular pump. All channels (approximately 25 to 60) of one disk are connected in parallel. A rotor and a stator disk are combined into a pumping stage. Thus, a turbomolecular pump consists of several in-series-connected individual pumping stages which are in turn formed by parallel connection of many elementary molecular pumps.

Fig. 29 b
Rotor of the BALZERS - PFEIFFER turbo-molecular pump TPU 510.

Independent of whether the pumped gas is treated as a continuum [236] or as a collection of individual particles (e. g. the Monte-Carlo method) [237], the theory of turbo-molecular pumps provides the following correlations of eqns. (39) and (40). The maximum compression K of a turbomolecular pump is an exponential function of the geometrical dimensions G, the circumferential speed u and the molecule mass M of the gas used:

$$K \simeq [\exp. \sqrt{(\frac{M}{RT})}] \, u \, G \qquad (39)$$

R = ideal gas constant
T = absolute temperature

As can be seen from eqn. (39), the compression increases exponentially with the root of the relative molecule mass M for a given rotor speed. For heavy gases, it is considerably higher than for light gases. Equation (39) does not take into account the losses resulting from backstreaming. However, the principal correlation is maintained [236].   In the region of molecular flow, the compression of a turbo-molecular pump for a specific gas is independent of the pressure. However, when the pressure increases and the region of molecular flow is left, the compression K is strongly reduced.

The volume flow rate S of turbo-molecular pumps is in the first approximation, the product of a geometrical factor G specific to the pump and the circumferential speed u:

$$S \simeq u\,G \tag{40}$$

In practice, the flow rate for light gases is somewhat higher (20 %) compared to that of heavy gases. This is caused by the flow resistance of the inlet channels of the disks and the high-vacuum socket. Since the resistances increase with the root of the molecule mass, the volume flow rate for heavy gases is somewhat smaller than that for light gases. In the range of molecular flow, the volume flow rate is independent of the pressure. Beginning with about 10^{-3} mbar, the volume flow rate is reduced with increasing pressure.

The ultimate total pressure that can be attained by means of turbomolecular pumps mainly depends upon the partial pressures p_{part} of the various gases on the pre-vacuum side of the turbo-molecular pump. The partial pressure p_o of a gas on the high-vacuum side is calculated from the partial pressure p_{part} on the fore-vacuum side, divided by the compression K for this gas:

$$p_0 = \frac{P_{part}}{K} \tag{41}$$

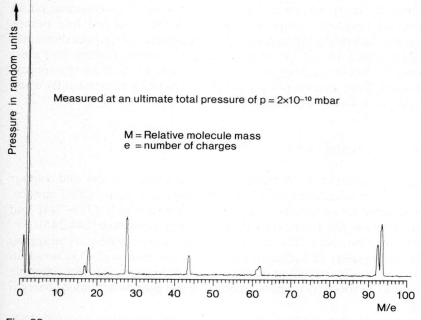

Measured at an ultimate total pressure of $p = 2\times10^{-10}$ mbar

M = Relative molecule mass
e = number of charges

Fig. 30
Typical residual gas spectrum of a turbo-molecular pump.

Since turbo-molecular pumps have for hydrogen the smallest compression (about 10^3) as a result of its low molecule mass, the hydrogen fraction is dominant in the ultimate total pressure. Fig. 30 shows these conditions at an ultimate total pressure of 2×10^{-10} mbar. The fractions of masses 17 and 18 (HO^+ and H_2O^+) can be even further reduced by careful heating, so that the hydrogen fraction in percent increases to such a level that it may exceed 95 %. When considering the residual gas spectrum, it is interesting that masses higher than 44 are not present. The heavy masses measured in the spectrogram are doubly and triply ionized metal vapours of the rhenium cathode of the quadrupole mass spectrometer (Balzers QMG 311) that was used.

Generally a vacuum system evacuated by a turbo-molecular pump and a rotary fore pump or by an oil diffusion pump and a rotary fore pump can practically be kept free of oil vapours (e.g. hydrocarbons) only if a dry roughing pump system is used [238]. When oil backstreaming is mentioned in connection with a high vacuum system, one thinks almost exclusively of the pumping fluid from the diffusion pump, or of the oil used for lubricating the ball bearings of a turbo-molecular pump, but not of the contribution of oil from the roughing pump even although the vapour pressure of these oils is often several powers higher. Turbo-molecular pumps are clearly superior to diffusion pumps concerning the oil backstreaming on the intake side. However, the parts of a turbo-molecular pump containing the drive elements and bearings are water- or air-cooled and may be contaminated with oil vapours from the warmer oil-sealed rotary fore pump during venting if no special measures are undertaken. The contamination can be prevented, however, using a catalyser trap between turbo-molecular pump and the oil-sealed roughing pump. This accessory for oil-sealed fore pumps reduces the oil backstreaming through a catalytic conversion of hydrocarbons into CO_2 and H_2O to about 10^{-3} of the original value. This then realizes to a great extent the concept of a dry roughing pump. The reaction products are transported by the pump itself. Regeneration of the catalyzer takes place automatically when the vacuum chamber is vented or pumped down [238].

6.2.1.5.1.1.4 CRYO PUMPS

The action of cryopumps is based on the fact that the gas and vapour molecules present in a vacuum chamber condense on a deep-cooled surface. Several review papers concerning cryopumping are available [239–244] and detailed analyses of the phenomena occuring have been published [244,245].

For less condensible gases, the pumping action is supported by cryotrapping (for example the trapping of hydrogen and helium by means of CO_2 or argon pre-condensed on to cryopanels at 20 Kelvin) and cryosorption (for example hydrogen sorption on specially prepared molecular sieves or on activated charcoal at 20 Kelvin).

Cryopumping surfaces cannot generally be exposed directly to a source of gas at room temperature because the heat load due to radiation would exceed that due to the condensation of gas molecules. Therefore, the cryogenic surface is protected on the side facing the gas source. As protection against thermal radiation, an optically dense baffle comprising liquid-nitrogen-cooled blackened shields is used often. There are cryopumps with bath cryostats operating between 4.2 and 2.3 Kelvin and with refrigerators operating approximately between 20 and 12 Kelvin. Refrigerator cryopumps are preferred. With the BALZERS cryopumps, the refrigeration capacity is provided by a refrigerator integrated into the pump as can be seen in Fig. 31. The refrigerator operates with a closed helium circuit (gaseous He); the high-pressure helium is supplied by a separate (low noise level) compressor via two flexible lines. The actual pump condensor is enveloped by an optically dense, blackened radiation screen which is cooled to about 80 K with liquid nitrogen. The cooling fluid is stored inside the pump in a tank fitted with an automatic level control system. The temperature of the pump condensor is measured with a hydrogen vapour pressure thermometer having an indication range between 14 and 30 Kelvin. The pressure gauge is equipped with two adjustable switch contacts for control operations. A remote measuring system is possible. The cryopump has a pumping capability for all gases. Hydrogen that is not condensible at 20 K is pumped by a solid-state adsorbent at nearly the same pumping speed as air; and argon is pumped by condensation at virtually the

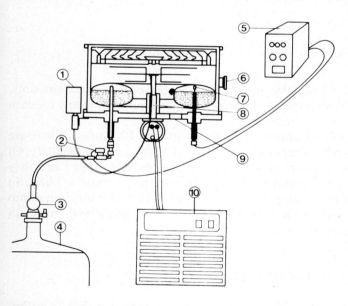

Fig. 31 a
Sectional drawing of a refrigerator cryopump.

1 = H₂ thermometer
2 = LN₂ valve
3 = LN₂ filling device FET 003
4 = LN₂ tank
5 = LN₂ controller

6 = gauge head connection
7 = LN₂ probe
8 = cryo generator
9 = fore-vacuum connection
10 = compressor unit

Fig. 31 b
BALZERS cryopump RKP with LN₂ shielding for installation within chamber.

nominal pumping speed. Though He and Ne are adsorbed only in small amounts, they are very rarely found as dominating or disturbing components in any vacuum system.

The working range of a cryopump is between 10^{-3} and 10^{-10} mbar. There are practically no problems with regeneration because of a high puming capacity. Of all high-vacuum pumps, cryopumps have the highest pumping speed and they also generate a very clean vacuum. Cryopumps are extremely versatile. They can be used both as the main pump for example, in conjunction with a sorption pump or with a twin stage rotary vane pump) and also as an additional pump in conventional vacuum systems.

6.2.1.5.1.2 HIGH-VACUUM PROCESS SYSTEMS

Independent of the type of PVD method used to deposit a film or a film system, the deposition process is carried out in a sealed chamber which is first exhausted to a pressure of the order of 10^{-5} mbar or even lower values. The glass chambers used formerly have been replaced, with the exception of glass recipients for special purposes, such as apparatus for electron microscopic preparation techniques, by those of metal. Cylindrical and cubic chambers made of stainless steel provided with various flanges and windows are used today. The walls of the chambers can be heated and cooled by water running through double-wall constructions or in brazed-on half-round pipes fitted on the outside. The vacuum chambers are evacuated by different pumping systems. The simplest consists of a diffusion pump with or without a liquid-nitrogen-cooled baffle and twin-stage rotary vane pump as fore and backing pump. When large quantities of gases have to be pumped away, a Root's blower should be used in addition.

Another pumping system may consist of a turbo-molecular pump, a Root's blower and a twin-stage rotary vane backing pump.

The cleanest vacua are obtained with a pumping system consisting of a cryopump, a Root's blower and a twin-stage rotary vane pump. It may be an advantage with all three types of pumping systems to use the catalyzer trap [238] in combination with the oil-sealed rotary fore pump. However, to obtain a very low level of organic residues in the residual gas, all other sources of contamination beside pumping fluid backstreaming must be considered, for example the desorption of organic compounds from the rubber seals (mostly Viton) used to close off the whole coating system. Careful cleaning, degassing and then proper handling of the formed seals minimizes their contribution to residual gas contamination.

It is interesting to note that pumping systems equipped with diffusion pumps are the cheapest but that the cryopumping systems are only slightly more

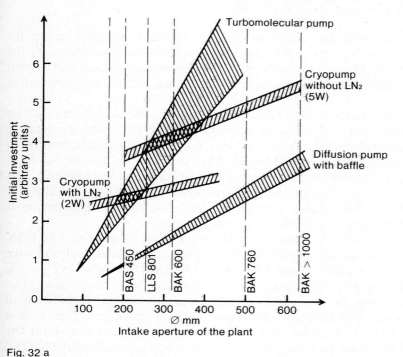

Fig. 32 a

Costs of acquisition of various types of high vacuum pumps as function of the size of the plants to be evacuated.

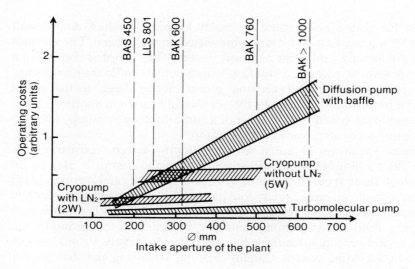

Fig. 32 b

Annual operating costs of various high vacuum pumps (220 days with 9 hours; electrical power, water and liquid nitrogen) in arbitrary units as function of the intake aperture of the plant. BALZERS plants are specially considered.
(calculated for Germany and Holland, March 1982)

expensive. Turbo-molecular pumping systems, however, although cheap for small plants become expensive for larger plants. These facts are shown in the graph presented in Fig. 32a. Considering also the operating costs, it can be seen in Fig. 32b that differences in the initial investment are partially compensated at least.

Various types of coating plants are available. Fig. 33 shows a classical bell jar coater evacuated by an oil diffusion pumping system. This coating system is fabricated for example by METALLUX (Italy). A cylindrical vacuum chamber in a horizontal construction is typical for the BAH 1600 coating system, shown in Fig. 34 which is fabricated for example by BALZERS AG (Liechtenstein). The horizontal plant is generally equipped with evaporation sources, but it is also best suited for supply with planar magnetron sputtering targets. Evacuation is performed generally by an oil diffusion pumping system.

Some examples of cubic coaters are shown in Figs. 35 and 36. The BAK coater series, for example BALZERS AG (Liechtenstein), includes small and large plants such as the BAK 600 and the BAK 1400. The BAK 760 shown in Fig. 35 is a medium-size compact high vacuum coater for large scale production. The standard plant forms the basis of a versatile modular system. Installation of various accessories and ancillary equipment makes it possible to turn such a coater into a special plant for very different coating applications. With the standard oil diffusion pump station which has a pumping speed of about 3800 l s^{-1} for air and 11.400 l s^{-1} for water vapour, a low ultimate pressure and short batch times can be achieved. The special design of stand and vacuum chamber enables

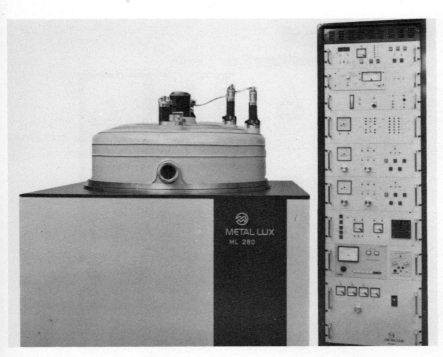

Fig. 33
High vacuum coating unit of the bell jar type (METAL LUX).

Fig. 34
High vacuum coating unit with horizontal cylindrical chamber BAH 1600
(BALZERS).

162

Fig. 35 a
Cubic high vacuum coating unit BAK 760 (BALZERS).

Fig. 35 c
Coating unit BAK 760 with refrigerator cryopump

Fig. 35 b
Coating unit BAK 760 with oil diffusion pump.

Fig. 36 a
Cubic high vacuum coating unit BAK 1200 (BALZERS).

Fig. 36 b
Pumping system of the BAK 1200 consisting of rotary vane pump, Roots
blower and refrigerator cryopump.

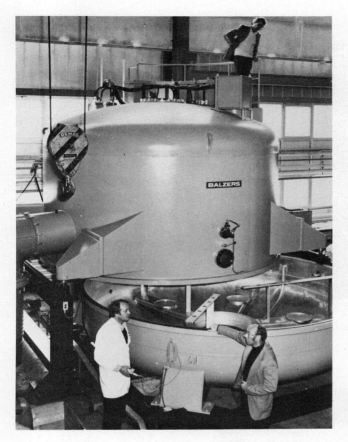

Fig. 37
Special coating unit to metallize large astronomic mirrors at
the Calar Alto observatory in Spain (BALZERS).

integration of the plant into a dividing wall which separates the vacuum chamber
from the pump station, thus meeting the requirements for installation in a
dust-free room.

A large batch coating system especially designed to coat astronomical telescope mirrors with Al films is shown in Fig. 37. The mirror substrate to be coated
is made of glass ceramics, its thickness is 60 cm and the weight is about 14 tons.

To coat large panes, two different methods are generally applied which
require different types of plants. In the first, the panes are inserted into the vacuum
chamber and coated by moving the vapour source, e. g. planar magnetrons, back
and forth along the surfaces of the panes. Fig. 38 shows such a plant made by
LEYBOLD-HERAEUS GmbH (Germany) used to coat architectural glass. In the
second method, coating is performed continuously by moving the panes over the

Fig. 38 a
Special plant to coat four window panes simultaneously by back-and-forth
moveable vapour sources (planar magnetrons) (LEYBOLD-HERAEUS).

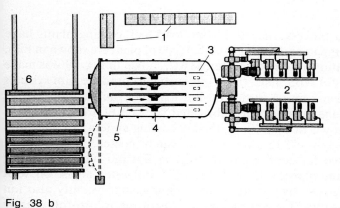

Fig. 38 b

Sectional drawing of the coating plant shown in Fig. 38 a.
1 = power supply   4 = cathodes
2 = pumping station   5 = glass panes in their holders
3 = cathodes starting position   6 = loading carriage

Fig. 39
Multi-station automatic coater MAC (OPTICAL COATING LABORATORIES
INC., USA) equipped with evaporation sources. The control system can
be seen on the left, and the six coating chambers and part of the lead
in vacuum system on the right.

vapour sources. For this process, special on-line load lock coating plants have
been made by various companies. Examples of this type of plant are shown in Figs.
39 and 40. The multi-station automatic coater, MAC, shown in Fig. 39 was made
by OPTICAL COATING LABS, INC. (USA); it is able to produce film systems
by evaporation and condensation. A further on-line load lock multistation coating
plant equipped with planar magnetrons fabricated by LEYBOLD-HERAEUS
GmbH (Germany) is shown in Fig. 40. With this coating system, architectural
glass having pane dimensions of 318 times 600 cm^2 can be coated.

Roll coaters are used for continuous coating on flexible substrates as for
instance paper, textiles and plastic foils of PTFE, PET, Polyester, and so on. Roll
coating can be performed by evaporation and by sputtering: recently also ion
plating was tried successfully. Typical process speeds of the roll-to-roll coating
apparatus are between 1 and 10 m s^{-1}. Various types of plants are commercially
available. A standard roll coater for the packaging industry made by LEYBOLD-
HERAEUS GmbH (Germany) is shown in Fig. 41.

Fig. 40 a
Multi-station coater (LEYBOLD - HERAEUS) to deposit multilayers on large architectural glass panes by sputtering.

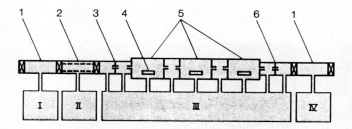

Fig. 40 b
Sectional drawing of the coating plant shown in Fig. 40 a.
1 = lock chamber   5 = coating chambers
2 = glow discharge chamber   6 = run out chamber
3 = run in chamber   I - IV = pumping stations
4 = vapour sources (planar
    magnetrons)

Fig. 41
Standard roll coater (LEYBOLD - HERAEUS).

6.2.1.5.2 FILM DEPOSITION BY EVAPORATION AND CONDENSATION IN HIGH VACUUM

Thin films of materials of such different properties as metals, halides, oxides and sulphides can be obtained in the crystalline or amorphous state by condensation of the vapour on a glass substrate. The mechanism involved in forming the film may be a pure physical e.g. simple condensation or may also involve chemical reactions.

Evaporated films were probably first made by Faraday [246] in 1857 when he exploded metal wires in an inert atmosphere. The deposition of metal films in a vacuum by resistance heating of platinum wires was performed by Nahrwold [247] in 1887 and only a year later this technique was used by Kundt [248] to produce films for measuring the refractive indices of metals. In the following period, evaporated thin films had only academic interest, although the vacuum evaporation of metal films by Pohl and Pringsheim in 1912 [249] was performed under better developed technological conditions.

The techniques of forming thin films by condensation at very low pressures have been developed in parallel with industrial development of techniques of producing high vacua in large volumes, using pumps of very high speed and suitable cleanliness.

Films obtained by condensation alone are usually made at pressures between 10^{-6} and 10^{-8} mbar and down to the lowest values that modern vacuum technology can attain. This is combined with fast deposition rates to produce deposits in which only few foreign gas atoms are incorporated.

Of great technical importance was the discovery and development of the reactive evaporation process by Auwärter 1952 [250] and Brinsmaid 1953 [251]. Such processes which involve a chemical reaction between evaporated constituents and the gas atmosphere are carried out mainly in the 10^{-4} mbar region with slow deposition rate. The reaction is often activated either thermally or by uv radiation or alternatively by ionic or electronic bombardment.

The industrial development of these evaporation and condensation techniques is due largely to the relative ease of obtaining either pure metal films or exact stoichiometric compound deposits of uniform thickness, because the laws which determine the phenomena at low pressure are better defined than close to or at atmospheric pressure. At very low pressures, the mean free path of the vapour atoms or molecules exceeds the usual distance between evaporation source and the substrate. Thus the substrate receives a flux of vapour which travels in straight lines with no or only very few gas/vapour collisions and the spatial distribution of the condensate obeys purely geometrical laws.

If, however, the residual pressure has such a value that the source to substrate distance corresponds to the mean free path of the vapour atoms, then only 37 % of the atoms travel without collision with the residual gas molecules, as mentioned in Section 6.2.1.5.1. At residual pressures above 10^{-3} mbar, the propagation of the evaporated atoms between source and the substrates is subject to an isotropic diffusion effect because of increasing frequency of intermolecular collisions.

The pressure of the atmosphere in which deposition is carried out optimally depends on the applied technology.

6.2.1.5.2.1 EVAPORATION

The number of atoms or molecules evaporating from a liquid or solid surface depends strongly on the temperature. As is generally known, the equilibrium vapour pressure would be obtained in a thermodynamically closed system. However, in practical evaporation, no equilibrium is obtained because the environment of the vapour source acts as a vapour sink. The evaporant atoms condense on all parts that are at lower temperature than the vapour source.

Systematic investigations of evaporation rates in vacuum have been performed carefully mainly by Hertz, Knudsen and Langmuir [252]. In these experiments it was found that a liquid has a specific ability to evaporate and cannot exceed a certain maximum evaporation rate at a given temperature even if the heat supply is unlimited. The theoretical maximum evaporation rates are obtained only if the number of vapour molecules leaving the surface corresponds to that required to exert the equilibrium pressure p_e on the same surface, with none returning to it.

From these experiments and considerations, the following equation for the molecular evaporation rate could be formulated:

$$dN/A\ dt = \alpha_e[(p_e - p_h)/\sqrt{(2\pi mkT)}] \quad cm^{-2}\ s^{-1} \tag{42}$$

dN = number of evaporating atoms
A = surface area
t = time
α_e = evaporation coefficient
p_h = hydrostatic pressure
m = atomic mass
k = Boltzmann's constant
T = temperature (Kelvin)

With $\alpha_e = 1$ and $p_h = 0$, the maximum evaporation rate is obtained:

$$dN/Adt = p_e/\sqrt{(2\pi mkT)} \quad cm^{-2}\ s^{-1} \tag{43}$$

Langmuir [252] showed that eqn. (43) is also correct for the evaporation from free solid surfaces. Multiplying by the mass of an individual atom or molecule yields the mass evaporation rate per unit area:

$$\Gamma = mdN/Adt = (m/2\pi kT)^{1/2}p_e \qquad (44)$$

$$\Gamma = 5.84\times10^{-2}\ (M/T)^{1/2}p_e \quad g\,cm^{-2}\,s^{-1} \qquad (45)$$

M = molar mass and p_e is given in torr.

The total amount of evaporated material m_{total} can be obtained through the double integral:

$$m_{total} = \int_t\int_A\ \Gamma\,dA\,dt \qquad g \qquad (46)$$

The mass evaporation rate Γ is for most elements of the order of $10^{-4}\,g\ cm^{-2}\,s^{-1}$ at $p_e = 10^{-2}$ mbar. Numerical values have been tabulated by Dushman [220]. Graphs of the equilibrium vapour pressure as a function of temperature for the elements are shown in Figs. 42a to 42e.

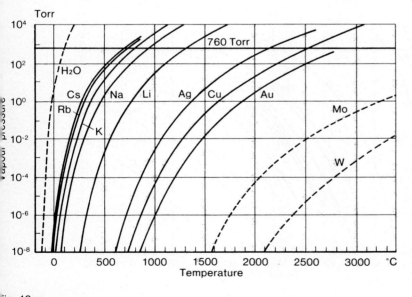

Fig. 42 a

Vapour pressures of the elements. I. group of the periodic system.

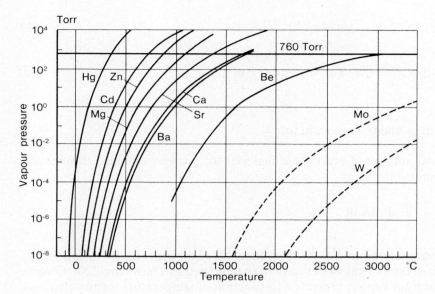

Fig. 42 b
Vapour pressures of the elements. II. group of the periodic system.

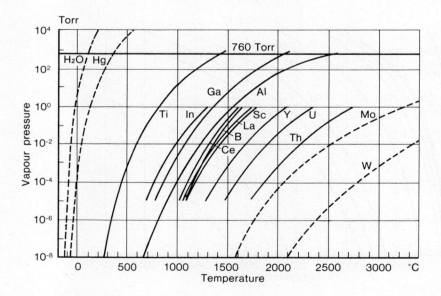

Fig. 42 c
Vapour pressures of the elements. III. group of the periodic system.

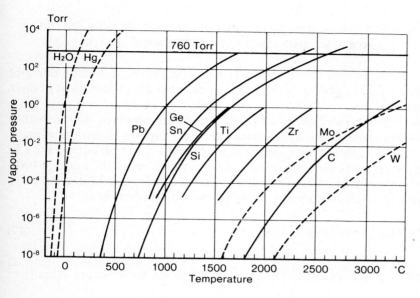

Fig. 42 d
Vapour pressures of the elements. IV. group of the periodic system.

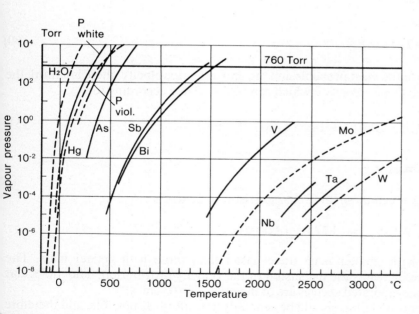

Fig. 42 e
Vapour pressures of the elements. V. group of the periodic system.

6.2.1.5.2.2   ENERGY, VELOCITY AND DIRECTIONAL DISTRIBUTION OF THE
VAPOUR ATOMS AND THICKNESS UNIFORMITY OF THE FILMS

Atoms and molecules effusing from a Knudsen cell have kinetic energies
which are distributed according to their temperature but are independent of their
molecular mass.

Asymmetrical Maxwellian distribution curves are obtained with a higher
energetic tail. As some of the atoms have relatively high energies, the average
energy $E_k = 3/2\ kT$ is three times greater than the most common energy $1/2\ kT$.
Considering the applied evaporation temperatures which are in the range of 1000
to 2500° C, the average energies of the vapour atoms lie between 0.1 and 0.2 eV.

According to kinetic theory, the random velocity distribution is:

$$\Phi(c^2) = (4/c_m^3 \sqrt{\pi})\, c^2 \exp(-c^2/c_m^2) \tag{47}$$

c_m is the most probable velocity. The relationship to the mean square velocity $\overline{c^2}$
is:

$$c_m^2 = (2/3)\ \overline{c^2} \tag{48}$$

and

$$\overline{c^2} = 3kT/m \tag{49}$$

therefore

$$c_m = (2kT/m)^{1/2} \tag{50}$$

In addition to the most probable and the mean square velocity, there is a third, the
arithmetic average velocity $\bar{c}$ which can be calculated according to:

$$\bar{c} = (8kT/\pi m)^{1/2} = 14.6\,(T/M)^{1/2} \quad cm\ s^{-1} \tag{51}$$

M = molar mass in g.

The ratio between these three characteristic velocities is:

$$\sqrt{\overline{c^2}} : \bar{c} : c_m = 1.225 : 1.128 : 1$$

Molecules with smaller mass travel faster than those with greater mass. The
temperature increases the dispersion of the velocity distribution. The molecular
velocities of evaporated species are of the order of 10^5 cm s^{-1}.

The deposited films should be generally as uniform as possible, and therefore
it is important to know the directional distribution of the evaporating species.
This, however, depends mainly on the applied evaporation source. There are

various types of evaporation sources in use, for instance the true point source, the small-area source, the ring source and the rod source. Many investigations have been done to optimize the film thickness uniformity on the substrates [253, 254,257-260]. In technical evaporations, the small-area source is because of its wide application the most important. The small-area source is characterized by a small planar surface of a few square centimeters from where the molten material evaporates to one side alone within an angle 2α. The formation of a meniscus with the walls of the source in the case where the evaporant wets the source material can generally be neglected. Small boats and crucibles can be classified into this type. In practice, however, such a source can be approximated as a point source. This approximation is much better fulfilled if the source area is kept small and the distance to the substrate is made large. In most technical equipment these requirements are sufficiently well realized. Using a computer program, Deppisch [255] has recently developed a method to determine the film uniformity if the directional vapour distribution was first measured. This elegant method will be explained in more detail because of its great practical importance.

Generally, the mass of material emitted from an evaporation source at a solid angle ω is:

$$dm = m(\omega)\,d\omega \tag{52}$$

For rotationally symmetric lobe-shaped vapour clouds, the dependence of the emitted mass on the solid angle ω can also be expressed by:

$$m(\omega) = m \cos^n \alpha \tag{53}$$

α denotes the angle between the direction of the emitted atom and the symmetry axis of the evaporation source.

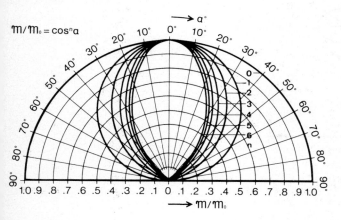

Fig. 43
Sectional drawings of calculated lobe-shaped vapour clouds of various cosine exponents [255].

The exponent n determines the exact shape of the vapour cloud. An increase of the n value decreases the emitted amount at larger emission angles. The emission of the source is confined towards the perpendicular direction.

Fig. 43 shows a diagram of vapour cloud distributions as a function of various n values. m_0 is the amount of material emitted under the condition $\alpha = 0$ degree

Transformation of eqn. (52) yields:

$$dm = m(n + 1/2\pi) \; \cos^n \alpha \, d\omega \tag{54}$$

This equation describes the solid-angle dependent emission of mass from sources with rotationally symmetric lobe-shaped vapour cloud characteristics. With $n = 1$, Knudsen-cell evaporator, it is identical with other published [253,254] calculations. With $n = 0$, point source or spherical source, there are slight differences with other published values which generally consider evaporation in the total space $\omega = 4\pi$, whereas here, evaporation is only considered in the half-space above the vapour source.

The thickness of a film condensing on a substrate element dA which is in the half-space with a radius s above the evaporation source can be calculated from eqn. (54).

The substrate element dA which may have an arbitrary spatial orientation is viewed from the source under a solid angle of

$$d\omega = (\cos\beta \, dA)/s^2 \tag{55}$$

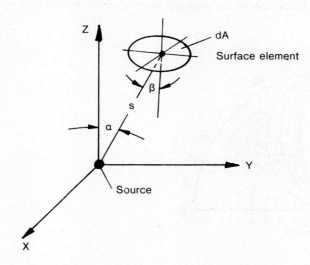

Fig. 44
Geometrical representation of an evaporation equipment.

As can be seen in Fig. 44, β is the angle between the normal to the surface at the center of the substrate element and the incident vapour beam, s is the distance between the center of the substrate element and the vapour source. The geometrical position of the substrate element is completely determined by the angles α and β and the distance s.

The evaporated material transported in the solid angle $d\omega$ condenses on the substrate element dA and forms a thin film. The relation existing between mass, density, volume and thickness is:

$$dm = \rho\, dV = \rho\, d\, dA \tag{56}$$

Substituting (56) into (54) for the thickness on the substrate element yields:

$$d = m/\rho\,(n+1/2\pi)\,(\cos^n \alpha \cos \beta / s^2 \tag{57}$$

Technical substrate holders, such as plane palettes or spherical segments generally have a preferred point, the center of rotation, whose relative position to the evaporation source remains unchanged during the rotational motion. This center of rotation is best suited as a normalization point for the film thickness d_o. The corresponding distance and the angles are given the suffix o:

$$d_o = m/\rho\,(n+1/2\pi)\,\cos^n \alpha_o \cos \beta_o / s_o^2 \tag{58}$$

In normalized representation from eqns. (57) and (58), eqn. (59) is obtained:

$$d = d_o\,(s_o/s)\,(\cos^n \alpha \cos \beta / \cos^n \alpha_o \cos \beta_o) \tag{59}$$

Equation (59) is the basic equation for the calculations of film thickness distributions. As well as calculations on plane and spherical substrate holders, thickness distributions can also be calculated on conical surfaces and on parabolic and hyperbolic surfaces.

Fig. 45 shows the geometrical arrangement typical of technical evaporations using a plane rotating substrate holder. Generally, the position of the evaporator is out of center by a section q. The substrates rotate with a radius r in a vertical distance h above the vapour source. In order to obtain the film thickness distribution for this geometrical arrangement, the parameters of eqn. (59) must be determined by considering the angle of rotation.

Thus the reaction of the distances s_o and s can be obtained from eqn. (60):

$$\left(\frac{s_0}{s}\right)^2 = \frac{1+(q/h)^2}{1+(r/h)^2+(q+h)^2-(2r/h)(q/h)\cos\varphi} \tag{60}$$

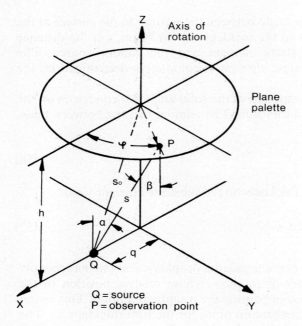

Fig. 45
Evaporator geometry with a plane palette as substrate holder.

Further, the angles α and β can be obtained from eqn. (61):

$$\cos\alpha = \cos\beta = \frac{1}{\sqrt{[1+(r/h)^2+(q/h)^2-(2r/h)(q/h)\cos\varphi]}} \quad (61)$$

Substituting (60) and (61) into eqn. (59) and normalization at the center of rotation $(r/h) = 0$ yield:

$$\frac{d}{d_0} = \frac{[1+(g/h)^2]^{(n+3)/2}}{[1+(r/h)^2-(q/h)^2+(2r/h)(q/h)\cos\varphi]^{(n+3)/2}} \quad (62)$$

Equation (62) describes the relation of the thickness of a random point to the thickness of the point on the axis of rotation under stationary conditions. In practice, however, the plane substrate holder will be rotated. This physical fact is solved mathematically by forming the intergral of eqn. (62) with the integration variable. The resulting thickness distribution on a rotating plane substrate holder is then:

$$\frac{d}{d_0} = [1+(q/h)^2]^{(n+3)/2} \frac{1}{\pi} \int_0^{\pi} \frac{d\varphi}{[1+(r/h)^2+(q/h)^2-(2r/h)(q/h)\cos\varphi]^{(n+3)/2}} \quad (63)$$

No closed solution of this integral exists for all possible n values of the lobe-shaped vapour cloud, therefore numerical calculation methods must be used. Closed solutions are possible with the exponent of the cosine-distribution $n = 1,3,5$.

With the following abbrevations:

$$k_1 = [1+(g/h)^2]^{(n+3)/2}$$
$$a = 1+(r/h)^2+(q/h)^2$$
$$b = (2r/h)q/h$$

eqn. (63) reduces to:

$$\frac{d}{d_0} = k_1 \frac{1}{\pi} \int_0^\pi \frac{d\varphi}{[a-b\cos\varphi]^{(n+3)/2}} \tag{64}$$

After integration of (64), the following expressions are obtained for the different evaporator characteristics:

$$n = 1, \quad \cos^1-\text{evaporator:} \quad d/d_0 = k_1 [a/(a^2-b^2)^{3/2}] \tag{65}$$

$$n = 3, \quad \cos^3-\text{evaporator:} \quad d/d_0 = k_1 [(a^2+1/2b^2)/(a^2-b^2)^{5/2}] \tag{66}$$

$$n = 5, \quad \cos^5-\text{evaporator:} \quad d/d_0 = k_1 [(a^3+3/2ab^2)/(a^2-b^2)^{7/2}] \tag{67}$$

The shape of the resulting thickness profile in the direction of the radius of the plane palette $d/d_o = f(r/h)$ is strongly influenced by the geometrical position of the vapour source and by the exponent n of the vapour cloud. How n can be determined will be shown later.

Fig. 46a shows the normalized thickness distributions as functions of various lobe-shaped vapour clouds with the position of the evaporation source on the axis of rotation ($q/h = 0$). It can be seen that the decrease in film thickness with increasing distance of the source from the axis of rotation is more marked at higher values of the exponent n of the vapour cloud characteristic.

A source position outside the axis of rotation produces not only a monotonic decreasing thickness distribution but also, depending on the characteristic of the source, to some extent peaks. This behaviour can be seen in Fig. 46b.

The application of eqn. (59) to the geometrical arrangement shown in Fig. 47 allows the calculation of the film thickness distribution on spherical segments for static as well as for rotating operation. This is a very important case because spherical segments are preferred in practice. They offer a better thickness

180

uniformity as compared with plane palettes of the same area. For clarity of representation the following abbreviations are used:

$$c = R/h + 1(1 - R/h)\cos\gamma$$

$$e = (q/h)\sin\gamma$$

$$f = 1 + (q/h)^2 + (2R/h)[R/h - 1 + (1 - R/h)\cos\gamma]$$

$$g = (2R/h)(q/h)\sin\gamma$$

$$k_2 = 1 - (R/h)(1 - \cos\gamma)$$

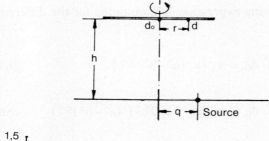

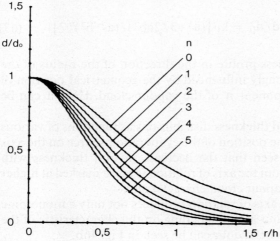

Fig. 46 a
Film thickness distribution on a plane palette as function of various cosine exponents n of the evaporator with the vapour source in the axis of rotation (q/h = 0) [255].

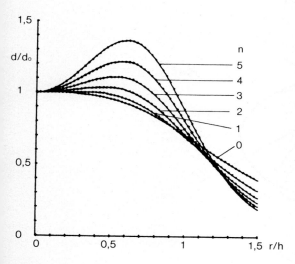

Fig. 46 b

Film thickness distribution on a plane palette as function of various cosine exponents n of the evaporator with the vapour source outside the axis of rotation (q/h = 0,75) [255].

Insertion of these abbreviations yields the following equations:

$$(s_0/s)^2 = 1+(q/h)^2/(f-g)\cos\varphi \tag{68}$$

$$\cos\alpha = k_2/[(f-g)\cos\varphi]^{1/2} \tag{69}$$

$$\cos\beta = (c-e)\cos\varphi/(f-g)\cos\varphi \tag{70}$$

Insertion of eqns. (68), (69) and (70) into eqn. (59) and normalization at the center of rotation results in the film thickness distribution of the stationary spherical segment:

$$\frac{d}{d_0} = k_1\, k_2^n\, \frac{c-e\cos\varphi}{[f-g\cos\varphi]^{(n+3)/2}} \tag{71}$$

In the same way as for the palette, the radial thickness distribution of a rotating spherical segment $d/d_0 = f(\gamma)$ can be obtained by integration of eqn. (71);

$$\frac{d}{d_0} = k_3\, \frac{1}{\pi} \int_0^\pi \frac{c-e\cos\varphi}{[f-g\cos\varphi]^{(n+3)/2}}\, d\varphi \tag{72}$$

$$k_3 = k_1\, k_2^n$$

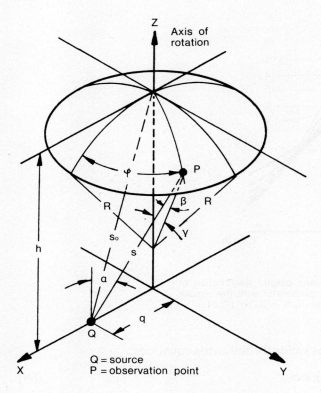

Fig. 47
Evaporator geometry with a spherical segment as substrate holder.

As with eqn. (63) or (64), eqn. (72) also allows a closed solution only for discrete n values. Such solutions are for:

$$n = 1, \ \cos^1\text{-evaporator:} \qquad \frac{d}{d_0} = k_3 \frac{cf-eg}{[f^2-g^2]^{3/2}} \tag{73}$$

$$n = 3, \ \cos^3\text{-evaporator:} \qquad \frac{d}{d_0} = k_3 \frac{cf^2+1/2cg^2-3/2efg}{[f^2-g^2]^{5/2}} \tag{74}$$

$$n = 5, \ \cos^5\text{-evaporator:} \qquad \frac{d}{d_0} = k_3 \frac{cf^3+3/2cfg^2-2ef^2g-1/2eg^3}{[f^2-g^2]^{7/2}} \tag{75}$$

Figs. 48a and 48b demonstrate the film thickness distributions on rotating spherical segments for different evaporators in two different geometrical positions. The curves in Fig. 48a are obtained with the source on the axis of rotation (R/h = 1, q/h = 0) and those of Fig. 48b are obtained with a source position out of center (R/h = 1.1, q/h = 0.5).

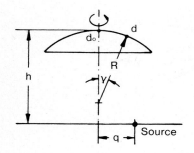

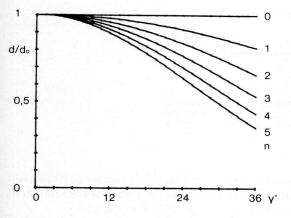

Fig. 48 a
Film thickness distribution on a spherical segment as function
of various cosine exponents n of the evaporator with the
vapour source in the axis of rotation (q/h = 0, R/h = 1) [255].

The cosine exponents n of the various evaporation sources have values between n = 1 and n = 6. Most evaporators have a vapour characteristic corresponding to n values between n = 3 and n = 5. Thus the closed solutions of eqns. (65), (66), (67) and (73), (74), (75) make it possible to obtain quickly a survey of the thickness distribution which can be expected.

The geometrical position of the evaporation source is easy to establish, but its exponent n characterizing the shape of the vapour cloud, must be determined experimentally. The processes inside an evaporation source that finally lead to vapour emission are too complex to allow the value of n to be calculated. It is generally even hardly possible to specify the evaporating area. The experimental determination of n is done by measuring the thickness profile at equidistant positions on the substrate holder and by comparing these values with those calculated using eqn. (64) or (72). Best fitting is approximated by a systematic variation of n in the calculation. The Gaussian method of least-squares can be

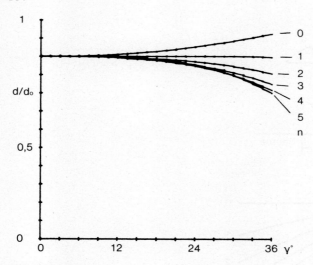

Fig. 48 b

Film thickness distribution on a spherical segment as function
of various cosine exponents n of the evaporator with the vapour
source outside the axis of rotation (R/h = 1,1, q/h = 0,5) [255].

applied as a decision criterion for optimum fitting. Practical experience has shown
that the best fit between experiment and calculation is obtained if the statement
for the solid-angle-dependent mass emission in eqn. (63) is replaced by:

$$\frac{m}{m_0} = (1-A)\cos^n\alpha + A \tag{76}$$

The symbol A describes an isotropic component of the distribution. Equation
(53) is then only a special case of eqn. (76) with A = 0 [256]. Further, the film
thickness distribution eqns. (64) and (72) for plane palettes and spherical segments
have to be modified by this statement (76).

As the result of this modification, the eqns. (77) and (78) are obtained for the
plane palette and the spherical segment respectively.

$$\frac{d}{d_0} = \frac{k_4(1-A)}{\pi}\int_0^\pi\frac{d\varphi}{[a-b\cos\varphi]^{(n+3)/2}} + \frac{k_4\,A}{\pi}\int_0^\pi\frac{d\varphi}{[a-b\cos\varphi]^{3/2}} \tag{77}$$

The abbreviation k_4 means $\left[\dfrac{1-A}{k_1} + \dfrac{A}{k_1^{3/(n+3)}}\right]^{-1}$

$$\frac{d}{d_0} = \frac{k_4 \, k_2(1-A)}{\pi} \int_0^\pi \frac{c - e \cos\varphi}{[f - g \cos\varphi]^{(n+3)/2}} \, d\varphi + \frac{k_4 \, A}{\pi} \int_0^\pi \frac{c - e \cos\varphi}{[f - g \cos\varphi]^{3/2}} \tag{78}$$

From eqns. (77) and (78), the vapour cloud parameters n and A can be also obtained by variation of the coefficients. Fig. 49 shows the measured vapour distribution curves of aluminium obtained by evaporations with a standard commercially available 270° bent electron beam gun. A 12 kW, 10 kV electron beam power supply and a quartz crystal thickness and rate monitor were used [256]. At an aluminium evaporation rate of 1.8 nm s^{-1}, the cosine exponent equals n = 2.3 with no isotropic component A = 0. With increasing rate the exponent increases and also the isotropic component appears. With rates of 10.5 and 81.4 nm s^{-1} the corresponding characteristic data are n = 4, A = 0.14 and n = 5.8, A = 0.14.

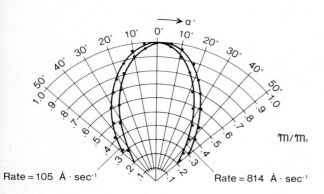

Rate = 105 Å · sec^{-1} Rate = 814 Å · sec^{-1}

Fig. 49
Measured vapour distribution of aluminium fitted with a function of the form $m/m_0 = (1 - A) \cdot \cos^n + A$ [256].

The isotropic component A in evaporation follows from the existence of a high-pressure region some distance (1 – 2 cm) above the melt which acts as an isotropic emitter.

The object of all efforts is to obtain uniform thickness distribution over the full area of the substrate holder under the conditions typical for the chosen evaporation source for a given coating plant.

Two principal approaches can be followed to attain this goal:

1. variation of the geometrical position of the evaporation source, and
2. the use of, or better, the additional use of static shutters between evaporator and substrates.

Optimization of the source position is done by evaluating eqns. (77) or (78) using the source parameters n and A. For this purpose, the geometrical ratios q/h and

R/h are used and are allowed to vary in a computer program under defined geometrical boundary conditions, such as dimensions of the plant, size of palettes or spherical segments, area to be coated, distance between source and substrates, etc. The computation is continued until the desired thickness uniformity lies within the tolerance.

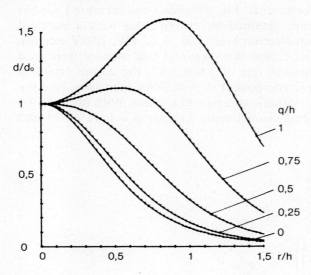

Fig. 50 a
Film thickness distribution on a plane palette with fixed cosine exponent of the evaporator (n = 3) as function of the source distance q/h from the axis of rotation [255].

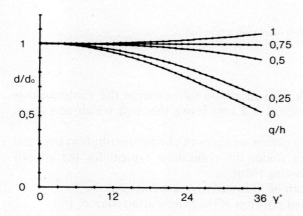

Fig. 50 b
Film thickness distribution on a spherical segment with fixed cosine exponent of the evaporator (n = 3) as function of the source distance q/h from the axis of rotation [255].

The variation of the thickness distribution on a plane palette and a spherical segment for various distances q/h of the source from the axis of rotation is shown in Figs. 50a and 50b.

In both cases, the source was assumed to be a $\cos^3$ – evaporator with no isotropic vapour component. From the results shown in Figs. 50a and 50b, it becomes obvious that, with spherical segments, films of equal thickness can be obtained over the whole area, while with plane palettes useful results are achieved only over a much smaller effective area.

In practice, the reproducibility of the thickness distribution is generally better than ± 2 %.

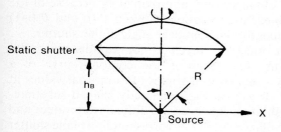

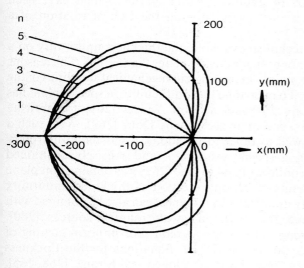

Fig. 51

Contour diagram of correction shutters calculated for sources of different cosine exponents n = 1, 2, 3, 4, 5 which are located in the axis of rotation of a spherical segment with a radius R = 500 mm. The built-in plane of the shutter is 350 mm above the vapour source [255].

There are some applications, however, for which the distribution uniformity that can be obtained by varying the geometry is still insufficient. This may occur particularly with sources having a high directional evaporation characteristic. In such cases, static shutters of suitable shape and size must be used. These partly cover the surface of the rotating substrates during deposition.

The boundary curve of such a correcting shutter can also be calculated with eqn. (78) for spherical domes. For that purpose, the minimum thickness $(d/d_o)_{min}$ on the dome must first be established. Then the fixed upper limit of integration π in eqn. (78) is replaced by a flexible limit θ $(0 \ldots \theta \ldots \pi)$. Equation (78) is integrated along the path of each substrate rotation at a special rotation angle θ as long as the current value of eqn. (78) corresponds to the specified limiting value $(d/d_o)_{min}$. This limiting value is numerically obtained only approximately because of the discrete integration step size. A linear interpolation between θ (i) and θ (i+1) results in proper approximation values for the boundary angle of the shutter.

This mathematical treatment must be repeated for equidistant radial positions of the spherical segment. The result is the contour curve of a corresponding shutter expressed in a sequence of triples of numerical values in spherical polar coordinates $P = P (R,\gamma,\theta)$ on the spherically shaped substrate-holder surface. Spherically shaped shutters, however, are difficult to produce and to adjust. It is therefore better to use plane shutters. A corresponding plane shutter is designed by perspective projection with the vapour source as projection center on a plane plate which is located below the spherically shaped shutter.

Fig. 51 shows the contours of static correcting shutters which have been obtained by calculation with a $\cos^n-$ evaporator on the axis of rotation of a spherical segment.

Despite the various skills technical available, the thickness uniformity on a larger number of substrates cannot be made clearly better than $\pm$ 1 %. The reasons for this fact are complex, but lie mainly in insufficient mechanical stabilities of the rotating and static installations. To overcome these difficulties, a special plant has to be constructed, as was done 20 years ago by Anders [261] to fabricate narrow-band interference filters, and more recently by Dönz [262]. With such a plant, having for example, as well as other installations, a self-centering rotary drive for the substrate holder, a special evaporator and precisely positioned correcting shutters, a thickness uniformity of $\pm$ 1 ‰ [262] was obtained on plane substrates. Theoretical considerations on optimization of thickness uniformity can be found in [263]. Generally uniformity in thickness can also be obtained with rotating shutters, see [264, 265, 266, 267], and with rotating sources [260]. However, this is beyond the scope of this monograph. For uniform coating of static large areas, ring sources can be employed. Equations for the thickness distribution from ring sources have been developed by Strong [268, 265], Thorndike [268], Fisher and Platt [269] and Holland and Steckelmacher [253, 258] and for curved surfaces by Behrndt [260]. It is generally assumed in the mathematical treatments that the evaporating ring of a specific radius is parallel and concentric to the substrate.

In 1935, Strong [268] coated telescope mirrors up to 2.5 m in diamter. Coils of tungsten loaded with aluminium were used as sources. An arrangement was formed consisting of three concentric circles with 4 coils close to the center, 12 coils and 24 coils on circles with 62.5 cm and 125 cm radius respectively. Later Strong [270] coated mirrors with 5 m in diameter using a total of 175 coils positioned on five concentric circles. The thickness uniformity was satisfactory in all instances and the mirrors were used in astronomical observatories.

6.2.1.5.2.3 EFFICIENCY OF ENERGY AND MASS

In thin-film production, the expenses of consumption of energy and material are important factors. For efficiency of energy usage, its effectivity η is generally used as a standard.

The thermal effectivity η_e of an evaporator is formed by the quotient:

$$\eta_e = \frac{E_2}{E_1} \tag{79}$$

in which E_1 is the input energy and E_2 denotes the effective energy. As with other processes in evaporation, only a portion of the input energy E_1 of the evaporator is consumed as effective energy E_2 for evaporation. The effective energy is the sum of the amount of energy required for heating the evaporation material, the heat of fusion and the heat of vapourization. Constituents of the lost energy are the heat capacity of the heated, possibly melted but non-vapourized coating material and the losses by thermal conduction and heat radiation of all parts of an evaporator. With evaporation by electron-beam heating, the losses produced by scattering during ionization and by reflected electrons have also to be considered. The losses caused by thermal conduction show a linear dependence on temperature T of the evaporant while the heat radiation increases as T^4. Since the specific evaporation rate is exponentially dependent on temperature T, good thermal effectivity is obtained when the vapour emitting surface is made very hot. Furthermore, the application of high heat transfer resistances, small heat radiation emitting areas and the use of heat reflectors increase the thermal effectivity. A qualitative measure for η_e is the specific expenditure of energy. As is reported in the literature [271, 272] the specific effective energy during evaporation of aluminium at 1500°C is 2.4 kWh kg^{-1}. However, in practical evaporation from hot crucibles with electron beam guns, the required energy ranges between 7 and 20 kWh kg^{-1}. In the case of aluminium evaporation from resistance-heated TiB$_2$–block sources, the required energy ranges between 50 and 100 kWh kg^{-1} [273]. Optimization of the thermal effectivity η_e is important because cheaper cooling systems can then be applied and the thermal loading of the plant walls and of the substrates can be kept lower. So the release of adsorbed and dissolved gases from the walls is reduced and heat-sensitive substrates such as most plastics receive lower intensities of damaging radiative heat from the hot parts of the evaporator.

It is interesting and frequently commercially significant to know the amount

190

of material of the evaporant which is lost because it condenses on the walls of the plant or on correcting shutters. The efficiency of the evaporated material is determined by the mass effectivity η_m. It is formed by the quotient of the mass deposited on the substrates m_s and the total mass emission of the evaporation source:

$$\eta_m = m_s / m \tag{80}$$

As an example, the mass effectivity of a spherical segment is investigated according to the calculations of Deppisch [255].

Equation (78) of Section 6.2.1.5.2.2 enables the determination of the normalized thickness profile $(d/d_o)_\gamma$ along the radius of the dome. The actual film thickness d in any position is obtained by multiplication with the thickness d_o of the normalization point:

$$d = d_0 \, (d/d_0)\gamma \tag{81}$$

Under consideration of the parameters n and A of the vapour cloud, d_o can be expressed by:

$$d_0 = m/\rho \, (n+1/2\pi) \, (\cos\beta_0/s_0^2) \, [(1-A)\cos^n\alpha_0 + (A/n+1)] \tag{82}$$

Making use of Guldin's rule, the mass covering m_s can be determined. As can be seen in Fig. 52, the cross section in the symmetry axis of the spherical segment

$$m_s = \rho \cdot 2\pi \cdot x_s \cdot A_s$$

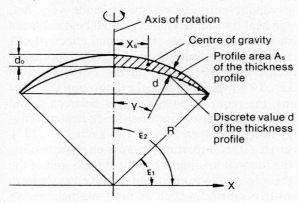

Fig. 52
Deposited film mass m_s on a spherical segment.

shows a sickle-shaped geometry of the deposited thin film. Considering one half of the symmetrical figure with the dashed profile area A_s, it becomes obvious that rotating this area A_s around the axis of rotation yields a body whose mass is identical with the required mass covering m_s of the spherical segment:

$$m_s = \rho \, 2\pi \, x_s \, A_s \qquad (83)$$

The symbol x_s is the distance between the center of mass of the profile area and the axis of rotation. For calculation of the profile area A_s and the distance of the center of mass x_s, it is useful to apply polar coordinates. The following equations are then obtained:

$$A_s = 1/2 \, (\pi/180) \, [R^2 \, (\varepsilon_2 - \varepsilon_1) - \int_{\varepsilon_1}^{\varepsilon_2} (R-d)^2 \, d\varepsilon] \qquad (84)$$

and

$$x_s = 1/3A_s \, [R^3 \, (\sin\varepsilon_2 - \sin\varepsilon_1) - (\pi/180) \int_{\varepsilon_1}^{\varepsilon_2} (R-d)^3 \, \cos\varepsilon \, d\varepsilon] \qquad (85)$$

Because we have thin films, the terms of eqns. (84) and (85) nearly equal in size are only small differences. Furtheron, the thickness profile is often not obtained in the form of a mathematical function but only as single discrete values.

Therefore the calculation of eqns. (84) and (85) must be performed with precise numerical integration methods, such as with Simpson's rule.

Specifying the amount m of the material evaporated from the source and knowing its density ρ, the values of d_o, x_s, A_s and thence, via the deposited mass on the spherical segment m_s, the desired mass effectivity η_m can be obtained.

The following η_m values have been obtained with Al films evaporated with an electron beam gun [255]. The evaporation source with a vapour characteristic of $m/m_o = 0{,}82 \cos^{3.5}\alpha + 0.18$ used in an optimized geometrical arrangement with R = 50 cm, h = 50 cm, q = 27.5 cm yields a mass effectivity of 32 %. Moving the source into the axis of rotation of the spherical segment increases the mass effectivity remarkably to about 57 %. The areal thickness or mass distribution, however, becomes considerably unsymmetrical in this case. Uniformity in mass distribution is obtained only by the additional use of a static shutter which again reduces the mass effectivity. Although the new mass effectivity is now 42 %, this value is clearly superior to that obtained with an optimized evaporation source alone.

6.2.1.5.2.4 EVAPORATON TECHNIQUES

There are various methods for performing evaporation [252, 257, 258]. However, for many materials there is only one optimum evaporation technique.

Principally, this concerns the correct choice of evaporation method, the evaporation source and the evaporation temperature. The technique to be applied depends primarily on the material used and the requried film purity, but of course in practice also on the existing plant and installations.

Direct resistance heating

The sublimation of, say wire or rod-shaped evaporation material clamped between two electrodes by heating up via direct flow of current can rarely be applied. Examples are elements such as C, Fe, Ti, Rh.

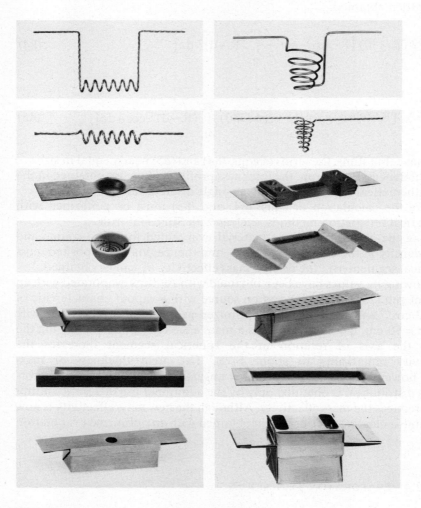

Fig. 53 a
Various types of resistance heated evaporation sources.

Fig. 53 b
Resistance heated metal boat during evaporation.

Indirect resistance heating

This method is most common. The evaporation material is placed in a container made of Mo, Ta, W or C which can be in the form of a boat, crucible, coil or strip. In some cases ceramic crucibles or inserts are also employed, made of Al_2O_3 and BeO, BN or BN/TiB_2. The container is heated by current flow and the material is evaporated or sublimed from this. Various types of resistance-heated evaporation sources are shown in Fig. 53. Undesired but possible chemical reactions can cause film contamination [274]. The formation and evaporation of vapourizable compounds of the boat material upon contact of the hot boat wall with reactive gases, as well as with some reactive and/or decomposable film material, is not always low [275], but can, however, often be avoided by appropriate choice of the evaporation source material and by special pre-treatment. A further evaporation method uses radiation heating. The radiation-heated vapour source consists generally of a resistance-heated spiral radiator of tungsten wire which is mounted above the evaporant surface in an open crucible. It can be used for the evaporation of easy volatilised materials.

To assist in choosing the type of source and which source material is best suited for a given evaporation material, reference should be made to the study of the information given in [252] and [258], and in the coating material catalogues of suitable companies.

Electron-beam heating

This mode of evaporant heating has recently been established as a universal method for producing highly pure films. The material is placed in a watercooled crucible where it melts in its own environment and there is practically no chance of unwanted reaction with the wall. The high-energy electrons produce very high

194

temperatures on the surface of the evaporation material so that even high melting metals and dielectrics can be evaporated. The versatility of electron beam guns makes possible the deposition of many new and unusual materials. There are easy ways of varying the power density so that excellent adaptation of consumption of energy can be performed. Even though the investment cost of this equipment is relatively high, in the long run this method will be cheaper due to reduction of the continuous expenses for other evaporation sources (e. g. boats can rarely be used more than once).

Of the various electron-beam heating methods [252, 257, 276], two proven types are mentioned here.

Deflection gun

As is shown in Fig. 54 the electrons emitted from a hidden hot filament and accelerated in the electrical field are deflected by, say, 270° magnetically with this type of gun and thus focused to the evaporation material. Small amounts of evaporated material from the hot filament cannot contaminate the films. A high voltage in the range between 6 – 10 kV accelerates the electrons in the direction to the anode that is the crucible. The size of the focal spot can be varied by variation of the Wehnelt potential and in addition there is an electromagnetic X-Y sweep. The first is particularly important when metals and dielectric materials have to be evaporated in the same charge. Since the power densities of some 10 kW cm^{-2} required for metals would destroy most of the dielectrics, a wobble modulation of the focused beam is inadequate, and dielectric materials must be evaporated by a soft evidently defocused electron beam. The required power density is about 1–2 kW cm^{-2}.

A further advantage of the electron beam gun lies in the possibility of using various crucible sizes and shapes. For special applications, for instance depositions at high rate, ceramic inserts can be used to minimize the loss of heat to the coolant water.

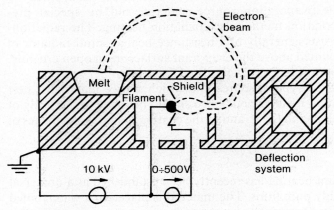

Fig. 54 a
Schematic representation of an electron beam evaporator with 270° beam deflection.

Fig. 54 b
Electron beam evaporation source with 270° electromagnetic
beam deflection BALZERS ESQ 110 with several types of crucibles. From left to right: The pot crucible, the grooved crucible,
the oscillating crucible and the four pot crucible.

Tele-focus guns

In the case of tele-focus guns, the electron emitter and acceleration system is usually at a great distance from the crucible, which can be of advantage for some applications.

Laser heating

Laser-induced vaporization has been studied widely [277-286]. On principle, it seems that lasers could be ideal power sources for evaporation. The light beam can easily be introduced into the plant and focused on a target where it produces the desired evaporation. However, the high initial costs, problems with splashing and other diffucilties have prevented application in technical evaporations.

High-frequency heating

In this method, the evaporation material is directly or indirectly heated up and evaporated inductively. Sometimes this method is applied for evaporating larger quantities of film material which become volatile at relatively low temperatures. The employment of this method is restricted however, because of the high apparatus investment and the very limited applicability.

Flash techniques

In the case of material combinations with various vapour pressures of the components like alloys and mixtures such as cermets, it is difficult to achieve a homogeneous film of a certain composition by evaporating from one single source. The method of controlled multi-source evaporation is a possible solution to overcome this difficulty, but is unfortunately complicated and expensive. Instead, satisfactory results have often been obtained by flash evaporation where the relevant material combination is supplied continuously to an evaporation source in small portions, the temperature of which is considerably higher than the evaporation temperature of all components of the combination, so that the varying vapour pressures play only a negligible role and a homogeneous film is generally obtained after condensation e. g. [252]. Homogeneity, uniform grain size and constant supply of the material mixture are decisive for a perfect film.

Fig. 55
Vibrating feeder for continuous feeding of small quantities of powders of alloys, cermets or mixtures.

Fig. 56
Wire feeder for evaporating large quantities of metal at high
evaporation rates.

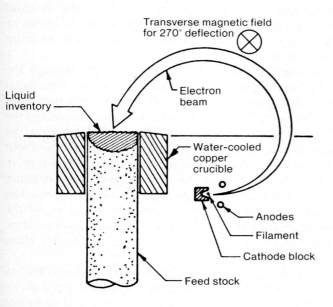

Transverse magnetic field
for 270° deflection

Electron
beam

Liquid
inventory

Water-cooled
copper
crucible

Anodes

Filament

Cathode block

Feed stock

Fig. 57
Schematic representation of a typical rod fed electron beam
evaporator.

Frequently, a vibrating feeder shown in Fig. 55 is used for the supply. Both resistance-heated boates and electron guns can be employed as evaporation sources.

Continuous evaporation

To operate continuous plants (e. g. roll coaters or multi-station coaters) and also to produce very thick films, the material must be supplied to the source under evaporation conditions. Re-filling of material is undertaken either by wire feeder, shown in Fig. 56, vibrating feeder or pill fill-up devices depending on type and form of the starting material. With electron guns, refilling of material can also be undertaken by lifting a rod supplier, as is shown schematically in Fig. 57. The use of the wire feeder in metal evaporation enables high evaporation capacity. The storage coil can hold 130 g of aluminium wire of 1 to 2 mm diamter. This quantity corresponds to a film thickness of about 50 μm on a substrate at a fixed position at a distance of 35 cm above the source. A high deposition rate for aluminium of up to 100 nm sec^{-1} can be achieved without spitting.

6.2.1.5.2.5 TRANSIT OF THE VAPOURIZED SPECIES THROUGH THE REDUCED ATMOSPHERE

In the classical model of physical vapour deposition, the atoms or molecules leave the vapourizing molten or solid surface with directions defined by the cosine law. They travel to the substrate without any interactions in the residual atmosphere and they impinge on the substrate, meeting only substrate atoms or atoms of their own kind. In reality, however, some events can occur to an atom between evaporation source and substrate.

The simplest events that can happen (to an atom on its flight through the reduced atmosphere) are collisions with inert residual gas atoms. The usual consequences are changes in direction and velocity. The atoms will no longer arrive in straight lines from the source. Depending on the residual pressure and the distance between source and substrate, evaporating atoms can arrive on the substrate surface from almost any direction. At higher pressures of about 5×10^{-2}mbar, a collision mechanism is responsible for the condensation type in which all surfaces of a substrate are coated. In this so-called high-pressure deposition, the deposition rate is controlled by the number of collision events. The coating thickness distribution depends primarily on the concentration of the vapour atoms in the volume near to the surface. This non-line-of-sight deposition is of great importance because of its use in decorative coatings of plastics, glass, ceramics and metals.

At higher pressures, as a consequence of these collisions the atoms lose so much energy that problems with adhesion may arise. In particular, metal atoms under such conditions form black, less dense coatings. However, to obtain films with properties close to those of the bulk material, deposition must be performed at very low pressures and/or condensation should be performed at higher substrate temperatures. Another event that can happen to a vapour atom is collision with reactive residual gas atoms or molecules in flight or shortly after condensation on the substrate. Such events may lead to chemical reactions if the required activation energy can be brought up. Generally chemical reactions during vacuum deposition are more likely to take place on the substrate surface than in the gas phase. This fact is applied to synthesize stoichiometric compound films of materials which show dissociation and fractionation during evaporation. The phenomena will be discussed later in the reactive deposition technique [250, 251]. Finally an evaporating atom or molecule can be ionized during the evaporation process and the formed ion can be accelerated to a biased substrate. For every atom, there exists a finite probability that it will ionize. This probability is a function of the ionization potential and the temperature. In a less dense vapour phase, it is possible to deviate substantially from thermal and electronic equilibrium since the atoms are far apart. The application of resistance heating for vapour generation yields relatively few ions, as is expected from thermal considerations alone. If, however, electron-beam heating is used deviations from the equilibrium population of ions can be gained. Electron-ion pairs are created by collisions of the vapour atoms with the high-energy electrons of the primary electron beam when it passes the vapour cloud directly above the source. The secondary electrons generated when the primary electrons bombard the vapour source provide, however, a much more powerful source of ion production, because their lower energy increases their cross section for reaction with neutral atoms. Their population is also considerably higher than that of the primary beam. These ions can be used to control the rate of evaporation [287] and when they are accelerated in direction to the substrate, their high impact energy increases the adhesion and density of the deposited film.

6.2.1.5.2.6 CONDENSATION AND FILM FORMATION

The early experiments and considerations of the condensation phenomena of vapours on solid surfaces are treated extensively by Holland [258] and Neugebauer in Ref. [252] and will therefore not be repeated here in full detail. In condensation, there exists a relation between a critical deposition rate and the substrate temperature. An incident vapour atom has a certain retention time on a surface which is proportional to the inverse substrate temperature and the binding forces. Some atoms are reflected, which means that an atom leaves the surface by a process which is the direct result of the collision of the atom against the surface. However, this effect will normally only be observed with substrates held at room temperature when the binding forces between the condensed atoms and the substrate are very weak and the energy of the arriving evaporated atoms is relatively low, for instance with Zn and Cd on glass. If, however, the number of the condensing Zn or Cd atoms per second is very high, few but large nuclei are formed. The condensation can be improved by pre-deposition with small amounts of other metals, such as Ag or Cr.

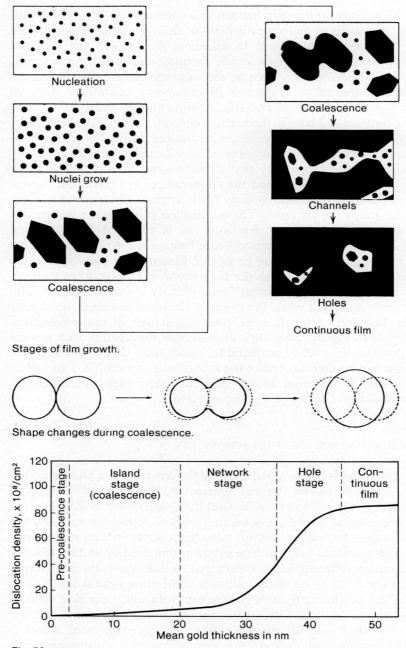

Nucleation

Nuclei grow

Coalescence

Stages of film growth.

Coalescence

Channels

Holes

Continuous film

Shape changes during coalescence.

Fig. 58 a

Schematic representation of the stages of metal film growth and approximate relationship between dislocation density and film thickness according to Pashley [318, 319].

Formation of a solid film by condensation is generally an irreversible process. A certain dwell time on the substrate and surface diffusion processes of the atoms are responsible for nuclei formation. Heterogeneous nucleation is important as the first step in the formation of a thin film by deposition from the vapour phase. Some theoretical models basing on thermodynamic and atomistic considerations have been developed in the last three decades, see [252,288]. Also, many experimental investigations have been performed on various substrate – film-material combinations, see [288,289].

Remarkable insights have been gained especially from electron microscopical in-situ experiments of nucleation and subsequent film growth under clean and controlled vacuum environment, see [290]. Many experiments have been performed with noble metals on mainly single crystalline substrates, but under other conditions also non-metals and compounds have been investigated, see [291].

With metal films at the earliest observable stage, very small (about 10 Å) three-dimensional nuclei form, and these may decorate cleavage steps on the substrate if the surface is generated by cleaving. The individual nuclei grow, without any appreciable increase in numbers, until neighbouring nuclei begin to touch. Liquid like coalescence phenomena then begin, and continue in one form or another until a continuous film is formed. The intermediate stages shown

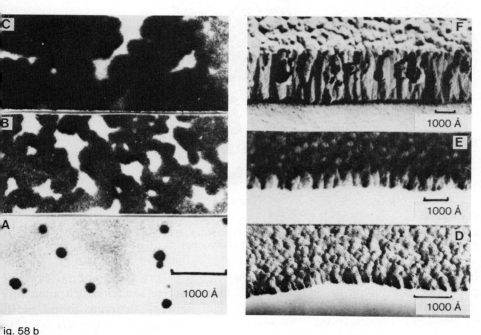

Fig. 58 b

Sequence of electron micrographs (direct transmission and Pt/C-replicas) of ZnS films of various mass thickness: A = 2 nm, B = 10 nm, C = 15 nm, D = 35 nm, E = 70 nm and F = 220 nm [291].

schematically in Fig. 58a are the island stage, the network stage, and the hole stage with secondary nucleation and growth in the holes. Finally, the continuous film is achieved. Although there have been relatively few investigations, it seems that compounds may behave similarly with no or less marked liquid-like coalescence, as can be seen in Fig. 58b.

As follows from electron diffraction, films of crystalline materials formed on amorphous substrates under normal conditions are generally polycrystalline in their structure. During further thickness growth, a columnar microstructure can often be observed and thicker film may develop a texture.

Most films deposited even at room temperature are in a non-equilibrium state and highly imperfect containing vacancies, dislocations, stacking faults and grain boundaries as can be seen in Fig. 58a, unless there is some mechanism for achieving equilibrium. The method of approaching equilibrium is by movement of atoms in and on the surface layers. The most important parameter controlling the mobility of atoms in a solid film is diffusion. Therefore if the condensation process occurs closer to the melting point of the film material, a better ordered solid film is formed. This can be achieved, for example, by increasing the substrate temperature. In addition to influencing surface mobility and ordering processes, the substrate temperature will also affect the grain size as can be seen in Fig. 59.

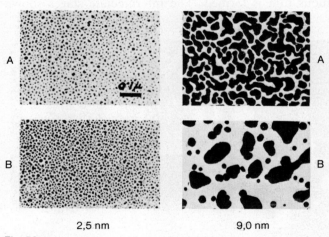

2,5 nm   9,0 nm

Fig. 59
Copper film deposited onto SiO-substrate at $p = 10^{-9}$ mbar before (A) and after (B) heat treatment at 500°C during 20 minutes.

Increasing film thickness generally also increases crystallite size. Glass-like polymeric films such as SiO_2 first also form discrete nuclei which soon link together into a polymeric three-dimensional network structure found in IR spectroscopic investigations. The properties of thin films are frequently found to differ appreciably from those of the bulk material. Numerous authors have pointed out that film structure and grain size as well as chemical changes can be responsible for these discrepancies.

6.2.1.5.2.7 EVAPORATION MATERIALS

There are about 70 elements and about 50 compounds of inorganic substances available which can be evaporated without great difficulty. When employing complicated methods, the number of compounds can perhaps be doubled. Alloys and mixtures are available in large quantities for relevant evaporation techniques. Other criteria of selection, dependent on the film application, limit the choice of material. Only a relatively small percentage of the large number of known inorganic materials can be evaporated under high vacuum without dissociation and condense as a definded thin film. In the other cases, technical tricks in the form of special evaporation techniques, are often necessary to produce stoichiometric films. Regarding technical standard, for optimum evaporation, the material is required in a relevant form and purity. Not only is the purity regarding solid contaminations important but also the gas content. The material should be pre-degassed, and it is often advantageous to be vacuum melted or vacuum sublimed.

In addition, it must be of the correct form. Powders are unsuitable, since the large surface causes them to desorb far too much gas. Grained material is frequently employed. The grain size must be adapted to the evaporation technique, e. g. for flash evaporation about 0.2 mm. Further, pressed and sintered pills are produced. Often metallic materials are offered granulated or as wire. Also, thick disks with the dimensions of the crucible have proved satisfactory especially for electron guns.

Chemical elements

Evaporation without contamination has today been solved by modern means such as new types of non-reactive boat material and inserts or electron beam guns for refractory metals most difficult to evaporate because of high evaporation temperatures. Condensation, problematic in a few cases, such as Zn, Cd, Ga, Sn, Sb, can be influenced by nucleation and the selection of evaporation speed (generally high) and substrate temperature (generally low). Reactions with the residual gas during condensation which cause undefined film products, difficult to reproduce, can be reduced by decreasing the residual gas pressure and the unwanted gas components, and by correct choice of substrate temperature and evaporation speed [292, 293].

Chemical compounds

A few simply composed stable compounds can usually be evaporated practically without dissociation, and condense to form stoichiometric compound films.

Compounds having complex anions dissociate almost without exception and show fractionating phenomena of their components.

Halides

Some simple halides can be evaporated practically without any dissociation as shown in Table 9a, the mass spectrum of MgF_2 vapour above a resistance-heated evaporation source [294]. Complex types decompose into the components and show a fractionation so that the films are no longer stoichiometrically composed, as can be seen for example in Tab. 9b for cryolite: $Na_3(AlF_6) \overset{4T}{\rightarrow}$ $3NaF + AlF_3$, [294,295]. NaF is found in the films and by recombination the desired $Na_3(AlF_6)$ is formed as well as various amounts of $Na(AlF_4)$ [296] as can be seen in Fig. 60.

Diffraction pattern	Compo-sition	Cryolite source temperatur (°C)	Evap. rate (Ås⁻¹)	Film thick-ness (Å)
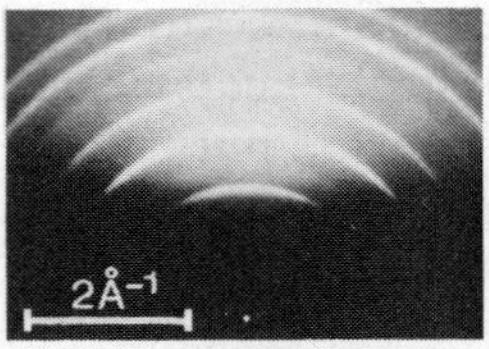	NaF	880	2	190
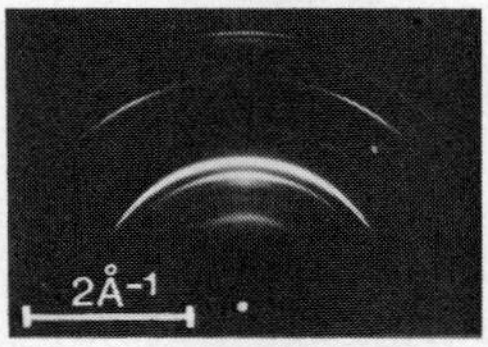	NaF + Na₃ (AlF₆)	970	15	280
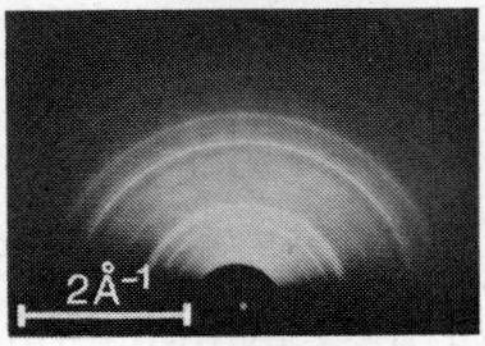	Na (AlF₄)	1020	25	1020

Fig. 60
Reflection electron diffraction patterns of films condensed from a cryolite source according to Pulker and Zaminer [296].

TABLE 9a

MASS SPECTRUM OF MgF₂ VAPOUR
Ion current of the fragments in arbitrary units

T(°C)	F⁺ 19	HF⁺ 20	MgF⁺⁺ 21.5	Mg⁺ 24	25	26	MgF⁺ 43	44	45	MgF₂⁺ 62	63	64
1045		0.4										
1070		0.5					0.8	0.1	0.1			
1100	0.1	0.4		0.2	0.03	0.03	4.3	0.6	0.6	0.4		
1120	0.2	0.5	0.35	0.8	1.0	0.1	17.0	2.3	2.2	2.0	0.3	0.3
1140	1.0	0.4	1.2	1.9	0.3	0.3	50.7	7.2	7.0	5.0	0.6	0.7
1165	2.8	0.7	3.7	4.5	0.6	0.6	117.0	14.9	15.2	11.7	1.5	1.6
1185	4.6	0.9	5.1	7.5	1.0	1.1	175.1	25.2	26.1	17.5	2.4	2.6

Before the evaporation of MgF_2 starts, a slight HF evolution is observed as a consequence of the water content in the crystals. It is known that at higher temperatures, a chemical reaction takes place according to

$$MgF_2 + H_2O \rightarrow MgO + 2HF, \qquad \Delta H = -44 \text{ kcal.}$$

In the region of 1070°C the evaporation of MgF_2 starts, as is evident from the appearance of the mass numbers 43, 44 and 45. The corresponding MgF^+ is formed by dissociative ionisation of MgF_2 vapour according to

$$MgF_2 + e^- \longrightarrow MgF^+ + F^{\cdot} + 2e^-.$$

In an analogous reaction Mg^+ is formed. This behaviour is like that of the alkali halides.

Higher aggregates like Mg_2F_4 and Mg_3F_6 which were found in small amounts (1 % dimer and 0.01 % trimer) in vapour effusing from a Knudsen-cell could not be detected in the evaporation from open boats.

Table 9b shows the mass spectrum obtained from the vapour beam above the cryolite source. During heating up of the boat, F^+ and HF^+ are also observed in the spectrum. Before melting, cryolite dissociates into NaF and AlF_3 and because of the higher vapour pressure of NaF fractionation occurs. The first mass numbers appearing in the spectrum belong to NaF. Ions Na^+ and Na_2F^+ (because NaF also forms a dimeric species in the vapour) are the main fragments in the NaF spectrum. The amount of NaF^+ is small and was therefore found only with higher vapour densities. At about 875°C, the AlF_3 component appears in the mass spectrum.

TABLE 9b

MASS SPECTRUM OF CRYOLITE VAPOUR
on current of the fragments in arbitrary units

T (°C)	F+	HF+	Na+	NaF+	AlF+	Na₂F+ / AlF₂
	19	20	23	42	46	65
810	2.4	0.9				
830	2.5	2.0	0.1			
860	2.7	3.9	9.0	0.2		1.2
870	2.8	4.7	16.0	0.3		1.8
875	3.5	5.5	45.8	1.3	1.1	19.7
900	3.7	5.5	124.0	3.2	1.7	34.6
935	5.0	5.6	236.1	6.7	3.5	75.8

Of the chemical compounds, simple fluorides give rise to fewest difficulties as regards evaporation. Technically important simple fluorides are: MgF_2, CaF_2, AlF_3, LaF_3, CeF_3, NdF_3, PbF_2 and ThF_4 [297].

Oxides

In the case of oxides, only from a few suboxides molecules can be evaporated having the same formula composition as the starting material, e. g. SiO, GeO, SnO and PbO. Evaporation of suboxides usually takes place from resistance-heated boats. By employing the evaporation via electron-beam heating, SiO_2, Al_2O_3, BeO and ZrO_2 for example can be evaporated with more or less strong dissociation. The dissociation during evaporation is partly concealed by recombination during condensation. By employing the method of reactive evaporation [250,251,298], completely oxidized films can be obtained when evaporating suboxides or even metals, e. g. TiO or Ti $+$ $\frac{1}{2}O_2$ or $O_2 \rightarrow TiO_2$. For this purpose, the suboxide or metal is slowly evaporated in an oxygen atmosphere of 1 to 3×10^{-4} mbar.

Sulphides

Dissociation is already strongly apparent when evaporating sulphides. However, during condensation a strong recombination is also evident. For example ZnS decomposes completely into Zn and S_2 according to mass spectrometrical analyses [299,294]. Nevertheless, stoichiometric films are usually obtained [300]. The dissociation of ZnS into its components supplies an explanation for the clear temperature dependency of the condensation coefficient of ZnS, which is observed even at low temperatures of 25° C to 200° C [301].

In the case of sulphides of Cd and Sb, larger deviations in stoichiometry can arise always with a metal atom surplus. The behaviour of selenides [302, 303] and of tellurides [304–306] is similar.

III – V Compounds

In the case of these compounds, dissociation into components is the rule. Nevertheless, using flash evaporation for example AlSb [307, 308] GaAS [309] and InSb [310] or specific recombination as with the so-called three-temperature method [311], e. g. (InSb, InAs, GaAs and also the compounds Bi_2Te_3 and CdSe [312, 313]), a stoichiometric film deposition can also be achieved. Moreover three-component films, such as $GaAs_xP_{1-x}$ can be produced in this way [314]. With the three temperature method, the single components of a compound are separately evaporated each from one source. The concentration of each component in the vapour phase is adjusted via the relevant source temperature. With a suitable substrate temperature, stoichiometric condensation and compound formation can now be achieved, for example in such a way that the Sb atoms which do not react with the In atoms to InSb on the substrate surface are re-emitted into the vapour phase.

Alloys and mixtures

With alloy evaporation, the vapour composition and thus also the film composition depend on the ratio of the vapour pressure of the alloy component and on the activity coefficients. In the case of varying vapour pressures of the alloy components, depletion of the easily volatile components in the evaporation source results [315]. More favourable conditions are obtained with flash evaporation. In rare cases, high-frequency induction heating is also successful, e. g. for AlSi films [316]. Mixtures of several compounds or cermets also evaporate in the ratio of the single vapour pressures of their constituents. However, the possibility of mutual chemical reaction of the materials must also be considered. Flash evaporation is also a suitable method in this case [317]. However, the evaporation temperatures of the single materials should not differ too greatly and thorough intermixing is necessary. Binary alloy films can practically always be produced by controlled evaporation of the components from two separate sources. Other possibilities of multi-source evaporation and the composure of complicated materials by synthesis in the film production phase are conceivable and have been carried out to some extent. The necessary apparatus, e. g. reliable rate meters, is currently available.

6.2.1.5.2.8 EVAPORATION PLANTS

Film deposition by evaporation and condensation is, with few exceptions, generally performed by evaporating the coating material from sources mounted at the ground plate and by condensing the vapour on substrates mounted in holders at the top of the chamber. The sidewards evaporation is very seldom applied, but an inverted arrangement to the simple bottom to top deposition direction is sometimes used when large and very heavy parts such as astronomical mirrors

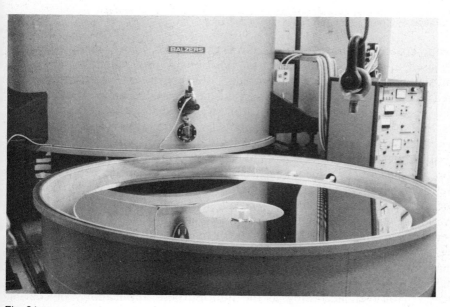

Fig. 61 a
Special coating unit to metallize large astronomical mirrors.

Fig. 61 b
Ring shaped glow discharge cathode and electrodes for the spiral shaped
Al-evaporators mounted at the top of the metallizing plant.

have to be coated, as shown in Fig. 61. As mentioned before, the evaporation
sources used are generally in eccentric positions to the palette or spherical
segment shaped and related substrate holders. Further static shutters can be used
to increase coating uniformity. A rod or ring-shaped gas discharge cathode is also
installed   because glow discharge treatment immediately before evaporation
degases the substrates and all installations, and generally increases the film
adherence. In many applications, film quality is improved by deposition on
heated substrates. For this reason, radiation heaters are often used. Infrared
radiators are specially proficient because they only heat the ir absorbing glass. If,
however, temperature-sensitive substrates have to be coated, the vapour source is
shielded by special shields to reduce the undesired heat of radiation from hot but
non-evaporating parts. Film thickness and deposition rate monitors are installed
to determine the speed of thickness growth and to detect arrival at the final
thickness. To interrupt deposition at this point, a moveable shutter is mounted
between source and substrates. This classical arrangement of installation is often
used in vertical bell jar plants, as well as in modern cubic coaters, as shown in Figs.
62 and 63. However, according to the multipurpose applicability of coating
plants, there are various special types in use. Some are shown in the following
figures.

Fig. 62 b
Classical installations of an evaporation plant.

Fig. 62 a
Classical installations of an evaporation plant.

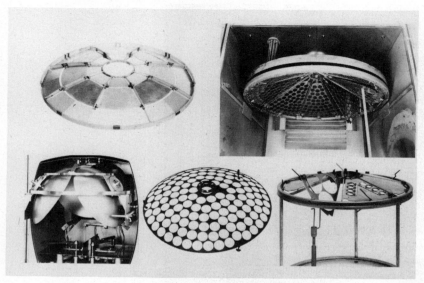

Fig. 63 a
Various types of substrate holders.

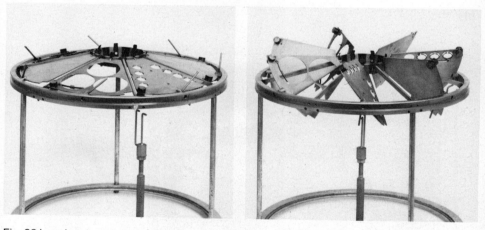

Fig. 63 b and c
Reversing calottes for coating both sides of a substrate in bell jar coaters.

With metallizing decorative plastic parts and automobile plastic reflectors, a good film quality is required but also important and decisive are low expenses of the coating process. Metallization of such parts is often done in horizontal cylindrical coaters with the evaporation sources in the cylinder axis and the substrates arranged around them on planetary axes as can be seen in Fig. 64.

Fig. 64
Horizontal cylindrical coating unit with spiral shaped resistance heated evaporators.

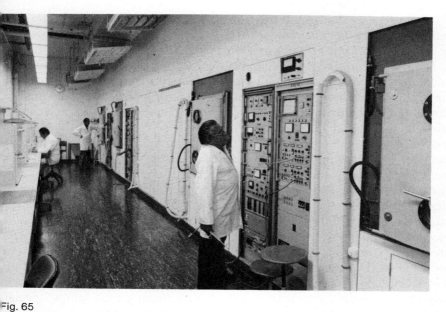

Fig. 65
Clean room in optical film production.

Fig. 65
Clean room in optical film production.

In the case of the metallization of plastic foils roll coaters are generally used. Today, both in semiconductor fabrication and also for the deposition of the highest quality optical films, clean and dustfree working conditions are required. Cubic coaters offer excellent possibility for installation in clean rooms, as can be seen in Fig. 65. The doors of such plants can be operated in the clean room but the pumping station with all other parts of the plant is in the service room and can be maintained here.

Coating large quantities of similar parts with the requirement of good and reproducible vacuum conditions is genereally performed in a continuous processing plant, which is shown schematically in its simplest form in Fig. 66. The coating chamber is always kept under vacuum and all necessary operations before and after coating like heating, glowing and cooling after deposition, are performed in separate chambers connected together and to the coating chamber by locks. In this way, in the coating chamber the conditions with respect to residual gas pressure and gas composition are very constant.

For some special applications as for instance electrode coating of oscillating quartz crystals, deposition of selenium films for electrical metal rectifiers and for dry copying systems, and for coating of decorative incandescent lamps, special plants are fabricated. Fig. 67 shows such a lamp bulb coating unit.

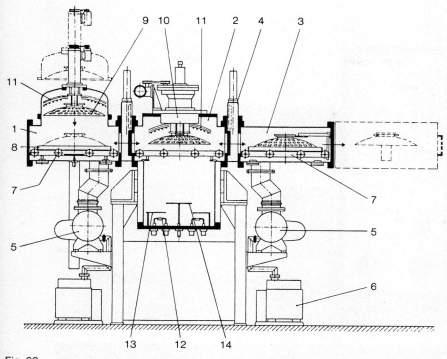

Fig. 66
Schematic representation of a simple continuous coating unit.

1 = lock chamber	6 = fore pump	11 = substrate heating
2 = coating chamber	7 = transport system	12 = evaporation sources
3 = lock chamber	8 = dolly	13 = shutter
4 = lock valve	9 = substrate holder	14 = glow discharge electrode
5 = turbo-molecular pump	10 = substrate holder drive	

6.2.1.5.3 FILM DEPOSITION BY CATHODE SPUTTERING

Cathode sputtering or more exactly the sputtering of (usually solid) materials by bombardment with positive noble gas ions is the oldest vacuum process for producing thin films. Sputtering, the cause of the erosion of the cathode in glow discharge, often an undesirable effect, was discovered more than 120 years ago by Grove 1852 in England [320] and Plücker 1858 in Germany [321] during gas discharge experiments.

Soon afterwards in 1877, metal sputtering was applied in the production of mirrors [322], and later it was used for decorating various articles with noble metal films. Around 1930 it was used for applying electrically conducting films of gold onto the wax masters of the Edison phonographs. It then became of less importance for the next 30 years compared with the rapidly developing deposition of films by evaporation and condensation in high vacuum.

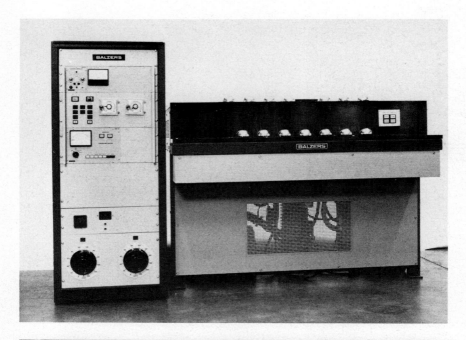

Fig. 67 a and b
Lamp bulb coating unit.
Parts of the lamp bulb are metallized with aluminium.

However, since about 1955 sputtering has undergone a renaissance. Intensive studies of the phenomena occurring during the sputtering process [323] and hence better control of the process together with the technical requirement for high quality films having very good adhesion and process specific properties have certainly contributed to the wide application.

6.2.1.5.3.1 GENERAL CONSIDERATIONS

Without going into detail, we will attempt to discuss the parameters and system components of significance for cathode sputtering, and thus give a general idea of the fundamental correlations of cathode sputtering and also of the opportunities offered by this process.

Fig. 68 shows a schematic drawing of a greatly simplified set-up for cathode sputtering. The process takes place in a vacuum chamber, which has been evacuated as well as possible before coating. In order to prevent contamination of the films to be produced by incorporation of less defined residual gas, the starting pressure should be 10^{-6} mbar or lower. The working pressure is then achieved with the working gas. The sputtering process itself takes place in a gas discharge which is ignited at a pressure between 10^{-3} and 10^{-2} mbar, depending on the special variant of the method. A requirement of the vacuum pumps is therefore an ultimate vacuum as high as possible and also a high and constant pumping speed in the mbar range. These requirements are met by diffusion pumps or even better by cryopumps and by turbo molecular pumps which are also usually applied in practice. In order to maintain the gas discharge, a gas inlet, for instance in the form

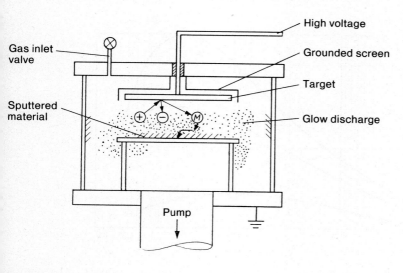

Fig. 68
Simple diode type sputtering system.

of a needle valve, has to be provided. The process, itself, runs on a flow-through principle. Noble gases, usually argon, are used as the working gas. However, for special applications, almost any other gases and gas mixtures can be used.

Two electrodes are installed in the chamber, one of them the so-called target serves as material source for the films to be produced and is at a high negative potential. A substrate holder is situated opposite the target, which can be earthed or applied to a floating potential. Furthermore, this holder can be heated or cooled. The positive ions produced in the gas discharge are then accelerated to the negative target. Upon bombardment of the target, they cause ejection of mainly neutral particles by impulse transfer phenomena. The ejected particles wander through the working gas and condense on the substrate. The energy range of the ions is usually between 10 and 5000 eV. A significant part of cathode sputtering is accordingly the bombardment of a solid surface with energetic particles. Therefore, the effect of such bombardment on the target will be considered in more detail.

As a result of the ion bombardment of solid materials depending on the energy and type of ions, various processes occur. In the case of ion energies of a few volts, only interchanges of sites or migration processes on the surface of the solid material are effected. With increasing energy, a threshold value is reached from which erosion of the material commences.

As shown in Fig. 69, the erosion process rises with increasing energy of the ions until a flat maximum, above which a preferential implantation of the bombarding ions takes place in the solid materials. This is an effect which, of course, is applied for doping solid state materials, usually semiconductors, with foreign atoms. For argon ions (Ar^+), the penetration depth in massive copper is

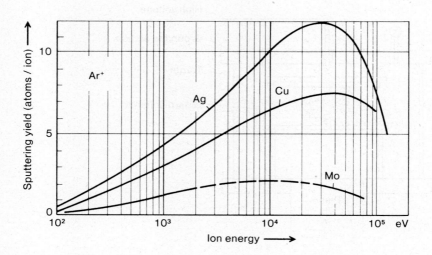

Fig. 69
Number of ejected atoms per incident ion as function of ion energy.

about 1 nm per 1 keV ion energy. The process of material erosion in the sputtering process is definitely determined by momentum transfer of the impinging ions on the atoms of the upperlayers of the lattice of the solid material, and has a faint resemblance to the behaviour of billiard balls. This was soon assumed [324], but could be confirmed only relatively late [325]. The process is schematically shown in Fig. 70.

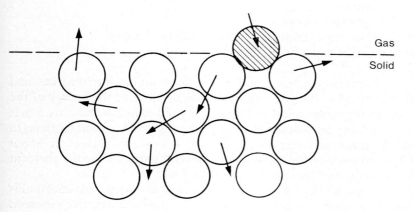

Fig. 70

Collision cascade in a solid material during ion bombardment with ejection of two atoms.

Even by 1942, Fetz [326] observed that the sputtering yield increases for flat angles of incidence of the ions and in 1954 Wehner [327] proved a preference of erosion in the forward directions if the surface is bombarded obliquely with low-energy ions of 1 keV.

In addition to other observations, such as the high particle energy of the sputtered-off atoms and also the sputtering-yield dependence on the crystal orientation and the high atom emission in the direction of densest packing in the crystal lattice discovered by Wehner 1955 [328], this behaviour clearly shows the error of an old idea that erosion takes place due to local overheating at the point of ion bombardment and associated thermal evaporation. In the case of monocrystal surfaces, it was further shown by Fluit 1961 [329] that the maxima and minima in the sputtering yield occur as a function of the angle of incidence of the ions. The minima arise at points of high transparency in the lattice, i. e. in directions where the bombarding ions penetrate deeply into the solid material. Sputtering experiments on various metals, e. g. made by Almen and Bruce 1961 [330], always result in a low ion incorporation in the case of materials with high sputtering yield, and high bombardment ions inclusion in the case of small sputtering yield. When employing technically sputtering, polycrystalline targets are used almost exclusively.

In the areas near to the surface of the solid target, various complicated processes occur simultaneously. These are:

knocking out of neutral atoms, compounds or fragmented species;
secondary electron emission;
ejection of positive and/or negative secondary ions;
temperature increase;
emission of radiation;
chemical reactions and dissociation;
implantation, solid-state diffusion, crystallographic changes;
reflection of incident and emitted particles.

All of these processes take place, both at the target and at the substrate, and determine the properties of the growing films. About 95 % of the energy of the bombarding ions on the target is lost as heat in the solid material and only 5 % is passed onto the secondary particles. The ratio of the sputtered neutral particles to the secondary electrons and secondary ions at 1 keV ion energy is about 100 : 10 :1. The frequency of the reflection as neutral particles for the incident argon ions (Ar^+) is 30 %, and for ions it is less than 1 ‰.

Classical sputtering is effected with a direct-current discharge (plasma) and is limited to conducting solids, since insulators quickly lose their negative potential required for sputtering with positive ions. In 1955, Wehner [331] suggested the use of high-frequency alternating voltages for sputtering non-conductors in order to compensate the positive space charge, accumulating during a semiwave, by electrons during the other half period. The high-frequency sputtering technique was developed from this, which is increasingly technically considered not only for sputtering insulators [332] and metals but also for etching them [333].

In addition to the diode methods, triode and tetrode as well as plasmas sustained by magnetic fields and special plasma guns are also applied.

The quality of sputtered films depends mainly on the quality of the target, on the temperature of the target and of the substrate, the plasma energy, the substrate potential, the vacuum conditions and the geometry of the equipment. A negative potential at the substrate, the so-called bias sputtering, has proven itself for reducing the inclusion of foreign gas.

6.2.1.5.3.2 SPUTTERING THRESHOLD AND SPUTTERING YIELD

As already mentioned, the ions incident on the cathode must have a minimum energy so that the sputtering effect can take place. Table 10 [334] shows the treshold energy (sputter limit) for a few metals with various bombardment ions.

It is surprising that the threshold energies are only slightly dependent on the ions mass but are, however, characteristic for the target material. It is remarkable that the various masses of the impact partners in this range have a hardly

TABLE 10

THRESHOLD ENERGIES OF ELEMENTS (eV)

	Ne	Ar	Kr	Xe
Be	12	15	15	15
Al	13	13	15	18
Ti	22	20	17	18
V	21	23	25	28
Cr	22	22	18	20
Fe	22	20	25	23
Co	20	25	22	22
Ni	23	21	25	20
Cu	17	17	16	15
Ge	23	25	22	18
Zr	23	22	18	25
Nb	27	25	26	32
Mo	24	24	28	27
Rh	25	24	25	25
Pd	20	20	20	15
Ag	12	15	15	17
Ta	25	26	30	30
W	35	33	30	30
Re	35	35	25	30
Pt	27	25	22	22
Au	20	20	20	18
Th	20	24	25	25
U	20	23	25	22

recognizable effect and that the threshold values found, 10 – 30 eV, are about 4 times the sublimation energy of the relevant target material. The association with the heat of sublimation illustrates that the impacts at such small energies are observed to be no longer independent of each other. The interaction lasts for quite a long time, so that neighbouring atoms are included before the primary impact is complete.

The sputtering yield is measured by the number of emitted atoms per incident ion and is an important value for distinguishing and characterizing a sputtering process. Above the sputtering threshold, the yield increases with increasing ion energy first exponentially and later achieves a flat maximum. With further increase of ions energy, the yield again decreases.

The sputtering yield, contrary to the sputtering threshold, is clearly dependent on the ion mass especially in the range of the maximum. For example, in the case of gold, for Xe^+ ions it is about 3 times higher than with Ar^+ ions.

Under otherwise similar conditions, the sputtering yield shows a periodicity as regards the position of the target material in the periodic classification of the elements and also as regards the mathematical interrelationship between the heat of sublimation and the crystal structure. The dependency of the sputtering yield on atomic number is shown in Fig. 71.

The mass of the target atoms enters into the energy transfer coefficient: $4 = 4m_1 m_2/(m_1 + m_2)^2$ (m_1 = mass of target atom, m_2 = mass of ion). The latter is of

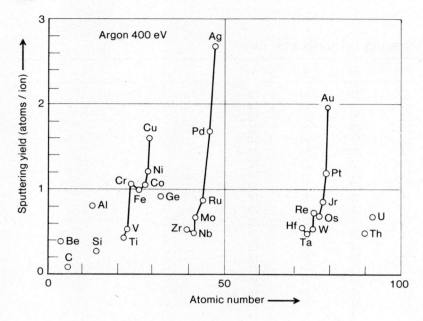

Fig. 71
Dependency of the sputtering yield on the atomic number of the target material.

significance for the theoretical treatment of the sputtering yield [335,336,337]. The sputtering yields of solid or liquid metals do not differ much, so that the target temperature has only an insignificant influcence on the yield as long as evaporation does not take place.

The sputtering yield values of various metals for the relevant conditions can often be found in tables.

The sputtering yield for alloys and compounds can generally not be ascertained from the values of the pure metals; however, estimation is usually possible.

Therefore, the yield is an important factor for sputtering; it determines the erosion rate of the target and, apart from the losses, also the growth rate of the films.

Investigations of the changes in yield with ion species have shown that noble gas ions give the best results. For economic reasons, argon is almost always used. In practice, the sputtering yields (Ar^+, $E \leqslant 2$ keV) are usually smaller than 5.

Sputtering yield is a function of many things such as: 1) relative masses; 2) energy of incident particles; 3) structure and crystallographic orientation of the target; 4) target material; 5) angle of incidence of the bombarding particle; 6) surface morphology; 7) sputtering pressure, etc. Relative sputtering yields are helpful in determining ease of sputtering, such as saying that beryllium will not sputter as readily as silver, but an absolute prediction of sputtering yield in a given

TABLE 11a

SPUTTERING YIELD OF VARIOUS ELEMENTS OBTAINED WITH NOBLE
GAS IONS AT 500 eV.

Gas	He	Ne	Ar	Kr	Xe	Reference
element						
Be	0.24	0.42	0.51	0.48	0.35	[349]
C	0.07	–	0.12	0.13	0.17	[349]
Al	0.16	0.73	1.05	0.96	0.82	[349]
Si	0.13	0.48	0.50	0.50	0.42	[349]
Ti	0.07	0.43	0.51	0.48	0.43	[349]
V	0.06	0.48	0.65	0.62	0.63	[349]
Cr	0.17	0.99	1.18	1.39	1.55	[349]
Mn	–	–	–	1.39	1.43	[349]
Mn	–	–	1.90	–	–	[350]
Bi	–	–	6.64	–	–	[350]
Fe	0.15	0.88	1.10	1.07	1.00	[349]
Fe	–	0.63	0.84	0.77	0.88	[351]
Co	0.13	0.90	1.22	1.08	1.08	[349]
Ni	0.16	1.10	1.45	1.30	1.22	[349]
Ni	–	0.99	1.33	1.06	1.22	[351]
Cu	0.24	1.80	2.35	2.35	2.05	[349]
Cu	–	1.35	2.0	1.91	1.91	[351]
Cu<111>	–	2.1	–	2.50	3.9	[352]
Cu	–	–	1.2	–	–	[353]
Ge	0.08	0.68	1.1	1.12	1.04	[349]
Y	0.05	0.46	0.68	0.66	0.48	[349]
Zr	0.02	0.38	0.65	0.51	0.58	[349]
Nb	0.03	0.33	0.60	0.55	0.53	[349]
Mo	0.03	0.48	0.80	0.87	0.87	[349]
Mo	–	0.24	0.64	0.59	0.72	[351]
Ru	–	0.57	1.15	1.27	1.20	[349]
Rh	0.06	0.70	1.30	1.43	1.38	[349]
Pd	0.13	1.15	2.08	2.22	2.23	[349]
Ag	0.20	1.77	3.12	3.27	3.32	[349]
Ag	1.0	1.70	2.4	3.1	–	[353]
Ag	–	–	3.06	–	–	[354]
Sm	0.05	0.69	0.80	1.09	1.28	[349]
Gd	0.03	0.48	0.83	1.12	1.20	[349]
Dy	0.03	0.55	0.88	1.15	1.29	[349]
Er	0.03	0.52	0.77	1.07	1.07	[349]
Hf	0.01	0.32	0.70	0.80	–	[349]
Ta	0.01	0.28	0.57	0.87	0.88	[349]
W	0.01	0.28	0.57	0.91	1.01	[349]
Re	0.01	0.37	0.87	1.25	–	[349]
Os	0.01	0.37	0.87	1.27	1.33	[349]
Ir	0.01	0.43	1.01	1.35	1.56	[349]
Pt	0.03	0.63	1.40	1.82	1.93	[349]
Au	0.07	1.08	2.40	3.06	7.01	[349]
Au	0.10	1.3	2.5	–	7.7	[355]
Pb	1.1	–	2.7	–	–	[353]
Th	0.0	0.28	0.62	0.96	1.05	[349]
U	–	0.45	0.85	1.30	0.81	[349]
Sb	–	–	2.83	–	–	[350]
Sn solid	–	–	1.2	–	–	[356]
Sn liquid	–	–	1.4	–	–	[356]

setup is [323] impossible. The relative sputtering yield is shown in Tables 11a and 11b. Another variable in sputtering yield is gas purity. If one is trying to sputter aluminium in a dc system and there is appreciable oxygen, in effect, one is trying to sputter aluminium oxide, which is an insulator, so that very low rates will result. One solution is to go to rf sputtering, though the sputtering rate compared with dc sputtering of aluminium will be lower.

One important sputtering parameter is gas pressure. At constant discharge potential, the cathode current density is proportional to the pressure, and up to some pressure the sputter deposition rate is also proportional to the pressure [340], though there seem to be exception to this in rf sputtering [341]. At pressures above 0.1 mbar, gas scattering causes the impinging ions to lose much of their energy and causes much of the sputtered material to be scattered back to the cathode surface and the deposition rate falls off rapidly [342]. In addition to affecting the deposition rate, gas scattering may affect the film thickness uniformity and other film properties.

TABLE 11b

SPUTTERING YIELD WITH ARGON IONS OF DIFFERENT ENERGIES IMPINGING ON VARIOUS MATERIALS [323]

Target voltage:	200	600	1.000	2.000	5.000	10.000	Reference
Material			Sputtering yields, atoms/ion				
Ag	1.6	3.4				8.8	[342]
Al	0.35	1.2		2.0			[342]
Au	1.1	2.8	3.6	5.6	7.9		[342]
C	0.05*	0.02*					
Cr	0.7	1.3					[342]
Cu	1.1	2.3	3.2	4.3	5.5	6.6	[342, 347]
Fe	0.5	1.3	1.4	2.0	2.5		[342, 348]
Ge	0.5	1.2	1.5	2.0	3.0		[342, 347]
Mo	0.4	0.9	1.1		1.5	2.2	[342, 348]
Nb	0.25	0.65					[342]
Ni	0.7	1.5	2.1				[342, 348]
Pd	1.0	2.4					[342]
Pt	0.6	1.6					[342]
Si	0.2	0.5	0.6	0.9	1.4		[342, 347]
Ta	0.3	0.6			1.05		[342]
Ti	0.2	0.6		1.1	1.7	2.1	[342]
W	0.3	0.6			1.1		[342]
Zr	0.3	0.75					[342]
			Sputtering yields:	molecules/ion			
LiF<100>	–	–		1.3	1.8	2.2	[339]
CdS<1010>	0.5	1.2					[354]
GaAs<110>	0.4	0.9					[354]
PbTe<110>	0.6	1.4					[354]
SiC<0001>		0.45					[354]
SiO_2			0.13	0.4			[346]
Al_2O_3			0.04	0.11			[346]

*Kr^+ ions

Gas composition can also affect sputtering yields in reactive sputtering. Often in the rf sputter deposition of oxides, a partial pressure of oxygen is used in the discharge to compensate for oxygen loss during the deposition. A few percent oxygen usually cause an appreciable loss in rf sputtering rate [343,345]. This is not the case with all oxides, however [344].

6.2.1.5.3.3 EJECTION OF OTHER PARTICLES AND EMISSION OF RADIATION

Secondary electrons

Owing to the high negative potential, electrons are accelerated away from the target. Their commencing energy can be about target potential. They essentially contribute to maintaining the discharge by impact ionisation of neutral gas particles. When hitting the substrate, their energy, usually equal to target potential, is transformed into heat. Consequently, they contribute considerably to heating up the substrate.

Secondary ions

The emission of positive ions, which are for example used in the SIMS method (secondary ion mass spectrometry) for surface analysis, is unimportant in the sputtering process. Owing to the high negative potential, positive ions do not come away from the target surface.

Significant effects can be expected from negative ions. However, when bombarding pure metal surfaces with noble gases, these do not arise. They become important with compounds and in reactive sputtering, for example, if a reactive gas is used for the gas discharge and the growing film produces a compound of the target material with the gas.

The effects of negative ions should be considered as an analogue to secondary electrons, i. e. as ion bombardment of the substrate.

Desorption of gases

The desorption of gases has to be considered as a source of contamination. The following are possible gas sources:

physically adsorbed gases on the target surface: in this case, desorption usually takes place as a result of local heating;
chemisorbed gases: these are usually sputtered off and present a source for negative ions;
included gases contained in the target material are sputtered off or are thermally desorbed;
decomposition products in the case of compound targets when the material forms gases or other volatile components.

Radiation

The emitted radiation from the target surface consists mainly of photons in the ultra-violet and visible spectral range. The radiation results from particles and entering ions excited by resonance or Auger electron transitions. Hence, it is characteristic for the target material and the entering ions, and can be used for process analysis.

The uv radiation can have a disturbing effect when using substrates sensitive to light, as in the case of photoconductors or special plastics, etc. This is similarly valid for x-rays, the maximum energy of which is given by the energy of the incident primary particles.

6.2.1.5.3.4 ION IMPLANTATION

Some of the primary ions can penetrate, neutralize and be incorporated into the target. This takes place even at relatively low ion energies. As a consequence of implantation, the crystallographic structure of the target in regions near the surface may be affected. The energy of the ions is transformed to heat, causing considerable temperature rise at the target. To avoid this, special cooling is required.

6.2.1.5.3.5 ALTERATIONS IN SURFACE FILMS, DIFFUSION AND DISSOCIATION

Composition
The bombardment of multi-component targets with ions changes the chemical composition of the surface as a consequence of the various sputtering yields of the different components. In most cases, the surface is enriched with the material showing the smaller yield until some sort of balanced condition has been achieved. After achieving this sort of equilibrium, the sputtered compound corresponds to the target composition provided that the process parameters remain unchanged. The range over which this alteration in composition stretches is the so-called «altered layer». In the case of metals, it is usually a few nm, for compounds such as oxides it can be considerably thicker and may be about 100 nm.

Diffusion
The diffusion can considerably interfere with the build-up of the equilibrium, and also strongly influence the thickness of the affected layer. Various processes can cause the diffusion:

ion-induced diffusion;
thermal-induced surface diffusion; and
diffusion resulting from ion implantation of the sputtering gas.

An Ag-Cu target can serve as example. With a target temperature of 80°C the equilibrium is achieved after 40 min. However, with a target temperature of 270°C the equilibrium is not achieved before 200 min, and the corresponding thickness of the «altered layer» is about 1 μm.

Chemical dissociation

The dissociation energy of chemical compounds is in the order of 10 to 100 eV. Of course, such a compound can be expected to decompose if bombarded with keV ions. Part of the volatile products thus arising are pumped away with the sputtering gas so that the growing film shows a deficit of these components. This is why stoichiometric compound films as oxides, nitrides, sulphides, etc. usually can only be prepared by sputtering by the addition of small amounts of O_2, N_2, H_2S, etc., to the sputtering gas, hence in a form of reactive sputtering which replaces the losses.

6.2.1.5.3.6 SPUTTERING RATE

In the frequently used energy range of 1000 – 3000 V, the sputtering rate generally only increases slowly with the energy of the ions, but increases quite strongly with current density, which determines the number of bombarding ions. The sputtering rate about 5×10^{-4} g cm^{-2} s^{-1} is small compared to an evaporation rate of 1×10^{-2} g cm^{-2} s^{-1} achieved with an electron-beam evaporation device.

According to Gurmin [338], the sputtering rate in the case of low to medium ion energies is given by:

$$\frac{dN}{A\,dt} = S\,\frac{I}{e} \tag{86}$$

with

$$S = \frac{3}{4\pi^2}\,\gamma\,\frac{4m_1 m_2}{(m_1 + m_2)^2}\,\frac{E}{E_0} = \beta\,V \tag{87}$$

S = sputtering yield, I = ion current, e = electronic charge, γ = function of m_2/m_1, m_1 = mass of target atoms, m_2 = ion mass, E = ion energy, E_0 = heat of sublimation, β = proportionality factor, V = voltage

To obtain a high sputtering rate the best parameters are: 1) high cathode potential; 2) high cathode current density; 3) heavy sputtering gas (krypton); and 4) low sputtering gas pressure. Of course, there are limits on each of these parameters and the process and product may be sensitive to one or more of the parameters.

For basic studies to obtain pure films and higher deposition rates, various schemes have been used to promote ionization in the sputtering gas, thereby extending the process to lower gas pressures. These schemes include the addition of auxiliary electron beams along with variously shaped electrodes giving rise to the triode and tetrode or other systems. Fig. 72 shows schematically a triode system. Lower sputtering pressure and greater overall efficiency can also be achieved by the use of an applied external magnetic field. The newest mode of sputtering to increase sputtering rate is based on the magnetron principle.

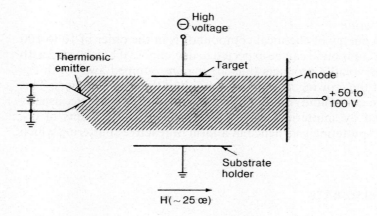

Fig. 72
Triode sputtering. Schematic view of thermionically supported glow discharge
adapted for sputtering.

The difference between conventional and magnetron processes lies largely in
the plasma environment. As will be shown later, the plasma is confined to the
surface of the cathode by a magnetic field created by permanent magnets located
under the target and by an electric field which is situated perpendicularly to the
surface. The electrons travel in spiral trajectories and can thus carry out many
ionizing collisions with the atoms of the sputter gas. An anode may be utilized to
prevent high currents from bombarding the substrate and causing excessive
heating or damage.

6.2.1.5.3.7 PARTICLE VELOCITY AND ENERGY

The mean particle velocity for most metal atoms having an atomic number
larger than 20 is between 4 and 8×10^5 cm s^{-1}, if sputtering is performed with
krypton ions of 1200 eV.

When using argon ions, the velocities are between about 3x and 6×10^5 cm s^{-1}
and are little influenced by changes in the ion energy between 600 and 1200 eV.

Heavy elements have the highest particle energy (e.g. U, $E \simeq 44$ eV) and
light elements the highest particle velocity (e. g. Be, $v = 11\times10^5$ cm s^{-1}).

The energy increases linearly with mass. For the lighter elements, it is about
10 eV and on average achieves 30 to 40 eV in the case of the heavy elements.
Materials with a high sputtering yield usually have low particle energy.

6.2.1.5.3.8 ANGULAR DISTRIBUTION

Distinct differences exist between monocrystalline and polycrystalline
targets, as mentioned above. For single crystals, clearly preferred directions of the

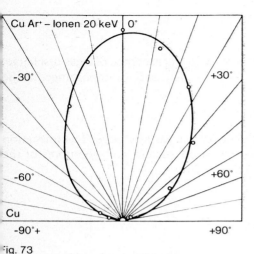

Fig. 73

Directional distribution, cosine law, of Cu atoms sputtered by Ar ions of 20 keV according to G. K. Wehner.

sputtered atoms can be observed. Polycrystalline solids show almost a cosine distribution as can be seen in Fig. 73. The distribution is influenced by the energy of the bombarding ions. The cosine distribution is better fulfilled with higher ion energies, e. g. > 1000 eV, than with low energy values. When sputtering is carried out from large, flat areas, the flux of ejected atoms will be nearly constant except near edges or other geometrical discontinuities where the electrical field curvature causes a focusing effect.

5.2.1.5.3.9 COMPOSITION OF THE SPUTTERED MATERIAL

In the case of a simple metal target, the sputtered species mainly comprises neutral, single metal atoms. There are only very small numbers of 2-atom aggregates. For manufacturing alloy films, the sputtering technique is very suitable because it is based on the mechanism of momentum transfer and the uniform erosion thus caused. The sputtering rate is almost independent of the temperature, and so the alloy target can be kept so cold that no interferring diffusion processes due to heat arise. Without mass transport in the target, after a short start-up time, the composition of the sputtered atom species remains almost the same. The composition in the film can, of course, change if the condensation coefficient and the directional distribution for each alloy component strongly differ.

The processes in the case of sputtering chemical compounds, such as oxides, are more complex than those of metals. As already mentioned, the impact energy often causes dissociation of chemical compounds, releases oxygen from oxides, and thus considerably alters the composition of the target, and the sputtered species may contain variable amounts of molecular fragments.

6.2.1.5.3.10 THE GAS DISCHARGE

Usually, the sputtering process takes place in the presence of a gas discharge. Therefore, in the following the aspects of the gas discharge of importance for the sputtering process will be discussed.

Direct current discharge

The most simple case is that of a direct voltage gas discharge. The formation of a gas discharge can best be discussed based on current/voltage curves of the phenomenon. Such a curve is presented in Fig. 74.

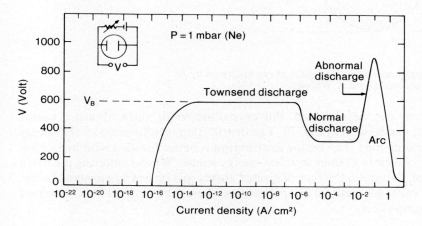

Fig. 74
The formation of a dc glow discharge.

By applying a dc voltage between both electrodes, first a very small current flow is effected because only a few ionized particles exist which can contribute to the current. Increase of the voltage has the effect that the charged particles gain sufficient energy in order to produce additional charged particles by impact ionization. This leads to a linear growth of the discharge current, the voltage thereby remaining constant, limited by the high output impedance of the voltage supply. This range of the discharge is frequently called a «Townsend» discharge.

Under certain conditions, an avalanche process can now take place. Ions strike the cathode and release secondary electrons there. These are accelerated in the field of the cathode and due to collisions with the residual gas atoms and molecules produce new ions, which are again accelerated towards the cathode and produce new secondary electrons. If a condition is achieved, where the number of produced electrons is sufficient to produce so many ions that they can again produce the same number of electrons, then the discharge is self-maintaining.

In this case, the gas starts to glow, the voltage falls and the current rises abruptly. This range is called «normal discharge». Since for most materials the number of secondary electrons produced per incident ion is approximately 0.1, about 10 to 20 ions must strike the given surface in order to release one secondary electron. The bombarded cathode surface adapts itself in this range to the discharge phenomenon so that the discharge is maintained. However, this leads to irregular bombardment of the cathode.

A uniform current intensity distribution can be achieved by further increasing the applied power, finally causing a rise of current and voltage. This range, the so-called «abnormal» discharge, is used for cathode sputtering.

In the case of the use of an uncooled cathode, with a current density of about $0.1 \ A \ cm^{-2}$, an additional thermionic emission of electrons takes place. This results in another avalanche effect and since the output impedance of the supply limits the voltage, a discharge with low voltage and high current density commences.

The so-called «breakdown voltage» V_b is decisive for the formation of a gas discharge and is dependent on:

the mean free path of the secondary electrons and
the cathode-anode distance.

This is a qualitative statement of Paschen's law which relates V_b to the product of gas pressure p and electrode separation d : $V_p = f(p \ d)$.

The formation of a gas discharge can be prevented by too low a pressure, or too short a distance between cathode and anode, since then the electrons do not collide with sufficient gas particles to produce sufficient ions for maintaining the discharge. On the other hand, too high a gas pressure can too strongly decelerate the produced ions by collision.

This means that the discharge can only be maintained in a certain pressure range.

As seen in Fig. 75, there are various dark and bright regions in a gas discharge and the voltage as well as the space charge change near the cathode.

Concerning the luminous phenomena, near the cathode there is a light region, the cathode glow. The glow arises from the neutralization of ions of the gas discharge and from the decaying of excited states.

The radiation is characteristic of the cathode material and the incident ions. A dark space arises in front of the cathode glow. In this region the secondary electrons are accelerated away from the cathode. The width of this zone corresponds approximately to the mean free path of the electrons.

In this range also, the positive ions are accelerated to the cathode. The velocity of the ions is very much smaller than that of the electrons; the ion density is high in the dark space. A consequence of this is the high space charge which is shown in Fig. 75. At the end of the dark space, the first collisions of electrons with neutral gas particles take place and thus the production of ions. This range of negative glowing is again characterized by a distinct emission of light. The Faraday dark space and the positive column adjoining the region of the negative glow present almost field-free spaces, as can be derived from the potential curve

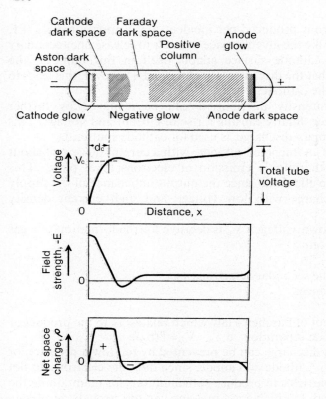

Fig. 75
Glow discharge at low pressure with its luminous and electrical characteristics.

and from the electric connection to the anode. In the case of cathode sputtering, the substrate holder is usually located in the region of the negative glowing. The cathode-anode distance should, however, be at least three to four times the width of the dark space.

High-frequency discharges

Considerable changes arise in the discharge if high-frequency electric fields of more than 50 kHz are applied. The method of high-frequency cathode sputtering differs from direct voltage sputtering in two main ways:

The electrons oscillate in the range of the negative glow and can gain sufficient energy to carry out more ionizing collisions. This results in the discharge being less dependent on the emission of secondary electrons, and the «breakdown» voltage is reduced.

The electrodes (target) no longer need to be electrically conductive since the hf voltages can be capacitively coupled. In principle, any material including insulators can therefore be used as target.

In practice, work is carried out in the frequency range 5 to 30 MHz, usually on the industrial frequency of 13.5 MHz.

With these frequencies, the ions are relative immobile and at first no ion bombardment of the surface is expected to take place. That this is not the case can be seen in Fig. 76a and 76b.

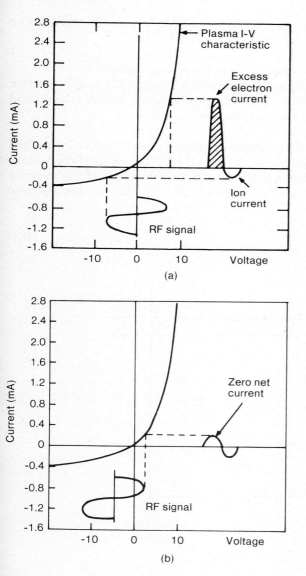

(a)

(b)

Fig. 76

The formation of a pulsating negative sheath on a capacitively coupled surface in a high frequency glow discharge.

232

The current/voltage characteristic curve of the discharge is similar to that of a rectifier with a high leakage current. At first, a high excess electron current arises. However, since owing to the capacitive coupling of the target no voltage compensation can take place, the target surface charges negatively until the voltage is called the dc potential and corresponds approximately to the peak value of the applied voltage.

In order to now achieve sputtering of an electrode, two conditions must be fulfilled:

the target must be coupled capacitively to the voltage supply;
the cathode surface must be small compared with the directly coupled surface which is usually formed by the system earth, and hence includes the ground plate and the surface of the vacuum chamber.

It is decisive that the majority of the voltage drop takes place in the vicinity of the cathode, in order to achieve sputtering of the cathode. The ratio of the voltage between the capacitively coupled small electrode and the range of the negative glow V_c to voltage V_d between the negative glow and the directly coupled large electrode is given by:

$$V_c / V_d = (A_d / A_c)^4 \tag{88}$$

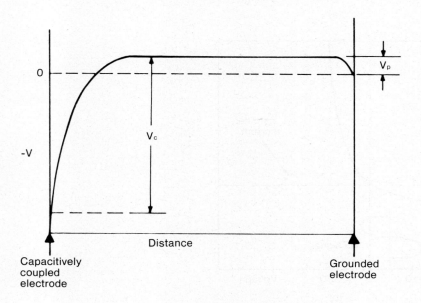

Fig. 77
Voltage distribution in an high frequency glow discharge from a small, capacitively coupled electrode (target) to a large, directly coupled electrode.

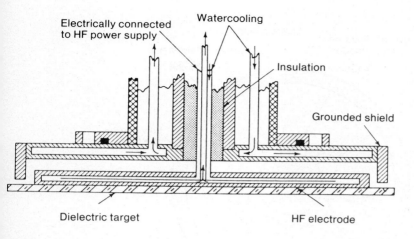

Fig. 78

Cross section of a typical HF-cathode including cooling system.

where A_d, A_c represent the corresponding surfaces. For the case $A_d \gg A_c$, the potential characteristic can be seen in Fig. 77.

Summarizing, it can be stated that the advantages of hf sputtering lie in the reduction of the breakdown voltage which can be applied in order to reduce the gas pressure and in the possibility of being able to sputter almost any material, hence also insulators. Fig. 78 shows a typical configuration of a hf cathode.

Discharge supported by a magnetic field

A normal gas discharge is a relatively inefficient source for ions, since only a few percent of the gas particles are ionized.

However, a higher portion of ionized particles is desirable for the sputtering process because firstly the required gas pressure can thus be reduced and simultaneously an increase of the erosion rate and growth rate, respectively, is possible.

This can be achieved by applying magnetic fields perpendicularly to the target surface. Due to these E B configurations, the electrons are forced in spiral paths parallel to the target surface. This step results in a considerable increase in ionization efficiency of the electrons. A schematic representation is shown in Fig. 79. Principally, there are two possible geometries, the cylindrical magnetron where the target is formed in the shape of a cylinder and the planar magnetron. Both kinds are shown in Figs. 80a, 80b and 80c. The magnetron set-up can be used for both dc and hf discharges. The process is similar to the cases discussed; however, usually irregular erosion of the target occurs owing to the inhomogeneous progress of the magnetic field and also altered potential conditions at the substrate and target.

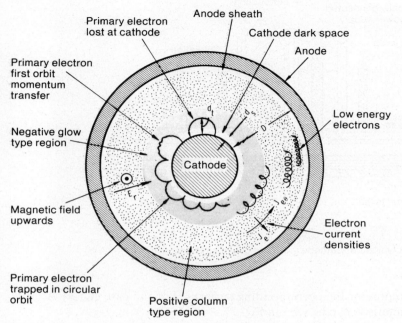

Fig. 79

Schematic representation of electron transport processes in a discharge between a cylindrical cathode and a coaxial anode in a uniform axial magnetic field.

The increase in rate, obtainable with such a device can be demonstrated in the case of a Cu-target:

sputtering in the planar magnetron set-up gives a rate of about 7 mg/s kW conventionally, about 0.7 mg/s kW is achieved.

The gas pressure in the plant can be reduced to the range of 10^{-4} mbar.

Potential conditions at the substrate

Since the properties of the growing film can be strongly influenced by particle bombardment and this bombardment – as far as ions are concerned – is controlled by the potential gradient in the vicinity of the substrate, these conditions are of great importance.

The substrates in the discharge can be considered as probes on a floating potential in a plasma, especially if they are insulating, such as glass. However, this is only an approximate consideration, because the gas discharge is not an ideal plasma. Fig. 81 shows the ideal characteristic of the I/V curve of such a probe. If the substrates are not earthed, they adopt a slightly negative potential. This is a

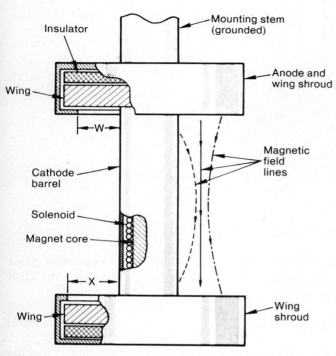

Fig. 80 a
Design of a cylindrical-post magnetron sputtering source.

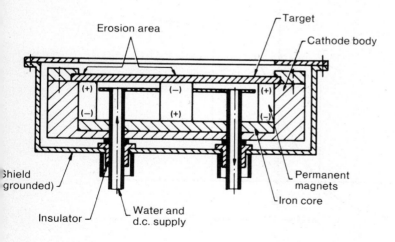

ig. 80 b
ross section view of a planar magnetron.

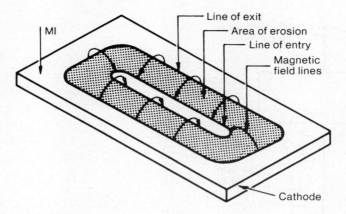

Fig. 80 c
Magnetic field and electron path in a planar magnetron.

consequence of the high electron mobility relative to the ions. On average, more electrons strike the substrate, hence it becomes negatively charged. In the ideal case, this potential V_f can be written:

$$V_f \;=\; -(1/2\,e)\,k_B\,T_e \ln\,(\pi m_e/2M) \tag{89}$$

Hence this voltage V_f is mainly dependent on the electron temperature T_e and the ratio of the electron mass m_e to the ion mass M. The substrate has the same potential as the plasma at V_p, i. e. in this case charged particles strike the substrate

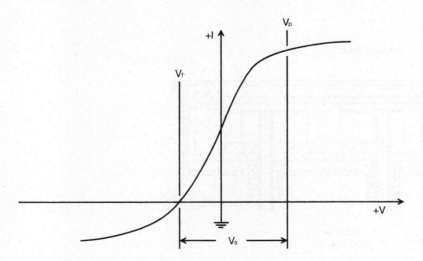

Fig. 81
Ideal I/V characteristic of a Langmuir probe in a gas discharge.

only statistically due to thermal movement. Under ideal condidtions

$$V_p = k_B T_I \ln (2/q_I) \tag{90}$$

can again be written, where T_I presents the temperature of the ions and q_I the ion charge.

The two relationships are, however, not strictly valid, but can nevertheless be applied in order to estimate the ions and the electron temperature. Two consequences result from this consideration for cathode sputtering:

the substrates are always at a potential slightly negative in relation to the plasma;
if the substrates are earthed, the substrate potential is reduced by an amount V_f.

Therefore, all insulating substrates should be applied to a floating potential; otherwise the surface adopts a potential V_f and the conducting substrate holder a potential of $V_s - V_f$. This causes an inhomogeneous discharge at the edges and hence causes inhomogeneity in the growing films.

In the case of hf discharges, the conditions are somewhat more complicated. As already discussed, the geometry plays an important role for the potential distribution. Frequently, external voltages are applied to the substrates in order to better control the ion bombardment, so-called bias voltages. Usually negative bias voltages are involved here. These take over the role of V_f, and V_p remains unchanged. In this case $- (V_p - V_o)$ arises as total negative bias voltage.

In a dc discharge, positive bias voltages make the substrate virtually an anode. This results in a high electron current and subsequent heating up of the substrate. In both dc and hf discharges, a further rise in plasma potential to positive values is observed, i.e. the actual substrate bias potential never exceeds the plasma potential. Positive voltages are applied only in exceptional cases. It can be statet as a further consequence of this consideration that substrates are almost always under the influence of a negative bias voltage which means that the substrate must also be regarded as a «target», it thus being subject to effective ion bombardment.

Contamination of the substrate can be reduced by sputter etching. However, the best method depends on the kind of contamination and the kind of the substrate. Often, treatment in an argon discharge is not sufficient and therefore a sort of chemical etching in a reactive gas atmosphere is more frequently used.

The sputter gas

Changes in gas pressure have a distinct effect on the discharge; generally, rise in pressure causes:

rise in the discharge current;
increase in back scattering;
deceleration of energy-bearing particles by collision.

The first two effects occur opposingly; the latter can be applied in order to minimize the particle energy at the substrate.

The gas flow, mostly defined by the pumping speed of the vacuum system, should be as high as possible in order to quickly pump down contaminants. Pressure gradients must be avoided in the vicinity of the discharge, since they cause inhomogeneity.

Gas contamination

The main sources of gas contamination causing impurity of the deposited films are:

remaining residual gas after pumping down (H_2O);
desorption from the walls as a result of ion bombardment;
adsorbed and included gases in the target which are released;
leaks;
contaminations in the working gas itself.

Most of these sources can be eliminated with good vacuum technology and appropriate pre-sputtering times.

In practice, these steps often reduce undesired inclusion of foreign gas. Gas inclusion in the film can be reduced by applying a negative initial potential to the substrate, in the so-called bias-sputtering. This method is based on the fact that the strength of bonding between film atoms is larger than the bonding to the contaminations. Hence the included foreign atoms are sputtered off favoured by the incident inert gas ions. Therefore, bias sputtering has proven more suitable with Ta, Mo, Nb than with Al or Mg, since the latter form bonds with oxygen which are too strong. Since 1962 [357], the bias technique has also been applied with hf discharges.

In principle, the asymmetric hf sputtering with power division between the two electrodes is equivalent, but generally the gas content is higher, the bias voltage being applied only for a half period. Table 12 shows optimum values of the bias voltage for various materials [358].

TABLE 12

OPTIMUM BIAS FOR LOW GAS INCORPORATION

Material	Bias voltage (V)	Material	Bias voltage (V)
Ag	35	W	150
Cu	35	Mo	150
Au	40	Ta	200
Ni	50	Nb	200
Pd	75	Ge	100
Rh	100	SiO_2	150
Pt	150	Corning 7059	100

Since inclusion of reactive gases in the condensed film can result in chemical stable compounds, it is obvious, analogous to the method of reactive deposition, to undertake sputtering in reactive gases [359,360] in order to produce oxides, nitrides, sulphides and carbides. The occurrence of negative ions of the species B, which bond to the compound AB, is characteristic for the process. The element B is introduced to the gas discharge as molecule B_n e.g. (O_2, N_2) or as compound BC e.g. (H_2O, H_2S, NH_3, CH_4), dissociated there – under the condition that the ionization energy of C is smaller than that of the sputtering gas – in B$^-$(and C$^+$) by field or collision ionization and travels to the anode where a positive ion film accumulates. The incident target atoms are ionized here by collision or via a dipole interaction, merge with B$^-$ ions to form an AB molecule and condense. Increased substrate temperatures promote the reaction. In the case of several AB forms – for example CuO, Cu_2O – the expected product is dependent on the cathode voltage and the Paschen product p d, albeit in an obscure way.

Oxides can be reactively produced with the addition of just a few percent of oxygen, but higher gas contents cause coating of the target with an oxide film and result in termination of the (direct voltage) discharge.

The necessary oxygen content depends on the oxygen affinity of the element and increases with sputtering rate. Reactive sputtering is also used with nitrides (Si_3N_4, Ta_xN). In contrast with oxygen, work must be undertaken almost always in pure, oxygen-free nitrogen. Nitrifying is simplified by using NH_3; even with 5 % partial pressure Si_3N_4 forms, but also the released hydrogen is partly embedded in the film. With N_2O, oxynitrides form. The sulphides of Cu, Mo, Cd and Pb [361] could be produced with H_2S, and with Ta and CH_3 tantalum carbide [362] is formed.

Reactive gas must also be partly added to hf discharges in order to maintain the stoichiometry of the sputtering dielectrics. Sputtering of SiO_2 in pure argon causes condensation of the compound SiO_x (x = 1.94 – 1.97). Therefore, a low admix of oxygen is required – causing reduction of the sputtering rate – or subsequent tempering at about 500°C.

6.2.1.5.3.11 THICKNESS UNIFORMITY AND MASS EFFICIENCY IN SPUTTERING

Thickness uniformity of sputtered films is also of great practical importance. Model calculations exist which determine thickness distributions in the substrate plane below the sputter cathodes not only for static but also for moving substrates [363].

In contrast to evaporation with sputtering, the source of material and the substrate plane are relatively narrow together over a large area. The cathode surface is therefore assumed to consist of a very great number of point sources of equal emittance. It is further assumed that with magnetron sputtering the energy of the sputtered species is so high that in collisions with residual gas atoms only little energy is lost and only small changes in direction will occur. Thus the gas

240

pressure can be neglected in the model. For the cosine exponent of a single point source, n = 1 is assumed.

The contribution dR of one source to the total rate is:

$$dR = \frac{m}{\rho A} \frac{n+1}{2\pi} \frac{\cos^n \alpha \cos \beta}{s^2} dA \tag{91}$$

m/A	=	sputtered target material per second per square centimeter
ρ	=	density of the target material
α	=	angle between the normal of surface of the source Q and the observed point P
β	=	angle between the normal of surface at the observed point and the tie line to the source Q
s	=	distance between elementary source and observed point
dA	=	surface element of the elementary source
n	=	cosine exponent of the elementary source.

Above a rectangular sputter cathode, the total rate at an observed point in the substrate plane is obtained by integration over all elementary sources, that is the erosion zone of the cathode:

$$R = \int_A \frac{m}{\rho A} \frac{n+1}{2\pi} \frac{\cos^n \alpha \cos \beta}{s^2} dA \tag{92}$$

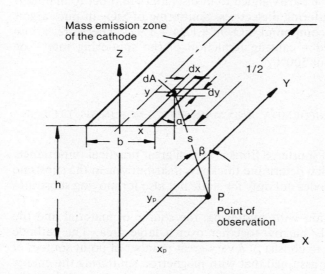

Fig. 82
Sputtering assembly with rectangular cathode [363].

In the case of a long planar magnetron, there exist two parallel erosion zones of the broadness b and the length l. At first the rate in the surface plane for one emission zone is calculated according to (92) and finally by superposition completed with the second emission zone. Under consideration of the geometrical sizes, schematically shown in Fig. 82:

$$s^2 = (x-x_p)^2 + (y-y_p)^2 + h^2 \qquad \text{and}$$

$$\cos\alpha = \cos\beta = h/s$$

and of the assumed cosine exponent n = 1, eqn. (93) is obtained from (92):

$$R = \frac{m}{\rho A \pi} h^2 \int_{-b/2}^{+b/2} \int_{-l/2}^{+l/2} \frac{dy\, dx}{[(x-x_p)^2 + (y-y_p)^2 + h^2]^2} \qquad (93)$$

As the next step, the inner integral of (93) is calculated and this intermediate result has now to be integrated according to X. Edge effects on the cathode are not considered. After integration the final result is:

$$R = \frac{m}{\rho A} \frac{1}{2} \left[\frac{b/2-x_p}{\sqrt{((b/2-x_p)^2 + h^2)}} + \frac{b/2+x_p}{\sqrt{((b/2+x_p)^2 + h^2)}} \right] \qquad (94)$$

Equation (94) describes the condensation rate of sputtered material from a long planar magnetron cathode at an arbitrary point X_p in the substrate plane. Since, however, such cathodes have two erosion zones of equal size both rate contributions are, under consideration of the parallel displacement of the coordinates, superimposed to the total rate.

The rate run R/R_o and with this, at constant rate, also the film thickness run d/do in the substrate plan transverse to the longitudinal axis of the cathode, are obtained by normalization of eqn. (94) at the symmetry point of the cathode. Equation (93) can be solved in a similar way also for other n values but the volume of calculation increases considerably with increasing n. Optimization in the thickness uniformity can be obtained by the use of shutters.

The material efficiency ε in sputtering is usually better than in evaporation. With a long planar magnetron cathode, e. g. PK 500 L (Leybold Heraeus), we obtain for the inner cathode area,

$$\varepsilon = \frac{m_s}{m_g} = \int_{-b/2}^{+b/2} R(x)dx \Bigg/ \int_{-\infty}^{+\infty} R(x)dx \qquad (95)$$

where m_s = deposited mass in the useful field, and m_g = total mass deposited on the infinite substrate plate. The desired deposited masses are obtained by calculation from the sputter rate distributions of the corresponding cathode type, since m is proportional to film thickness and this again is proportional to the condensation rate. Fig. 82a shows the mass efficiency factor ε as a function of the opening width b of the coating slit length at various target-substrate distances h. With a standard distance h = 5 cm and a minimum effective opening width of the aperture of 10 cm a mass efficiency of about 65 % is obtained [363].

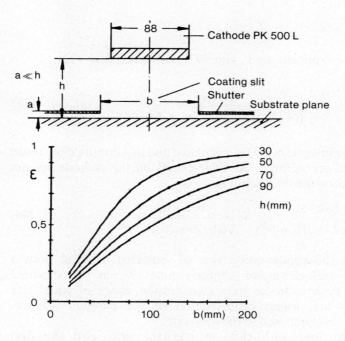

Fig. 82 a
Material efficiency factor of a planar magnetron cathode PK 500 L (LEYBOLD–HERAEUS) as function of the geometry of the sputtering assembly [363].

6.2.1.5.3.12 SPUTTERING MATERIALS

Usually, sputtering targets are made of polycrystalline materials. They are available as round disks, rectangular plates and rods. The choice of material is similar to that in the case of evaporation. Metals, semi-conductors, alloys, halogenides (usually fluorides), oxides, sulphides, selenides, tellurides, III–V and II – VI compounds and cermets are available as targets. Because of the field of employment of the sputtering technique, metals, semiconductors, alloys and oxides are most frequently used, but also nitrides (e.g. AlN, BN, HfN, Si_3N_4, TaN,

TiN, etc.) silicides (e. g. Cr_3Si, $MoSi_2$, $TiSi_2$, WSi_2, ZrSi, etc.), carbides (e.g. CrC_2, HfC, SiC, TaC, TiC, WC, etc.) and borides (e.g. CrB_2, HfB_2, MoB_2, TaB_2, TiB, WB, etc.) are sputtered.

The high power density in sputter processes nearly always demands effective water cooling in order to avoid undue target heating and hence limited diffusion, dissociation and even melting and evaporation of the target material. Normally, target disks are assembled on the support (screws, soldering, adhesion with conductive epoxy resins); in the case of very cheap materials, the target can be produced from one piece. However, for reasons of insufficient thermal conductivity this is not always acceptable so that many cheap materials have also to be mounted on metal supports. Owing to specific material bonding of the target with its socket, with just a few exceptions, continuous operation at full load is possible. Destruction especially of dielectric target disks due to inner mechanical stress occurs very rarely.

Compound films such as oxide, nitride and carbide can be produced by reactive sputtering as mentioned above. Electrically non-conducting substances are processed by rf sputtering. Whether the noble gas ions required for sputtering originate from a gas discharge plasma in the plant, from special ion guns or from other set-ups does not have any influence on the choice of material.

Fig. 83
Planar magnetron sputtering unit BAS 450 PM (BALZERS).
The processing chamber has a diameter of 44,5 cm and a hight of 50 cm, it contains 3 planar magnetrons, quartz lamp substrate heaters and substrate holders.

6.2.1.5.3.13 SPUTTERING PLANTS

Typical for most sputtering plants with few exceptions is the remarkable smaller distance of 6–10 cm between source and substrate compared with that of 40 – 50 cm in high-vacuum evaporation plants. Various technical designs of sputtering plants have been used to fabricate films with this coating technology. Modern development of magnetron sputtering systems has replaced most of the older technologies.

Coating systems using planar magnetron deposition assume three general configurations at present. The first design is a batch metallizer with planetary motion of the substrates. It is used where one or two layer films must be deposited on simple- or complex-shaped parts that require planetary motion for proper coverage. Such a batch coater is usually either an older evaporative system that has been converted to sputtering or, better, a completely new system. The second type of design is the batch metallizer with barrel-type coating jig. It is used for complete coverage of small parts that cannot be readily jigged. The third and most important type of design is the linear chamber lock to lock system. This type is used for one- or two-side coating without planetary motion, to obtain good film property control and high throughput. It is also used when more than one cathode is required. Typical sputtering plants can be seen in Figs. 38 and 40 and in Figs. 83 to 85.

Fig. 84
Load lock sputtering systems LLS 801 (BALZERS).
Chamber diameter 80 cm, 5 work stations with planar magnetrons, rotary drum substrate holder.

Fig. 85 a and b
On line sputtering system DZA 750 V (BALZERS).
The plant is used to produce In-Sn-oxide coatings on glass.

6.2.1.5.3.14 COMPARISON EVAPORATION AND SPUTTERING

Evaporation/condensation and sputtering/condensation are quite different methods of producing thin films. A comparison of their characteristics is given in Table 13.

TABLE 13

THIN FILMS PRODUCED BY EVAPORATION AND SPUTTERING

EVAPORATION	SPUTTERING
1. Vapour production phase	
thermal process	*ion bombardment with momentum transfer*
therefore: relatively low kinetic energy of atoms (at $1500°C$, $E = 0,1$ eV) directional distribution according cosine-law (point or small area source) no or only few charged particles with alloys, fractionated evaporation with compounds, perhaps dissociation	therefore: relatively high kinetic energy of the atoms ($E = 1-40$ eV, changed energy distribution) directional distribution according to cosine-law mainly at higher bombarding energy a number of positive as well as negative ions between 10^{-3} and 10^{-1} per incident ion, in addition reflected primary ions with alloys fairly uniform sputtering of the components dissociation with compounds possible.
2. Transport phase	
evaporated species travels in high or ultrahigh vacuum, therefore no or few collisions (mean free path large compared with evaporator-substrate distance)	sputtered species travels at relatively high pressures of the working gas (10^{-3} to 10^{-4} mbar) energy-reducing collisions (mean free path smaller than cathode-substrate distance) charge-transfer processes strong change in direction (isotropic) high readiness to chemical reactions (presence of excited, ionized and dissociated vapour species including electrons).

3. Condensation phase

no influence on the substrate surface by incident atoms	strong influence on the substrate surface by incident ions and high energy neutral particles (roughening, penetration, imperfections, temporary local charges on the surface, chemical reactions with residual gases).
no change in the conditions of nucleation	strongly changed conditions of nucleation (formation of centres of simplified nucleation)
low incidence of residual gas atoms or molecules (impact number $\sim 10^{13}$ cm^{-2} s^{-1}), therefore low gas incorporation (pure films), no or only little chemical reactions with residual gas and no significant temperature changes of support and film	high incidence of working and residual gas atoms, molecules, ions (impact number $\sim 10^{17}$ cm^{-2} s^{-1}), therefore higher gas or foreign material inclusion (impure films possible); large tendency to chemical reactions (activation, ionization); higher temperature changes of substrate and film possible due to high kinetic energy of the impinging particles.

6.2.1.5.4 FILM DEPOSITION BY ION PLATING

Ion plating is a combination of the evaporation process and the sputtering technique. The method is relatively young compared with the other two PVD techniques. It was developed first by Berghaus, Brit. Patent Spec. No510993 (1938) and DRP No 683414, Klasse 12i, Gruppe 37 (1939), but at this time practically unused and overlooked. However, it later was rediscovered by Mattox, USA 1963, and other researchers [364,365]. Plant installations are similar to that used in evaporation. The source to substrate distance is generally in the range of 40 cm. The substrate holder, however, is electrically insulated and the substrates are biased negatively so that an electric field exists between the source and the substrates. If the gas pressure is high enough and the voltage gradient is adequate, a glow discharge is generated usually in an argon atmosphere. With a simple ion plating process, evaporation is performed in the presence of the gas discharge. In collisions and electron impact reactions, coating material ions are formed and accelerated in the electric field so that condensation and film formation take place under the influence of ion bombardment. It can involve ions of the working gas, of the film material vapour or of a mixture of both. In addition, the higher energy neutral particles of vapour and gas are also very important for the deposition process. This complex action is typically for ion plating. A large number of process variations is possible and different components can be combined to more complex triode and tetrode ion plating systems. It is furthermore remarkable that also high-speed sputtering cathodes are used as vapour sources in special ion-plating arrangements. Bias sputtering and some types of plasma CVD also fall under the definition of ion plating.

6.2.1.5.4.1 CHARACTERISTICS OF ION PLATING

The principle set-up of a diode ion plating system is shown in Fig. 86. We then ask what happens during deposition to a metal vapour atom on the path between evaporation source and substrate. Under the present conditions, the following can occur:

collisions with inert and reactive atoms, or ions of the working gas,

chemical reactions with the excited reactive atoms or ions either in the gas space or on the substrate surface, and

ionization resulting from charge exchange process during the collisions with ions or from electron impact and subsequent acceleration to the substrate.

The occurrences during collisions of the particles in the gas space cause:

changes in speed,

changes in the direction of flight,

changes in the chemical composition, charge-exchange phenomena and impact ionization with electrical acceleration of the metal vapour ions formed.

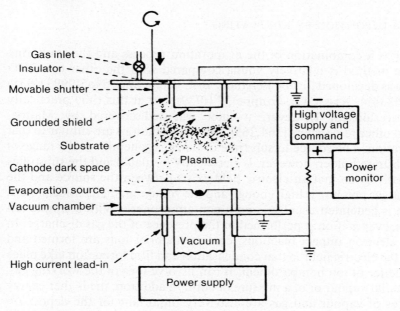

Fig. 86
Schematic representation of a diode type ion plating system.

This results in random enveloping deposition with particles of relatively high energy. The throwing power is a typical effect in ion plating. The chemical reaction with activated gas atoms, or with gas ions, is an important process for the reactive deposition. It is well known that many compounds are not volatile in stoichiometric molecules so that reactive deposition in the presence of reactive gases is an important method for producing stoichiometric films (e. g. oxide, nitride, carbide, boride and their mixtures). The presence of a plasma or of ions favours the occurrence of chemical reactions. Compound formation usually occurs on the substrate surface and only rarely in the gas phase.

The grade of ionization of the vapourized film material atoms depends on:

the material,
the applied evaporation or other vapour generation technique, and
special installations and coating conditions.

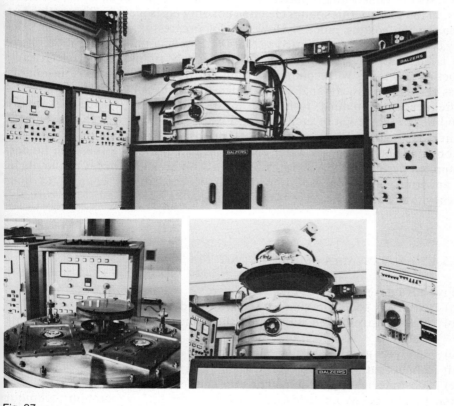

Fig. 87
Diode type ion plating plant with two differentially pumped 270° electron beam guns and a high voltage insulated, water cooled, rotating substrate holder.

With respect to the evaporation technique used, an increasing portion of ions in the vapour is observed in the direction of resistance heating (boats), heating with electron beams (E-guns) and high-frequency heating. The portion of ions is further increased by collision in the plasma.

The positive ions produced in various ways are accelerated towards the substrate in the electric field formed by the substrate applied to negative high voltage and the earthed parts of the equipment. The portion of the film material ions is usually very low (up to 2 %). However, there are many high energy neutral vapour particles which resulted from the fact that many film material atoms were ions for a short time, as such underwent an acceleration especially in the cathode dark space, lost their electric charge during further collisions, but in the main retained the high energy gained. The resulting particle energies are between 1 and 100 eV according to the applied potential.

The ions and high energy neutral particles incident on the substrate surface expend most of their energy in heating the substrate. With current densities of $\sim$ 0.5 mA cm^{-2}, this can result in heating up of the substrate by a few hundred degrees. At the same thime, sputtering of the surface films of substrate material occurs. In ion plating, this sputter cleaning not only takes place before coating but also during the coating process. However, for progressive film formation, the energy conditions must be adjusted so that the film material condensation rate is much larger than the sputtering rate. It is relatively simple to coat electrically conducting substrates with conductive films. However, in the case of insulators such as glass and plastics – analogue the high-frequency sputtering technique – a high-frequency ion-plating technique must be applied. There exist many possibilities in the design of such a plant. Figure 87 shows a technical set-up for the direct-current ion-plating technique. The installation is equipped with two electron guns and contains a water-cooled, insulated, rotating substrate holder. The applied negative high voltage is between 1 and 5 kV. To coat insulators in such an arrangement, the substrate has to be capacitively coupled to high frequency to maintain a negative bias, for example.

6.2.1.5.4.2 ADVANTAGES OF ION PLATING

Five advantages should be mentioned here:
a) extremely good adhesion between film and substrate;
b) high density of the film (often the same density as that of the bulk material;
c) relatively uniform coating of substrates of complicated shape;
d) a high coating rate, especially when using electron guns;
e) substrate heating accompanying the coating process.

Ion plating opens up fields of application for the evaporation technique previously only possible with sputtering, without having to forfeit the advantage of high rates of condensation.

As with sputtering, ion cleaning of the substrate can be undertaken. Generally, the high energy of the ions influences the film properties positively. Ion plating offers more opportunities than the sputtering technique as regards ion yield and ion energy. The question now arises of what explanation can be given for the excellent quality of ion plated films and the advantages of their production?

a) A clean surface is frequently a prerequisite for good adhesion. Improvement in adhesion occurs if the film material can penetrate the surface of the substrate material, e.g. by diffusion or other mechanisms. If diffusion is possible between film and substrate material, this is favoured by increased substrate temperature. In a similar way, imperfections of a crystal have the effect of lattice vacancies in areas near the surface of the substrate. The formation of intermediary films with limited thickness and continuous change in the chemical composition has, for example, a counterbalancing effect on varying coefficients of expansion and therefore reduces undesired mechanical stresses thus arising. Bombardment with ions and high-energy neutral particles causes sputtering of surface films and film material. With reduced sputter rate, this sputter cleaning usually takes place during the whole film formation. Because of this continuous bombardment of the growing film, particularly during the initial phase of film formation, the partial re-sputtering of film material, in addition to sputtering of substrate material and the back scattering resulting from collision processes in the gase phase in front of the substrate, cause a physical intermixing of substrate and film material (pseudo-diffusion). The limited intermediary film thus formed favours the adhesion and other film properties [366]. Also, implantation effects of the high-energy film material particles can have a positive effect on the adhesion of the growing film. Implantation depths of 20 – 50 Å [367] have been observed, which apparently lead to increased anchoring of the film. The mechanisms mentioned improve the adhesion mainly in the case of thick films particularly.

b) Regarding the densitiy of the films, it should be mentioned that continuous bombardment of the growing film with high-energy charged or neutral atoms or molecules can cause changes in the film structure and result in density values near those of the compact material. This result is obtained both when bombarding the films with ions of the same type [368,369,370] as well as with inert gas ions [368,369,371]. During film growth, partial sputtering especially of the loosely adhering film material affects the crystal growth and hence influences the structural formation and the topography of the surface of the film. The reason for the frequent formation of a prismatic columnar structure during evaporation generally results mainly from geometric shadowing effects once a rough surface has formed, e. g. due to a more rapid growth of a few crystals in crystallographically preferred directions in the initial stages of film formation.

As a consequence of particle bombardment during film growth in ion plating, the surface diffusion is increased, the nucleation conditions are changed and also geometric projecting regions, such as tips, are preferentially sputtered. The atoms sputtered forward fill the topographic valleys and the atoms firstly sputtered away from the rough surface return partially as a result of gas scattering and ionization processes and thus make the film surface smooth [372]. These microstructural modifications are shown in Fig. 88. In this manner, often instead of the normal prismatic growth a very densely packed spiky structure of a fine-grained crystalline, practically isotropic film micro-structure and a relatively smooth surface are obtained [368,370,373].

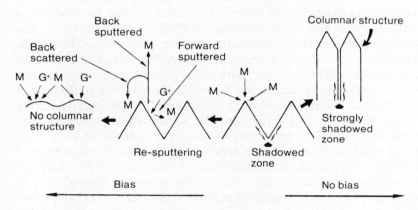

Fig. 88
Microstructural modifications obtained by ion bombardment during film growth.

c) Frequently, with ion plating, especially with higher working gas pressures, an enveloping good adhering film forms.

Two effects mainly contribute to this power. The first, rather weaker, effect is based on the fact that a small portion of the film material vapour is ionized. The vapour ions formed follow the electric field lines which end on all sides of the substrate applied to high voltage so that the front, side and rear parts and also (not too narrow) borings are coated. However, the literature shows that the ionization degree of many plasmas is very low with values between 0.1 % to 2 % [374,375,376,377]. The second, much stronger effect is therefore the scattering in the gas space. The high-energy particles do lose energy during collisions, but different from the low-energy vapour atoms in the other PVD methods, they retain sufficient energy in order to form very good adhering, dense and enveloping films. There are, as yet, very few published investigations regarding the subject of parameters for a uniform enveloping coating. Regarding gold coating, we refer to the work of Chambers and Carmichael [378]. Generally, the uniformity of the coating increases with rising gas pressure in the discharge and decreasing evaporation rate.

d) Three main evaporation methods are found in the literature regarding ion plating:
    evaporation of the material from resistance-heated crucibles or boats;
    evaporation by electron beam guns; and
    evaporation by inductive heating, which apparently produces the largest number of ions in the vapour of all three methods [373]. If high evaporation rates are required, the electron gun is currently the most favourable technical evaporator.

e) The energy transferred during ion bombardment is mostly transformed into heat and can be used for heating up the substrate. In many cases, a substrate temperature is required up to about 300°C because this results in desorption processes of interferring gases and in an increase of the chemical reaction ability in the case of reactive coating, and generally better developed, dense, crystalline films are formed. If substrate heating is undesirable, ion energy and ion density must be lowered and highly effective water-cooling is required. Even in this case excellent film quality is obtained.

5.2.1.5.4.3 APPLICATIONS OF ION PLATING

A main use of ion plating is to coat steel and other metals and alloys with special films, such as for tribological applications and with extremely hard films and film combinations, to increase surface hardness and life time of various tools, but also to create a durable decorative effect on metal parts.

Using mainly high-frequency ion plating, insulators can also be coated. SiO_2, TiO_2 and other oxide films have been deposited by reactive high-frequency ion plating onto glass substrates, CR 39 and other plastics [379,380].

In further investigations, ion beams (Ar, O_2) directed to the substrate surface have been applied prior to and during deposition of, for example BK 7 glass with Ag, Cu and Au and also MgF_2 films, ThF_4/AlF_3 and MgF_2/ZnS multilayers have been deposited at 50°C substrate temperature with success as regards low film stress, good adhesion and abrasion resistance [381]. One problem associated with this technique is connected with the possible change in material composition due to sputtering, forming an altered layer integrated into the film structure as deposition proceeds. This change in stoichiometry might explain the residual small absorption observed for some materials, such as ThF_4/AlF_3 multilayers deposited under ion bombardment. The interaction of low-energy ion beams with

254

surfaces has recently been reviewed [385]. In glow-discharge ion plating, the substrate surface is bombarded prior to and during deposition. Uniform etching and structural changes of the surface of, mainly, plastics such as CR 39 by the ion bombardment were revealed by electron spectroscopy for chemical analysis and transmission electron microscopy. Ion-plated silicon oxide films formed on such a modified surface showed clearly better adhesion than vacuum evaporated films; however, excessive ion bombardment degraded the adhesion quality of the films by too strong degrading of the plastic surface. Depth profiling by Auger electron spectroscopy of properly deposited films showed that the intermixing (pseudo-diffusion) layer, which seems to improve film adhesion, is formed by the ion plating [382]. Ion plating onto plastic-foil substrates of polyethylene terephthalate (ICI Melinex) was demonstrated to give considerable advantages. These are that adhesion is increased and film properties are obtained which would normally require higher temperatures than the substrate can with-stand. In order to coat large areas of sheet plastic, it is necessary to achieve high deposition rates without creating high substrate temperatures. This is done commercially by pulling the sheet substrate tightly over a very smooth cooled drum and transferring the plastic from an initial roll to a final roll over this rotating drum. Conventional resistance-heated sources are generally used, though induction and electron beam heating have also been applied.

Simple diode-type sputtering results in damage to the plastic from both heat and electron bombardment. This can, however, be avoided if a planar magnetron source is used [383].

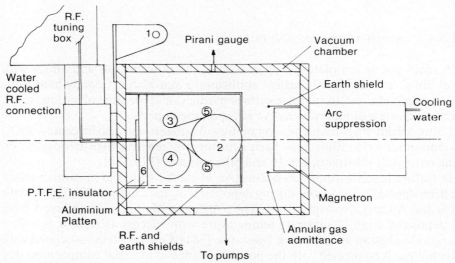

Fig. 89 a

Schematic representation of the roll-to-roll coating apparatus detail and vacuum system showing the magnetron source in position.

1 = motor, 2 = main drum, 3 = wind-on roll, 4 = wind-off roll, 5 = pinch rolls
6 = watercooled rf electrode [383].

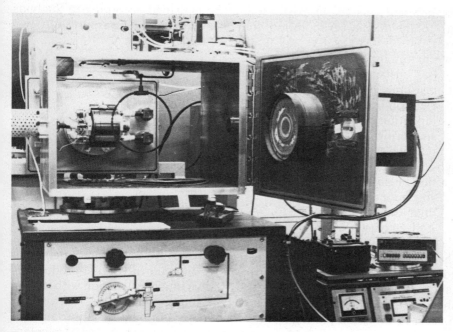

Fig. 89 b
The roll-to-roll coating apparatus with open chamber [383].

Figure 89a and b show an apparatus which allows plastic foils passing over a rotating drum to be subjected to a high-frequency discharge either before, or simultaneously with, the deposition of material from a planar magnetron source. The material may be deposited directly from the source, or reactive processes may be used with carefully chosen reactive gas inlet points. The high-frequency power is applied to the roll, which is isolated. The storage rolls and the drum are mounted within a high-frequency cage. The motor drive shafts are also insulated and only the surface of the drum, over which the plastic foil is rolled, is exposed to the coating chamber. The rest of the plant is protected by a closely spaced earth shield, as can be seen in Fig. 89a. In this ion plating unit, plastic substrates were continuously coated with $In_{90} Sn_{10}$ oxide. A precisely controlled sputtering rate of the metal alloy and an exactly determined mass flow rate of oxygen into the system enables the production of films which are 90 %-transmissive highly transparent and have about 400 Ω cm^{-2} electrical conductivity [383,384]. The introduction of the high-frequency bias to the substrate enhanced the reaction process considerably. It can be concluded that the high degree of ion plating variants which exist in instrumental set-ups and evaporation media are responsible for the many new terms that have been generated to designate a particular system more specifically.

The different versions in existence may be classified according to Spalvins [386] as to the media, evaporation source and mode of transport into two categories:

low vacuum ion plating
    (self-sustained glow discharge plasma)
    resistance evaporation
    electron-beam evaporation
    induction evaporation
    sputtering
    reactive evaporation and
high-vacuum ion plating
    (ion beams)
    single ion beams
    dual ion beams
    ionized cluster beam.

The extensive variation in the instrumental set-ups and requirements has precluded the manufacture of a standard ion-plating unit in volume production. Each laboratory or industrial operation uses its own unit construction with its own specialized instrumental features to best meet its requirements.

In order to bring ion plating from the laboratory into industrial production, great efforts were and are still necessary for optimizing equipment and parameters. In this respect,

special installations
economic process parameters and
basic knowledge of correct operation of the installations are significant pre requisites for wider employment of this relatively new technology of thin film production.

For further information, the proceedings of IPAT conferences (Ion and Plasma Assisted Techniques) are particularly recommended.

6.2.1.5.5 REACTIVE DEPOSITION PROCESSES

6.2.1.5.5.1 GENERAL CONSIDERATIONS

Mixtures of elements which have large differences in vapour pressure but which form vapour species which are the same as, or a multiple of, the mixture formula of the initial material, can produce satisfactory multi-element depositions by evaporation from a single source. Unfortunately, however, a great number of combinations of elements do not fulfill the above criterion.

The ionic bombardment in sputtering of chemical compound targets often produces a vapour which is dissociated, where the highly volatile and reactive components, such as oxygen, disappear rapidly and the compound films deposited are therefore often non-stoichiometric in composition. Thus, for instance, direct evaporation as well as direct sputtering of Al_2O_3 results in oxygen deficient films [387]. To overcome this difficulty, Auwärter [250] and Brinsmaid [251] suggested that the deposition be carried out in the presence of a partial pressure of oxygen or generally, depending on the compound processed, of some other desired reactive gases.

Reactive deposition has developed into a powerful technology over the last 30 years. It is historically interesting to note that in early studies, Soddy [388] found in 1907 that calcium vapour is highly reactive to most gases and that Langmuir [389] in 1913 investigated the formation of tungsten nitride by vapour-phase reaction of the elements.

The factors responsible for the efficiency of a reaction are numereous and very complex. They include at least the chemical nature of the reacting species, the stability of the reaction gas or compound species, the free energy of formation of the compound, its dissociation pressure and dissociation temperature, the substrate temperature, condensation behaviour of the reactants on the substrate, sometimes influenced by the substrate temperature and the partial pressure of the reactants, and so on.

The reaction probability on collision between the reactants can be enhanced by activating one or even both [250,390,391]. This means that it is an advantage both for the occurence and also for the yield of a reaction when the reactants are ionized, dissociated or in electronically excited states.

Activation can be performed by uv radiation, electron impact, substrate heating, or all together in the presence of a gas discharge plasma. Plasma-assisted reactive evaporation is therefore a very effective technology, along with reactive sputtering and reactive ion plating. The use of plasmas brings these and other quite different and independent coating technologies closer together [200,446].

Concerning the reaction site, we can state that a reaction between gases and the vapour species may occur at one or more of three possible sites:

on the substrate or film surface,
in the vapour phase, or
on the evaporant or target surface.

Generally special conditions and arrangements are used which aim at avoiding reactions at the vapour source surface because these may lower the evaporation and the sputter rate and can cause contamination, for example by formed volatile source material compounds. In most cases, the reaction takes place mainly on the substrate surface. Gas-phase reactions only tend to come into action at higher gas pressures to a given degree, depending on further conditions.

With all reactive depositions, it is a prior condition to have an adequate supply of reactants which can undergo collisions and react following the chemical reaction to form the desired compound film. There have been numerous theoretical treatments and reviews of reactive evaporation [390–397] of reactive sputtering [398–415] and to a smaller extent of reactive ion plating [416–422].

6.2.1.5.5.2 REACTIVE EVAPORATION

For the deposition of a stoichiometric oxide film by reactive evaporation, a relatively high O_2 partial pressure and a slow metal-atom condensation rate are required, so that completely oxidized metal-oxide films can be formed. The partial pressure of the reactive gas component is usually few 10^{-4} mbar. The significant technology of reactive evaporation [250] is applied in all cases where direct evaporation of a chemical compound is not possible because of thermal dissociation or too low a vapour pressure. In practice, oxide films are usually produced using sub-oxides or metallic starting materials. However, basically it is also possible to produce sulphides and nitrides or other compounds in this manner.

When evaporating under reactive conditions, care must be taken that the gas components consumed during the chemical reaction are continuously replenished. This is frequently carried out through a gas inlet valve controlled using the pressure measured in the chamber. The average mean free path at 10^{-4} torr is still only 50 cm, which is somewhat more than the distance from boat to substrate; reactions from collision in the gas space between gas molecules and metal atoms are only possible to a small percentage, although the collision probability is often more than 50 %. The reaction takes place to a far greater extent on the substrate surface which is exposed to the incident metal atoms and gas molecules.

Therefore formation of a metal-oxide film by reactive evaporation takes place in the following stages:

1) The substrate surface is exposed to the metal vapour atoms and gas molecules which impinge at a certain rate.

$$\frac{d\,N_{Me}}{A\,dt} = N_L \frac{\rho_{Me}}{M_{Me}}\,d'_{Me} \qquad \text{metal atoms cm}^{-2}\,\text{s}^{-1} \qquad \text{and} \qquad (96)$$

$$\frac{d\,N_{O_2}}{A\,dt} = (2\pi m kT)^{-1/2}\,P_{O_2} = 3{,}51.10^{22}\,(M_{O_2}\,T)^{-1/2}\,P_{O_2}\,\frac{\text{molecules}}{\text{cm}^2\,\text{s}} \quad (97)$$

A	= surface of the substrate (cm^2)
N_L	= Loschmidt number
ρ_{Me}	= density of the metal film (g cm^{-3})
M_{Me}	(or M_{O_2}) = molar mass of the metal (or O_2) (g mol^{-1}).
d'_{Me}	= condensation rate of the metal (cm s^{-1})
T	= gas temperature in Kelvin
P_{O_2}	= partial pressure of O_2 in torr

2) A portion of these atoms and molecules is adsorbed on the substrate surfaces, and another portion is either reflected or, after a short dwell on the surface, again desorbed. The ratio of the actually adsorbed quantity to the number of incident particles is given by the condensation coefficient σ.

There are two processes responsible for the fact that a portion of the incident atoms and molecules leave the surface again. An incident particle can rebound immediately if its translational energy is not absorbed by the substrate upon incidence. Fast desorption takes place if either a very small adsorption energy exists or the substrate has a high temperature.

The adsorption of oxygen by most metals is, however, strong and even if the adsorption energies are considerably reduced because the metal surface is, for example, already partially oxidized, the relevant dwell time of the adsorbed oxygen is usually longer than the rate of growth of the film to be evaporated if the surface temperature is not too high.

It can therefore be concluded that a condensation coefficient smaller than 1 usually indicates deficient energy accommodation. Quantitative measurements of the accommodation coefficients of gases on solid surfaces have seldom been carried out.

The condensation coefficients of metal vapours are frequently near to unity: large rates of incidence are chosen in the vacuum evaporation, so that they correspond to pressures which are much larger than the equilibrium pressures at the usual substrate temperatures.

3) In the adsorbed phase, in which the particles are mobile because of surface diffusion, the chemical reaction of the metal-oxide formation takes place via dissociative chemisorption of the oxygen. This means that, since the reaction takes place in the adsorption phase and the various reaction partners can have different condensation coefficients, the ratio of the rates of incidence is a necessary but not adequate criterion. According to Ritter [293,298], the chemisorption rate of oxygen is the critical step for the completion of the reaction:

$$\left(\frac{d\,N_{O_2}}{dt}\right)_{ads} = \left(\frac{d\,N_{O_2}}{dt}\right)_{incid} \frac{a\,(1-\theta)^2}{a-\theta}\ \sigma\ \exp\ \left(\frac{-E}{RT}\right) \qquad (98)$$

θ = degree of coverage with oxygen
a = number of nearest neighbours of a surface site
σ = condensation coefficient of the O_2 molecules
E = activation energy

During simultaneous condensation with oxygen, a condensation coefficient of unity is found for the incident metal Me or a metal suboxide MeO particle. Consequently, the O:Me ratio in the film is:

$$\left(\frac{N_O}{N_{Me}}\right) film = 2\frac{(d\,N_{O_2})\,ads}{(d\,N_{Me})\,inc} = 2\left(\frac{d\,N_{O_2}}{d\,N_{Me}}\right)_{inc} \frac{a(1-\theta)^2}{a-\theta}\ \sigma\exp\left(\frac{-E}{RT}\right) \quad \text{and} \quad (99)$$

$$\left(\frac{N_O}{N_{Me}}\right) film = 2\frac{(d\,N_{O_2})\,ads}{(d\,N_{MeO})\,inc} +1 = 2\left(\frac{dN_{O_2}}{dN_{MeO}}\right)_{inc} \frac{a(1-\theta)^2}{a-\theta}\ \sigma\exp\left(\frac{-E}{RT}\right)+1 \quad (100)$$

Results from experiments on reactive evaporation by Ritter [424] and Pulker [425] on Si [425] and SiO [424], are shown in Fig. 90. Ritter [293,298] considered the degree of oxidation of the films N_O/N_{Si}, as determined by chemical microanalysis, compared with the ratio of the number of incident O_2 molecules and the condensing vapour species Si (x) or SiO (o) respectively.

With the aid of eqns. (99) and (100), graphs were calculated, also included in Fig. 90, which show a functional relationship between the degree of oxidation and the ratio of the rates of incidence. The degree of oxidation of the films was inserted to calculate the degree of coverage with oxygen θ, where two oxygen atoms per Si atom were assigned a θ value of 1 (corresponding to the maximum oxidation stage SiO_2) and pure SiO films were treated as Si films with $\theta = 0.5$. The number (a) of nearest neighbours was taken as four in each case.

The product $\sigma \exp (-E/RT)$ was ascertained by calculation from the experimental data. A value of 0.6 in the case of Si and 0.2 for SiO fitted the experiments best. An analysis of the product $\sigma \exp (-E/RT)$ showed that for both reactions the activation energy is smaller than 1 kcal mol^{-1}. A value as low as 1 kcal mol^{-1} was found also for the formation of Al_2O_3 films with reactive

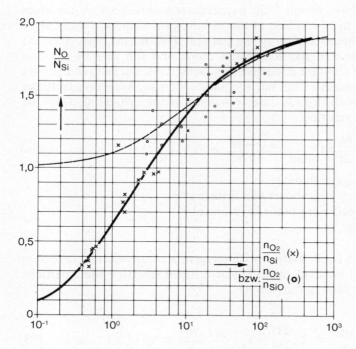

Fig. 90

Dependence of the degree of oxidation of Si (x) and SiO (o) condensates on the ratio of the simultaneously incident O_2 molecules on the substrate surface to the Si atoms and SiO molecules. The graphs were calculated by the use of the equations (99) and (100), [293, 298].

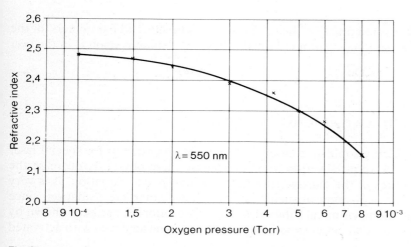

Fig. 91

Dependence of the refractive index of TiO₂ films on the oxygen gas pressure during evaporation [293, 298].

evaporation of Al. However, a different definition was used for σ [426]. This agrees well with the fact that the chemisorption of oxygen on metals usually takes place with only small activation energy [423,427]. The temperature dependence also supports this fact. A decrease of the reaction was observed with increasing temperature of the substrate surface. Ritter [293,298] explains this observation with a reduction of the condensation coefficient σ of oxygen.

More recent investigations have shown that reactive evaporations performed under usual conditions but with partial ionization of the reactive gas enhance the chemical reaction [390,428]. It can be concluded from this observation that a further lowering (decrease) of the activation energy is of advantage for complete oxidation. From Fig. 90, it follows that for a specific chemical reaction during evaporation, the ratio of the number of wall collisions of oxygen molecules to the number of condensing film material atoms should be kept as high as possible. However, there are limits imposed here by consideration of , for example, film hardness. During the reactive evaporation of TiO and condensation of its vapour on hot glass substrates at 300° C, the hardness and the refractive index of the TiO₂ films formed decrease if the O₂ pressure of 10^{-4} is increased to 10^{-3} torr [293,298]. Referring to Fig. 91, it seems that the film components lose too much energy with the increasing frequency of collisions in the gas space at high pressures and hence cannot form a compact film.

Summarizing, it can be stated that the oxide film formation with reactive evaporation is influenced by

the rate of incidence of the metal atoms or sub-oxide molecules respectively, and the oxygen molecules,
the condensation coefficients of the relevant reaction partners and
the substrate temperature.

In film deposition practice, the need thus arises to select all parameters for the relevant process so that the film obtained fulfills the demands upon it in the best possible way.

6.2.1.5.5.3 ACTIVATED REACTIVE EVAPORATION

Films for highly sophisticated optical applications, such as for special laser mirrors, should have no absorption losses because these obviously decrease the reflectivity and lower the damage threshold for high power radiation. It is very important in the deposition of such films to achieve complete oxidation, since traces of metal atoms or metal suboxides cause absorption. It has been shown by various researchers that better stoichiometric films can be achieved with activated

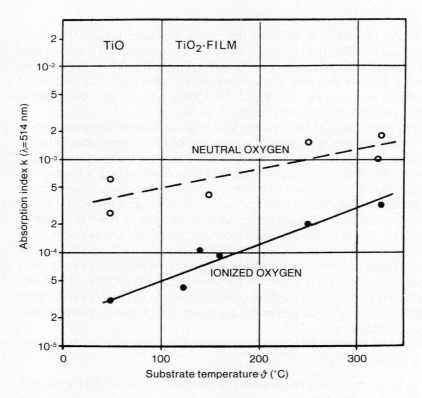

Fig. 92

Absorption index k (λ= 514 nm) versus substrate temperature; starting material TiO, oxygen pressure 1,5. 10^{-4}mbar, deposition rate 0,3 nm. sec^{-1} [396].

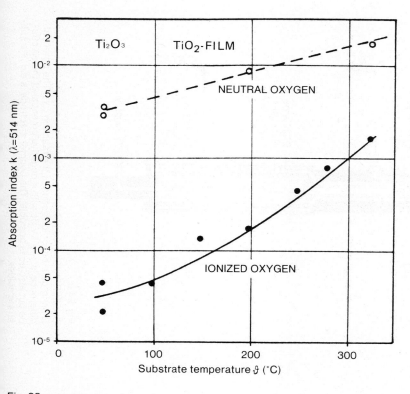

Fig. 93

Absorption index k (λ = 514 nm) versus substrate temperature; starting material Ti₂O₃, film thickness 400 nm, oxygen pressure 1. 10⁻⁴ mbar, deposition rate 0.5 nm. sec⁻¹ [396].

oxygen gas containing ions and excited molecules, than by the use of ordinary neutral oxygen gas [390,395,396,397,429]. This holds especially true for the fabrication of fully oxidized compound films.

In the case of TiO_2 films, the partial ionization of oxygen reduces the absorption by a factor of 10 for TiO starting material, or by a factor of 100 for Ti_2O_3 or Ti_3O_5 starting materials compared with the values found with ordinary oxygen. This can be seen in Figs. 92 and 93 [395,396]. It was found that under these conditions negative oxygen ions produced the best oxidizing effect [395]. They were therefore used to produce all the films.

The oxygen ions can be produced by different ion sources, which are shown in Figs. 94a – 94c. In the case described here, the Ebert source was used. The ions were generated in a glow discharge burning between a hollow cathode and the evaporation source as well as the walls of the coating chamber. The oxygen pressure in the coating chamber was 1 – 5×10⁻⁴ mbar, that in the discharge tube about 2 mbar. The current density at the orifice of the ion outlet nozzle is about

264

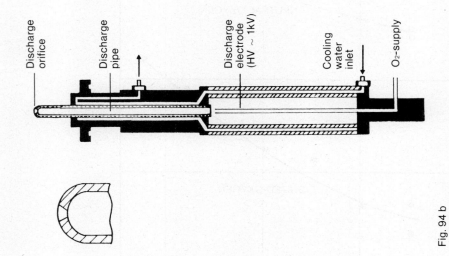

Ion outlet
nozzle

Hollow
cylinder
electrodes

O₂-supply

HV-supply

Fig. 94 a

Heitmann – Source [390].
Discharge tube for ionization of the reaction gas. The potential
difference is 1 kV ac, the discharge current is 600 mA which resulted in a peak value current density of 40 A.cm².

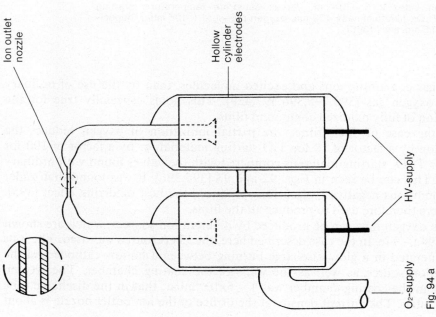

Discharge
orifice

Discharge
pipe

Discharge
electrode
(HV ~ 1kV)

Cooling
water
inlet

O₂-supply

Fig. 94 b

BALZERS – Source [431].
Discharge tube for ionization of the reaction gas. The potential
difference is ∼ 1 kV dc with a negative potential of the electrode but also ac is possible. The counter electrode is e.g. the substrate holder.

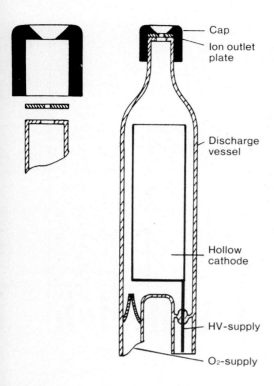

Cap

Ion outlet plate

Discharge vessel

Hollow cathode

HV-supply

O₂-supply

Fig. 94 c
Ebert – Source [395, 396].
Hollow cathode discharge tube for the ionization of the reaction
gas. The discharge current is 400 mA and the corresponding
current density at the ion out-let is 100 A.cm⁻². The amount of
ions is about 6 %.

100 A cm^{-2} and the degree of ionization of the oxygen is about 6 % for 400 mA
discharge current. The arrangement used and the results obtained with different
types of oxygen are shown in Figs. 95 and 96. At lower substrate temperatures,
Ti_2O_3 is a better starting material than TiO because a higher evaporation rate is
possible and the film refractive index is higher. At elevated substrate temperatures, there is a strong increase in the absorption as can be seen in Figs. 92 and 93.
In this case, TiO is the preferred starting material because of the smaller
absorption in the obtained TiO_2 deposit.

The measured refractive indices of differently prepared TiO_2 films as a
function of wavelength are shown in Fig. 97 [395]. As can be seen, the dispersion
curves of TiO_2 deposited from the two different starting materials are nearly
identical at high substrate temperatures. The difference with unheated glass
substrates seems to be caused by the partially changed evaporation conditions,
higher O_2 pressure and slower deposition rate for the TiO starting material as

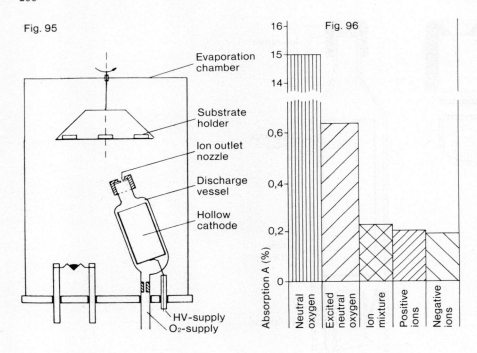

Fig. 95

Fig. 96

Fig. 95
Schematic representation of the arrangement for reactive evaporation with ionized oxygen [395].

Fig. 96
The absorption at 514 nm of TiO_2 films of 400 nm thickness for different types of oxygen. The starting material was Ti_2O_3 [395].

compared with Ti_2O_3. These parameters have a great influence on film properties, especially at lower substrate temperature. The data agree well with those obtained by other authors [430]. Measured absorption coefficient values for other oxide films are shown in Fig. 98 [396]. Comparable results have also been obtained in a similar arrangement with the Balzers source [429]. Substantial higher densities and refractive indices as well as lower optical losses of evaporated oxide films were obtained by Macleod et al. [475] when bombarding the growing film with ions. This also results in an improvement in the resistivity against water vapour sorption of single films and multilayers on atmosphere.

Other compound films especially nitrides or carbides can be fabricated with high rates by a special activated reactive evaporation technique, ARE, developed by Bunshah, see [391], in which the evaporant vapour species and the reaction gas are ionized or activated by impact of secondary electrons.

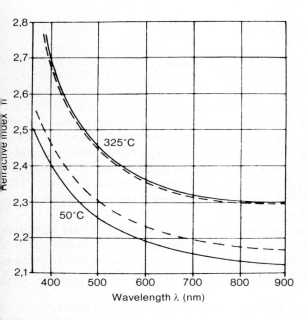

Fig. 97
The dispersion graphs of TiO$_2$ films for the two starting materials: TiO ———, and Ti$_2$O$_3$ — — — [395].

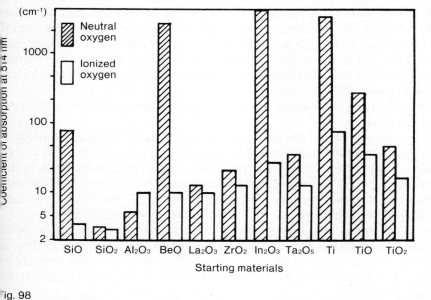

Fig. 98
Activated reactive evaporation with different starting materials, the ion current density
is approximately 0,3 mA.cm^{-2} [396].

6.2.1.5.5.4 REACTIVE SPUTTERING

With reactive sputtering, one has to distinguish between a compound film synthesis under the action of a diluted reactive gas atmosphere starting from pure metal, alloy or multi-element targets on the one hand, and the making up for the lost constituent when using compound targets on the other. The main difference is found in the dependence of the deposition rate on the partial pressure of the reactive gas. Generally the reactive gas component, which can be for example O_2, H_2O for the formation of oxides, N_2, NH_3 for nitrides, $O_2 + N_2$ for oxynitrides, H_2S for sulfides, C_2H_2, CH_4 or other hydrocarbons for carbides and HF or CF_4 for fluorides, is added to the sputtergas – mainly Ar – in small amounts, between 10^{-5} and 10^{-4}mbar. In all gas discharge sputter arrangements, the added gas is in the activated states and therefore highly reactive.

Reactions in the gas phase, ignored in reactive evaporation, are here also generally negligible; the heat of reaction liberated cannot be dissipated in a two-body collision. Conservation of momentum and energy lead to heterogeneous reactions on the substrate surface but are unfortunately also possible on the target surface [398].

When the sputtering rate is high and the amount of reactive gas is low, it is very likely that the compound formation occurs on the substrate, and thus the stoichiometry of the film depends on the relative impingement rates of metal vapour species and reactive gas at the substrate. Where only one compound exists between the target material and the reactive gas, reactive sputtering is a straightforward process provided sufficient reactive gas is present to from the compound. However, more compounds can often be formed depending on the conditions prevailing during sputtering. If, however, the reactive gas partial pressure is increased and/or the metal sputtering rate is decreased, it may suddenly happen that the rate of compound formation on the target exceeds the removal rate of compounds. The rate decrease by increasing reactive gas concentration has three possible causes [403]:

Formation of compounds on the target, which then have lower sputtering rates than the metal:

The compounds formed often have higher secondary electron emission than metals, so that more of the energy transferred by the ions is used to produce and to accelerate secondary electrons. The increased secondary electron emission, in the case of constant-current power supplies, automatically decreases the cathode voltage for a fixed power setting. It is therefore better to maintain a constant voltage. In this case the abrupt rate decrease becomes more smoothed out.

A further reason is less efficient sputtering by reactive gas ions than by inert ions. A comparison with compound targets has shown that these targets show a much more gradual decrease in sputtering rate with increasing partial pressure of the reactive gas, and this seems mainly to be related to the less efficient sputtering ion concentration.

Reactive sputtering has shown itself to be a very versatile technique capable of straightforward synthesis of broad classes of materials [410,412,432–440,

477–479]. Reactive sputtering allows precise control of coating structure and composition, leading to rapid property optimization and, in some cases, coating performances better than those previously achieved with conventional evaporative techniques. New materials with complex compositions not synthesizable by traditional approaches can also be made by reactive sputtering, which provides a further route to improve coating performance [410].

Beside gas discharges also ion beams produced in ion guns can be applied in reactive sputtering. Recently reactive ion beam sputtering was used to deposit oxide films for optical applications. With this technique a beam of $\sim 1000\,\text{eV}$ positive ions of an argon oxygen gas mixture, neutralized by the equal number of electrons, bombards a target (metal or oxide), and the sputtered particles form a deposited film of, for instance, SiO_2, TiO_2 or Ta_2O_5 [476]. The ion current density in the process is typically 1 to 1.5 mA cm^{-2}. The obtained films are stoichiometric, mechanically stable, very dense and excellent adherent. Additionally a second ion beam at lower energy can be applied for predeposition treatment (sputter cleaning) and simultaneously for mechanical film stress modification.

6.2.1.5.5.5 REACTIVE ION PLATING

Reactive ion plating is a process for obtaining a chemical compound film, mainly by direct synthesis from the elements, e. g. $Ti + O_2 \rightarrow TiO_2$ or $2Ti + N_2 \rightarrow 2TiN$. As with reactive sputtering, the less reactive gases can also be activated so that nitride, carbide, boride and mixed compound films can be produced. This is done in the classical process by burning a gas discharge in either a mixture of noble gas and reactive gas or in a reactive gas alone, and by depositing an evaporated or sputtered vapour species in the presence of the discharge onto negatively biased substrates [368, 441].

In all plasma-assisted processes, the chemical reactions in the plasma region are normally very complicated, but this complexity is considerably reduced with single atomic coating material species and elemental reactive gases. Depending on the conditions in the discharge, many atoms and molecules are electronically excited and only a few are ionized. In any atom, a set of quantum mechanical energy levels corresponding to a set of electronic states [442] exists. In any molecule [443], each energy level consists of rotational and vibrational energy in addition to electronic energy.

These excited states of atoms and molecules can easily be detected and recorded using an optical spectroscope [444]. An analytical treatment of such spectra allows to detect the various chemical reactions in the discharge and can often help in optimizing the reaction conditions. Compared with reactive sputtering, classical reactive ion plating is generally a very fast process and results in deposits of excellent quality. With the recent process variants, it has also become possible to coat temperature-sensitive substrates because the energy, activation and dosage of the bombarding ions can be accurately adjusted and controlled, and can be kept very low.

Examining the newer literature, it becomes difficult to define reactive ion plating since each of the variants of ion- or plasma-assisted processes can principally be modified and so transformed into an ion plating process simply by negatively biasing the substrates. Alternatively, it becomes possible in other ways using special techniques to accelerate the condensing vapour species in the direction of the substrates.

A still increasing number of reactive process variants is under development. From the various possible reactive gas processes, the most frequently used ones are listed in Table 14, following Zega [445]. In many cases it will be necessary to develop both system and process to accomplish a given task. It is therefore impossible to give an all-purpose system design or to foresee all the problems which may arise.

6.2.1.5.6 PLASMA POLYMERIZATION

Plasma-polymerization reactions leading to solid films on solid substrates had already been performed by the middle of last century, at the same time as gas discharge experiments were being undertaken in many laboratories. The significance of polymer-forming plasmas such as glow discharges of acetylene, ethylene, styrene, benzene, etc., is that a considerable portion of the molecules of the starting material leave the plasma phase and deposit as a solid polymer. However, at this time, little development in polymerization was done and the process was used more as a laboratory curiosity. The situation changed rapidly in the late 1940's. Technical applications were found. The insulating organic films were first used in the electronics industry. A little later polystyrene coatings on Ti foils were used as dielectrics for a nuclear battery. Furthermore, from 1970 the process was used also to fabricate optical coatings. There are many review articles and books on this topic [446–459]. Studying the literature, it becomes obvious that the early emphasis was to obtain a product and fundamental considerations received minimal attention. On the other hand, we now know that plasma processes leading to desired polymer-film formation are rather complex and are therefore difficult to investigate. Plasma polymerization can appropriately be defined as a reaction between low molecular-weight vapourized organic materials as the result of the action of electrical discharges, to produce polymeric coatings on various substrates.

An unfortunately unsolved problem in plasma polymerisation is the discovery of the exact reaction mechanism. Often, even the type of the predominant active species is unknown. Considering the possiblities essentially, only two types of polymerization mechanisms have been found operable; these are shown schematically in Fig. 99 [458]. The first is conventional plasma-induced polymerization. The second is plasma-state polymerization. In conventional polymerisation, the conversion of the monomer occurs by reactive species which are activated in the gas discharge. For the occurrence of this mechanism the monomer must have a readily polymerizable structure of typically one or

TABLE 14

FREQUENTLY USED REACTIVE PROCESSES FOR COMPOUND FILM DEPOSITION [445]

	RIP	ARE	HCD	BARE	PUSK*	HRRS
	Reactive ion-plating	Activated reactive evaporation	Hollow cathode discharge deposition	Biased activated reactive evaporation	Accelerated plasma deposition	High-rate reactive sputtering
Vapoure source	High-voltage E-beam heated crucible		Low-voltage E-beam heated crucible		Low-Pressure arc-eroded target	Magnetron sputtering target
Atmosphere	$Ar+O_2, N_2, C_xH_y$	O_2, N_2, C_xH_y	$Ar+O_2, N_2, C_xH_y$	O_2, N_2, C_xH_y	O_2, N_2, C_xH_y	$Ar+O_2, N_2, C_xH_y$
Activation Process	- Discharge around substrate - Additional electrons	- Secondary	- Low energy primary electrons			- Medium energy electrons in penning discharge
Advantages	- High deposition rates - High reactivity		- Very high reactivity		- Orientation independent	- Easy scale-up - Easy monitoring - Low directivity - Semi-continuous Plant possible
Disadvantages	- Typically «Batch» processes		- Rate control?	- High voltage E-guns	- Additional ignition system needed	- Limited rate - Low material yield

* Plazmenii uskoritel

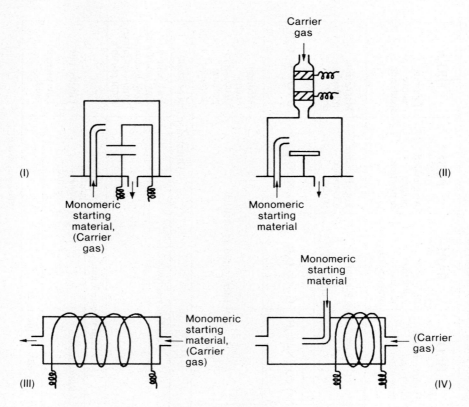

Fig. 99
Schematic representation of the mechanism of glow discharge polymerization [458].

more triple bonds or olefinic double bonds in the carbon chain or carbon ring system. The plasma-enhanced reactivity of such bonds leads to addition of other monomers forming the polymer.

In contrast with conventional plasma-induced polymerization, plasma-state polymerization is an atomic process in which reactive species are generated by the action of electron impact and high-energy ion/atom collisions which occur in the plasma. For reactive monomer formation in this process the starting gases need not be chemically unsaturated multiple bonds containing compounds. As a consequence, however, the final molecular structure of the polymer can generally be quite different from that of the original gas. A further difference of this mechanism is that polymer intermediate and gaseous by-products are formed as result of the plasma reactions. There is also some uncertainty here, depending on the special conditions and probably also on the materials used, what type of reactive species is predominant. Free radicals are regarded by some researchers to be the predominant active species while others consider cations. Evidence exists to support either point of view. The free radical mechanism is supported by the observations of high radical populations in the films formed [460–468].

An ionic mechanism is supported by evidence relative to an electrical field. It could be shown [469–471] that a floating electrode receives a very low coating rate compared to a power electrode.

Today plasma polymerization is generally assumed to be a combination of the induced and state mechanisms. Clarification of this basic problem is urgently desired and would put the entire field on a more predictive basis, which would further serve as a basis for evaluating the role of some parameters such as pressure of the monomer, typically between 10^{-1} and 10^{-4}mbar; power level, often between 10 and 300 W frequency being generally between 0 (dc) and the microwave range (gigahertz), as well as type and amount of carrier gases and reactor design, etc., which should then enable optimization of desirable film properties of the polymers.

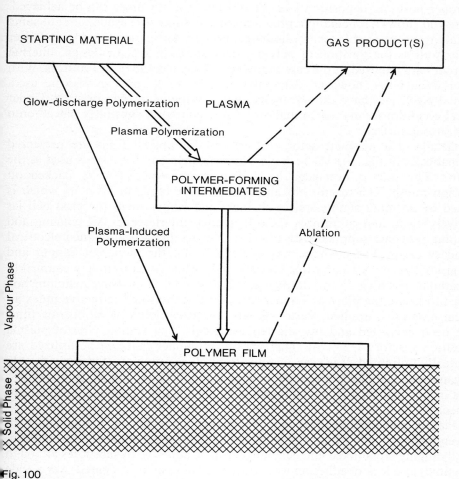

Fig. 100

Schematic representation of some typical arrangements of electric discharge used in plasma polymerization [458].

It is typical for all polymer deposition by glow-discharge polymerization that the film is formed on surfaces which directly contact the glow. Some polymer deposition may also occur on surfaces in non-glow regions, but the deposition rate is then orders of magnitude smaller. The surface where the polymer film is formed can be an electrode surface or a substrate surface suspended in the glow region. It is important to mention that plasma polymerization is system-dependent, so that the deposition rates are dependent on the ratio of surface to volume of glow discharge. Therefore the other operational parameters such as system pressure, flow rate, and discharge power are insufficient parametes for a complete description of the process. Many factors influence glow-discharge polymerization in an interrelated manner, so that a single factor cannot be taken as an independent variable of the polymerization process. Nevertheless, with substrates in proper position, deposition rates of 100 nm min^{-1} or more can be achieved. Figure 100 shows a schematic representation according to Yasuda [458] of some typical arrangements of electric discharge, flow of starting monomer and carrier gas, as well as the location of polymer deposition. In reactor design, internal electrodes are required in the case of low-frequency electric power sources. With higher frequencies, however, external electrodes or a coil can also be used. Internal electrodes have the advantage that any frequency can be used. Under typical conditions, polymer formation occurs mainly on substrates placed onto the electrode surface.

Details of a plasma-reactor coating unit for optical films are presented schematically in Fig. 101 [472]. A glass bell jar of 46 cm diameter is used as the reactor. The unit is evacuated by a conventional mechanically backed oil diffusion pump. The reactor has an active volume of about 80 liters which is formed by an inner stainless steel chamber and which keeps the glass bell jar relatively clean and minimizes radio-frequency interference. All gauging and pumping ports are supplied with metal screens so that a well defined electrical boundary exists. The chamber has inlet valves for the monomer vapour and background gases. A 13.56 MHz power supply which is capacitively coupled to the cathode disk (Ø = 15 cm) through an adjustable LC-impedance matching box is used to excite the plasma. To measure film thickness and refractive index an optical monitor is applied. Various kinds of plasma-polymerized organic films have been deposited and investigated using this or a similar type of plasma deposition reactor [472–474]. The types of monomer that can be utilized are practically unlimited. It is even possible to produce polymers starting with only inorganic materials, for example, from a mixture of CO, H_2 and N_2. It is quite obvious in this case that from the initial reactant a host of reaction products are formed.

Plasma-polymerized films have distinctive properties in comparison with conventional polymer films. The films are fundamentally pinhole free, uniform, usually highly crosslinked and generally strongly adhesive to substrates although sometimes also lack of adhesion was observed. They can be prepared over a wide range of thickness. The films are mostly very hydrophobic, unless some source of oxygen or nitrogen was present in the reaction mixture. Plasma-polymerized films

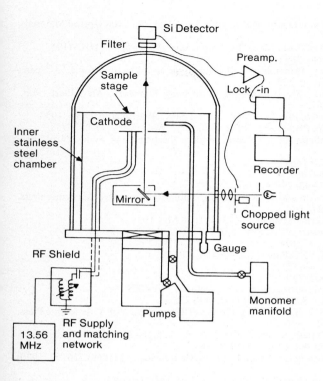

Fig. 101

Reactor to deposit plasma polymerized organic films for optical applications [472].

often exhibit excellent resistance to chemical attack, but, depending on deposition parameters, also sometimes exhibit residual chemical activity. This results in an increase in the mass of the polymer on exposure to the atmosphere. By producing a large number of active sites in plasma treatment, the film surface may be modified to form a tough cross linked shell which can act as a barrier to vapour and liquid solvent attack.

REFERENCES

[1]   H. D. Taylor, British Patent No. 29561 (1904).
[2]   W. Geffcken, Glastechn. Ber., 24 (1951) 143.
[3]   W. Geffcken, German Patent, filed in 1939.
[4]   a) H. Schröder, Z. Techn. Phys., 22 (1941) 38; 23 (1942) 196.
      b) H. Schröder, Z. Naturforsch. 4a (1949) 515.
[5]   H. Schröder, Glastechn. Ber., 26 (1953) 91.
[6]   E. K. Hussmann and R. Schnabel, in S. Musikant and J. Dupuy (Eds.), Proc. SPIE, Vol. 400, New Optical Materials, 1983, paper 20.
[7]   a) J. Koch, British Patent No. 667.784, Publ. (1952); Appl. in Denmark, July (1948).
      b) H. Anders, Dünne Schichten für die Optik, Wiss. Verlagsges. m. b. H., Stuttgart, 1965 p. 131.

276

[8] P. Mazzoldi, in S. Musikant and J. Dupuy, (Eds.), Proc. SPIE, Vol. 400, New Optical Materials, 1983, paper No. 18.

[9] S. Wein, Silver Films, Rep. No. PB 111236, Off. Techn. Serv., Washington, D. C. (1953).

[10] Nat. Bur. Stand. (U.S.), Circ. No. 389 (1931).

[11] W. Blum and H. Hogaboom, Principles of Electroplating and Electroforming, 3rd edn., McGraw-Hill, New York, 1949, p. 225.

[12] C. Davidoff, in N. Hall (Ed.) Metal Finishing Guidebook Directory, Met. Plast. Pub., Hackensack, N. J., 1977, p. 241.

[13] S. Wein, Copper Films, Rep. No. PB 111237, Off. Techn. Serv., Washington, D. C., 1953.

[14] S. Wein, Gold Films, Rep. No. PB 111332, Off. Techn. Serv., Washington, D. C., 1953.

[15] F. Pearlstein, in F. A. Lowenheim (Ed.), 3rd edn., Modern Electroplating, Wiley, New York, 1974, pp. 710 – 747.

[16] N. Feldstein, Plating, 61 (1974) 146.

[17] R. M. Lukes, Plating, 51 (1964) 969.

[18] N. Feldstein, US Patent No. 3.993, 799, (1976).

[19] F. A. Lowenheim, in J. L. Vossen and W. Kern (Eds.), Thin Film Processes, Academic Press, New York, 1978, pp. 212 – 221.

[20] W. Geffcken and E. Berger, German Patent No. 736411, 28 May 1939.

[21] C. H. Cartwright and A. F. Turner, Phys. Rev., 55A (1939) 1128.

[22] M. Banning, J. Opt. Soc. Am., 37 (1947) 688.

[23] W. Geffcken and H. Schröder, FRG Patent No. 906426, (1951).

[24] H. Schröder, FRG Patent No. 1063773, (1957).

[25] H. Schröder, Opt. Acta, 9 (1962) 249.

[26] H. Schröder, 7th Int. Congr. Glass, Bruxelles, 1965, Section 1, 7.1 – 7.5.

[27] H. Bach and H. Schröder, Thin Solid Films, 1 (1967/68) 255.

[28] H. Schröder, in G. Hass and R. E. Thun, (Eds.), Physics of Thin Films, Vol. 5, Academic Press, New York, 1969, pp. 87 – 140.

[29] H. Schröder, 10th Int. Congr. Glass, Kyoto, Japan, 1975, 8 – 118; 8 – 130.

[30] H. Dislich, Angew. Chem., Int. Ed. Engl., 10 (1971) 363.

[31] K. S. Mazdiyasni, R. T. Dollof and J. S. Smith II, J. Am. Ceram. Soc., 52 (1969) 523; 53 (1970) 91.

[32] H. Dislich, P. Hinz and R. Kaufmann, FRG Patent No. 1.941.191, (1969).

[33] H. Dislich and A. Jacobson, Angew. Chem., Int. Ed. Engl., 12 (1973) 439.

[34] H. Dislich and A. Jacobson, FRG Patent No. 1.494872, (1965).

[35] H. Dislich, Angew. Chem., Int. Ed. Engl., 18 (1979) 49.

[36] H. Dislich and E. Hussmann, Thin Solid Films, 77 (1981) 129.

[37] H. Dislich and P. Hinz, Kinetics and Mechanism of Polyreactions, Vol. 1, IUPAC-Macro, Budapest, 1969 Preprints 1/39.

[38] I. Keesmann, Z. Anorgan. Allgem. Chem., 346 (1966) 30.

[39] G. Bayer, O. W. Flörke, W. Hoffmann and J. J. Scheel, Glastechn. Ber., 39 (1966) 242.

[40] American Optical Company, US. Pat. No. 2.474.061, 1943; US. Pat. No. 2.466.119, 1944.

[41] W. Noll, Chemie und Technologie der Silicone, 2nd, edn., Verlag Chemie, Weinheim, 1968.

[42] H. Schmidbaur, Angew. Chem., 77 (1965) 206; Angew. Chem. Int. Ed. Engl., 4 (1965) 201.

[43] Houben-Weyl-Müller, Methoden der organischen Chemie, Vol. VI/2 Sauerstoffverbindungen I, Part 2, Thieme Verlag, Stuttgart, 1963.

[44] H. Dislich, Glastechn. Ber., 44 (1969) 1.

[45] Schott u. Gen., US Pat. No. 2.366.516.

[46] Leitz, German Patent No. 937.913, (1939).

[47] H. A. Tanner and L. B. Lockhart, J. Opt. Soc. Am., 36 (1946) 701.

[48] J. J. Licari, Plastic Coatings for Electronics, McGraw-Hill, New York, 1970.

[49] B. D. Washo, IBM J. Res. Dev., 21 (1977) 190.

[50] I. Langmuir, J. Am. Chem. Soc., 39 (1917) 1848; Proc. Roy, Soc. (London), A 170 (1939) 1.

[51] K. Blodgett, J. Am. Chem. Soc., 56 (1934) 495; Phys. Rev., 57 (1940) 921; J. Phys. Chem., 41 (1937) 975.

[52] K. Blodgett and I. Langmuir, Phys. Rev., 51 (1937) 964.

[53] K. Blodgett, J. Am. Chem. Soc., 57 (1935) 1007.

[54] H. Mayer, Aktuelle Forschungsprobleme aus der Physik dünner Schichten, Oldenbourg, München, 1950.

[55] P. S. Vincett, W. A. Barlow, F. T. Boyle, J. A. Finney and G. G. Roberts, Thin Solid Films, 60 (1979) 265.

[56] V. L. Ginzburg, J. Polym. Sci. C, 29 (1970) 3.
[57] A. D. Adler, J. Polym. Sci. C, 29 (1970) 73.
[58] J. Dresner, R. C. A. Rev., 30 (1969) 322.
[59] J. Desner and A. M. Goodman, Proc. IEEE, 58 (1970)1868.
[60] A. Aviram and M. A. Ratner, Chem. Phys. Lett. 29 (1974) 277.
[61] a) C. W. Pitt and L. M. Walpita, Electron. Lett., 12 (1976) 479.
      b) C. W. Pitt and L. M. Walpita, Thin Solid Films, 68 (1980) 101.
      c) C. W. Pitt and F. Grunfeld, in R. R. Jacobsson, (Ed.), Proc. SPIE, Vol. 401, Thin Film
          Technologies, 1983, paper No. 16.
[62] L. Esaki and R. Tsu, IBM J. Res. Dev., 14 (1970) 61.
[63] P. K. Weimer, Proc. IRE. 50 (1962) 1462.
[64] G. G. Roberts, K. P. Panda and W. A. Barlow, Electron. Lett. 13 (1977) 581.
[65] H. Müller, Präparation von Techn.–Physikal. Objekten für die Elektronenmikroskopische
      Untersuchung, Geest u. Portig KG, Leipzig, 1962.
[66] W. Kern and C. A. Deckert, Chemical Etching, in J. L. Vossen and W. Kern, (Eds.), Thin Film
      Processes, Academic Press, New York, 1978, p. 401 – 496.
[67] W. S. De Forest, Photoresist Materials and Processes, McGraw-Hill, New York, 1975.
[68] W. M. Feist, S. R. Steele and D. W. Readey, The Preparation of Films by Chemical Vapour
      Deposition, in G. Hass and R. E. Thun (Eds.), Physics of Thin Films, Vol.5, Academic Press,
      New York, 1969, p. 237 – 322.
[69] W. Kern and V. S. Ban, Chemical Vapour Deposition of Inorganic Thin Films, in J. L. Vossen
      and W. Kern, (Eds.), Thin Film Processes, Academic Press, New York, 1978, p. 257 – 331.
[70] J. M. Blocher jr., G. E. Vuillard and G. Wahl, (Eds.), Proc. 8th Int. CVD Conf., The
      Electrochemical Soc., Inc., 10 South Main St., Pennington, NJ 08534, (1981) and all the
      Proceedings of preceding CVD Conferences.
[71] J. W. Mellor, A Comprehensive Treatise on Inorganic and Theoretical Chemistry, Vol. 6,
      Longmans, Green & Co. Inc., New York, 1957.
[72] L. Mond, US Patent No. 455,230, June 30 1891.
[73] L. Mond, C. Langer and F. Quinke, J. Chem. Soc., 57 (1890) 749.
[74] J. M. Blocher, jr., Thin Solid Films, 77 (1981) 51.
[75] J. C. Viguié and J. Spitz, J. Electrochem. Soc.; Solid-State Science and Technol., 122 (1975) 585.
[76] W. Geffcken, Schott und Gen., German Patent No. 742463, Swiss Patent No. 223.344 and
      German Patent No. 742.463, (1944).
[77] D. V. Vorobieva and I. F. Polurotova, Izv. Akad. Nauk. SSSR, Neorgan. Materialy, 7 (1971) 266.
[78] O. V. Vorobieva and E. S. Bessonova, Steklo i keram., 21 (1964) 9.
[79] J. C. Manifacier, J. P. Fillard and J. M. Bind, Thin Solid Films, 77 (1981) 67.
[80] R. Pommier, C. Gril and J. Marucchi, Thin Solid Films, 77 (1981) 91.
[81] G. Blandenet, M. Court and Y. Lagarde, Thin Solid Films, 77 (1981) 81.
[82] J. Kane, H. P. Schweizer and W. Kern, J. Electrochem. Soc., 122 (1975) 1144.
[83] J. Kane, H. P. Schweizer and W. Kern, J. Electrochem. Soc., 123 (1976) 270.
[84] D. E. Carlson, J. Electrochem. Soc., 122 (1975) 1334.
[85] J. C. Manifacier, M. de Murcia and J. P. Fillard, Mater. Res. Bull., 10 (1975) 1215.
[86] R. Kalbskopf, Thin Solid Films, 77 (1981) 65.
[87] M. Yokizowa, H. Iwasa and I. Teramoto, Jap. J. Appl. Phys., 7 (1968) 96.
[88] D. R. Harbison and H. L. Taylor, in F. Vratny, (Ed.), Thin Film Dielectrics, Electrochem. Soc.,
      New York, 1969, p. 254.
[89] C. C. Wang and K. H. Zaininger, Proc. Electron Compon. Conf., 1969, p.345.
[90] S. Pakswer and P. Skoug, Proc. 2nd Int. CVD Conf., J. M. Blocher jr. and J. C. Withers, (Eds.),
      Electrochem. Soc., New York, 1970, p. 619.
[91] M. Balog, M. Schieber, S. Patai and M. Michman, J. Cryst. Growth 17 (1972) 298.
[92] J. R. Szedon and R. M. Handy, J. Vac. Sci. Technol., 6 (1969) 1.
[93] Y. Nakai, Electrochem. Soc. Extended Abstract No. 84, Spring Meeting, 1968, p. 215.
[94] J. A. Aboaf, J. Electrochem. Soc., 114 (1967) 948.
[95] M. T. Duffy and W. Kern, RCA Rev., 31 (1970) 754.
[96] M. Mutoh, Y. Mizokami, H. Matsui, S. Hagiwara and N. Ino, J. Electrochem. Soc., 122 (1975)
      987.
[97] M. Matsushita and Y. Yoga, Electrochem. Soc. Extended Abstract No. 90, Spring Meeting, 1968,
      p. 230.
[98] N. N. Tvorogov, Zh. Prikl. Khim. (Leningrad), 34 (1961) 2203.
[99] L. A. Ryabova and Y. S. Savitskaya, J. Vac. Sci. Technol., 6 (1969) 934.

278

[100] D. Peterson, Non-Vacuum Deposition Techniques for use in Fabrication Thin Film Circuits, No. Noobsr 91336, Final Rep., 1967.
[101] M. T. Duffy, C. C. Wang, A. Waxman and K. H. Zaininger, J. Electrochem. Soc., 116 (1969) 234.
[102] C. C. Wang, K. H. Zaininger and M. T. Duffy, RCA Rev. , 31 (1970) 728.
[103] L. A. Ryabova and Y. S. Savitskaya, Thin Solid Films, 2 (1968) 141.
[104] I. Weitzel and K. Kempter, Electrochem. Soc. Extended Abstract No. 76 – 2, 1976, p. 642.
[105] W. Kern and R. C. Heim, J. Electrochem. Soc., 117 (1970) 562.
[106] W. Kern and R. C. Heim, J. Electrochem. Soc., 117 (1970) 568.
[107] A. Mayer, N. Goldsmith and W. Kern, Surface Passivation Techniques for Compound Solid State Devices, Techn. Rep. AFAL-TR-65-213 Air Force System Command, Wright-Patterson Air Force Base, Ohio, 1965.
[108] T. Kamimori and M. Mizuhashi, in Proc. 8th Int. CVD Conf., J. M. Blocher jr., G. E. Vuillard and G. Wahl, (Eds.), The Electrochem. Soc. Inc., 10 South Main St., Pennington, NJ 08534, 1981, p. 438.
[109] L. Ben-Dor, R. Druilhe and P. Gilbart, J. Cryst. Growth, 24/25 (1974) 1972.
[110] K. L. Hardee and A. J. Bard, J. Electrochem. Soc., 123 (1976) 1024.
[111] J. MacChesney, P. B. O'Connor and M. V. Sullivan, J. Electrochem. Soc., 118 (1971) 776.
[112] M. Sullivan, J. Electrochem. Soc., 120 (1973) 545.
[113] P. K. Gallagher, W. R. Sinclair, R. A. Fastnacht and J. P. Luongo, Thermochim. Acta, 8 (1974) 141.
[114] D. R. Mason, J. Electrochem. Soc., 123 (1976) 519.
[115] W. Kern, RCA Rev., 32 (1971) 429.
[116] A. E. Feuersanger, Proc. IEEE, 52 (1964) 1463.
[117] Y. W. Hsueh and H. C. Lin, Ann. Rep. Conf. Electr. Insul. Dielectr. Phenom. 1974, p. 515.
[118] J. P. Dismukes, J. Kane, B. Binggeli and H. P. Schweizer Proc. 4th Int. CVD Conf., G. F. Wakefield and J. M. Blocher, jr., (Eds.), Electrochem. Soc., Princeton, NJ 1973, p. 275.
[119] T. Matsuo, Jap. J. Appl. Phys., 12 (1973) 1862.
[120] A. W. Fischer, J. A. Amick, H. Hyman and J. H. Scott jr., RCA Rev., 29 (1968) 533.
[121] W. Kern and R. C. Heim, Electrochem. Soc. Extended Abstracts No. 5, Spring Meeting, 1968, p. 234.
[122] A. W. Fischer and J. A. Amick, RCA Rev., 29 (1968) 549.
[123] K. Strater and A. Mayer, in R. R. Haberecht and E. L. Kern, (Eds.), Semiconductor Silicon, Electrochem. Soc. New York, 1969, p. 469.
[124] D. M. Brown and P. R. Kennicott, J. Electrochem. Soc., 118 (1971) 293.
[125] A. S. Tenney, J. Electrochem. Soc., 118 (1971) 1658.
[126] E. A. Taft, J. Electrochem. Soc., 118 (1971) 1985.
[127] A. S. Tenney and J. Wong, J. Chem. Phys., 56 (1972) 5516.
[128] D. M. Brown, M. Garfinkel, M. Ghezzo, E. A. Taft, A. S. Tenney and J. Wong, J. Cryst. Growth, 17 (1972) 276.
[129] G. Wahl, Proc. 5th Int. CVD Conf., J. M. Blocher, jr., H. E. Hintermann and L. H. Hall, (Eds.), Electrochem. Soc., Princeton, NJ, 1975, p. 391.
[130] F. C. Eversteijn, Philips Res. Rep., 21 (1966) 379.
[131] D. Peterson, IEEE Trans. Compon. Parts, CP-10, (1963) 119.
[132] D. B. Lee, Solid State Electron., 10 (1967) 623.
[133] H. Teshima, Y. Tarui and O. Takeda, Denki Shikenjo Iho, 33 (1969) 631.
[134] A. Cuccia, G. Shrank and G. Queirolo, in R. R. Haberecht and E. L. Kern, (Eds.), Semiconductor Silicon, Electrochem. Soc., New York, 1969, p. 506.
[135] E. Arai and Y Terunuma, Jap. J. Appl. Phys., 9 (1970) 691.
[136] P. C. Parekh, D. R. Goldstein and T. C. Chan, Solid State Electron. 14 (1971) 281.
[137] T. Wong and M. Ghezzo, J. Electrochem. Soc., 118 (1971) 1540; 119 (1972) 1413.
[138] J. Wong, J. Electrochem. Soc., 119 (1972) 1071, 1080; 120 (1973) 122.
[139] R. B. Fair, J. Electrochem. Soc., 119 (1972) 1389.
[140] M. Ghezzo and D. M. Brown, J. Electrochem. Soc., 120 (1973) 110.
[141] E. Tanikawa, O. Takayama and K. Maeda, in Proc. 4th Int. CVD Conf., G. F. Wakefield and J. M. Blocher, jr., (Eds.), The Electrochem. Soc. Princeton, NJ, 1973, p. 261.
[142] J. Wong, J. Electron. Mater., 5 (1976) 113.
[143] H. Nagai and T. Niimi, J. Electrochem. Soc., 115 (1968) 671.
[144] T. Yashiro, J. Electrochem. Soc., 119 (1972) 780.

[145] K. Sugiyama, S. Pac, Y. Tahashi and S. Motojiama, J. Electrochem. Soc., 122 (1975) 1545; Proc. 5th Int. CVD Conf., J. M. Blocher, jr., H. E. Hintermann and L. H. Hall, (Eds.), The Electrochem. Soc., Princeton, NJ, 1975, p. 147.
[146] C. F. Powell, J. H. Oxley and J. M. Blocher, jr., Vapour Deposition, Wiley, New York, 1966.
[147] A. J. Perry and N. J. Archer, in Materials Coating Techniques, AGARD Lecture Series No. 106, AGARD-NATO Publication, ISBN 92 – 835 – 1357 – 6, March 1980, No. 4.
[148] J. J. Crosby, USAF Rep., ASD-TRD-62-907, December 1962.
[149] W. M. MacNevin, US Patent No. 2, 867, 546, Jan. 6, 1959.
[150] C. Berger, US Patent No. 2, 921, 868, Jan. 19, 1960.
[151] V. Norman and H. B. Prestridge, US Patent No. 3, 202, 537, May 1; 1962.
[152] F. E. Drummond, US Patent No. 2, 332, 309, Oct. 19, 1943.
[153] J. J. Lander and L. H. Germer, Am. Inst. Mining Met. Engrs. Inst. Metals Div., Metals Technol. 14 (6) Techn. Publ. 2259 (1947); Metal Ind. (London), 71 (1947) 459, 487.
[154] B. B. Owen and R. T. Webber, Am. Inst. Mining Met. Engrs., Techn. Publ. 2306, January 1948.
[155] H. E. Nack, J. J. Bulloff and J. R. Whitacre, US Patent No. 3, 050, 417, Aug. 21, 1962.
[156] J. H. Oxley, M. F. Browning, N. D. Veigel and J. M. Blocher, jr., Ind. Eng. Chem. Prod. Research and Developm., 1 (2) (1962) 102.
[157] G. P. Conard and E. J. Jablonowski, 10th Quarterly Rep. on Magnetic Materials Research, ASTIA Doc. AD-255689, May 14, 1961.
[158] E. J. Jablonowski, Cobalt, 14 (1962) 28.
[159] W. C. Fernelius, (Ed.), Inorganic Syntheses, Vol. II, McGraw-Hill, New York, 1946, p. 238.
[160] Gmelin, Handbuch der anorgan. Chemie, No. 58, Part A, Verlag Chemie, Berlin, 1932.
[161] A. E. Wallis, Brit. Patent No. 620,287, March 22, 1949.
[162] L. J. Novak and H. J. Homer, US Patent No. 2,859,132, Nov. 4, 1958.
[163] M. V. Sullivan, US Patent No. 2,759,848, Aug. 21, 1956.
[164] W. E. Tewes, T. E. Zava and T. B. Hoover US Atomic Energy Comm. Rep. K-1533, 1962.
[165] E. C. Marboe, US Patent No. 2,430,520, Nov. 11, 1947.
[166] D. Beischer, Z. Elektrochem., 45 (4) (1939) 310.
[167] H. E. Carlon and J. H. Oxley, Am. Inst. Chem. Engs. J., 11 (1965) 79.
[168] V. Norman and T. P. Whaley, US. Patent No. 3,175,924, March 30, 1965.
[169] J. H. Simons and R. W. McNamee, C. D.Hurd, J. Phys. Chem., 36 (1932) 939.
[170] P. Pawlyk, US Patent No. 2,704,729, March 22, 1955.
[171] H. S. Williams and W. A. Weyl, Glass Ind., 26 (1945) 324.
[172] W. A. Weyl, Glass Ind., 28 (1947) 231.
[173] Pilkington Bros. Ltd., German Offenlegungsschrift 2,626,118, Dec. 30, 1976.
[174] L. Brewer, Papers 6 and 7 in The Chemistry and Metallurgy of Miscellaneous Materials: Thermodynamics, McGraw-Hill, New York, 1950.
[175] L. H. Marshall, US Patent No. 1,893,782, Jan. 10, 1933.
[176] D. A. Petrov, V. A. Butov and N. G. Gilyadova, Zhur, Neorg. Khim., 4 (1959) 1970.
[177] H. J. Homer and O. Cummins, US Patent No. 2,916,400, Dec. 8, 1959.
[178] C. E. Waring and W. S. Horton, J. Am. Chem. Soc., 67 (1945) 540.
[179] T. V. Sathyamurthy, S. Swaminathan and L. M. Yeddanapalli, J. Ind. Chem. Soc., 27 (1950) 509.
[180] H. W. Schultz, Met. Progr., 76 (1959) 74.
[181] Western Electric Company, Brit. Patent No. 589,966, July 4, 1947.
[182] J. J. Lander, US Patent No. 2,516,058, July 18, 1950.
[183] X. X. Metal Finishing, 47 (10) (1949) 79.
[184] J. J. Ward, A. D. Coon and J. H. Oxley, paper presented at the Conf. Applications of Fundamental Thermodynamics, Universitiy of Pittsburgh, Nov. 1964.
[185] J. J. Cuomo, in Proc. 3rd Int. CVD Conf., F. A. Glaski, (Ed.), American Nuclear Soc., Hinsdale, Ill., 1972, 270.
[186] A. E. van Arkel, Metallwirtschaft, 13 (1934) 405.
[187] N. V. Philips' Gloeilampenfabrieken, German Patent No. 495,751; Oct. 20, 1926; Brit. Patent No. 280,697; Oct. 29, 1926.
[188] Gmelins Handbuch der anorgan. Chemie, 8. edn. Verlag Chemie, Berlin, No. 65, 1942.
[189] R. S. Cattle, Brit. Patent No. 1,209,518, Oct. 21, 1970.
[190] M. I. Ermolaev and Y. Y. Gukova, USSR Patent No. 281,785, Sept 14, 1970.
[191] R. J. H. Voorhoeve and J. W. Merewether, J. Electrochem. Soc., 119 (1972) 364.
[192] T. Matcovich, E. Korostoff and A. Schmeckenbecker, J. Appl. Phys. Suppl., 32 (3) (1961) 93S.
[193] J. S. Mathias, W. Freitgag and C. J. Kriessmann, USAF Rep. AFCRL – 970, Nov. 1961; ASTIA Doc. AD – 275310.

280

[194]   F. C. Eversteijn, P. J. W. Severin, C. H. J. van der Brekel and H. L. Peek, J. Electrochem. Soc., 117 (1970) 925.
[195]   W. Kern and R. S. Rosler, J. Vac. Sci. Technol., 14 (1977) 1082.
[196]   C. H. J. van den Brekel, in Proc. 8th Int. CVD Conf., J. M. Blocher, jr., G. F. Vuillard and G. Wahl, (Eds.), The Electrochem. Soc. Inc., 10 South Main Str., Pennington, N. J. 08534, 1981, p. 116.
[197]   A. E. T. Kuiper, C. H. J. van den Brekel, J. de Groot and G. W. Veltkamp, in Proc. 8th Int. CVD Conf., J. M. Blocher, jr., G. F. Vuillard and G. Wahl, (Eds.), The Electrochem. Soc. Inc., 10 South Main Str., Pennington, N. J. 08534, 1981, p. 131.
[198]   F. Kaufman, Adv. Chem. Ser. (80) (1969) 29.
[199]   F. K. McTaggart, Plasma Chemistry in Electrical Discharges Elsevier, Amsterdam 1967; and J. R. Hollahan and A. T. Bell, Eds., Techniques and Applications of Plasma Chemistry, Wiley, New York, 1974.
[200]   T. Bell, in Proc. 8th Int. CVC Conf., J. M. Blocher, jr., G. E. Vuillard and G. Wahl, (Eds.), The Electrochem Soc., Inc., 10 South Main Str., Pennington, N. J. 08534, 1981, p. 185.
[201]   A. R. Reinberg, Int. Round Table Surf. Treat. Plasma Polymer., IUPAC, Limoges, France, 1977.
[202]   A. R. Reinberg, US Patent No. 3,757,733 (1973).
[203]   R. S. Rosler, W. C. Benzing and J. Baldo, Solid State Technol., 19 (6) (1976) 45.
[204]   H. Kobayashi, A. T. Bell and M. Shen, J. Appl. Polymer Sci., 17 (1973) 885.
[205]   M. H. Brodsky, M. A. Frisch and J. F. Ziegler, Appl. Phys. Lett., 30, (1977) 561.
[206]   W. E. Spear, P. G. Lecomber, S. Kinmond and M. H. Brodsky, J. Appl. Phys. Lett. 28 (1976) 105.
[207]   W. E. Spear and P. G. Lecomber, Philos. Mag., 33 (1976) 935.
[208]   H. Katto and Y . Koga, J. Electrochem. Soc., 118 (1971) 1619.
[209]   B. Mattson, Solid State Technol., January (1980) 60.
[210]   D. R. Secrist and J. D. McKencie, J. Electrochem. Soc., 113 (1966) 914.
[211]   P. Turner, R. P. Howson and C. A. Bishop, Thin Solid Films, 83 (1981) 253.
[212]   M. J. Rand, J. Vac. Sci. Technol., 16 (2) (1979) 420.
[213]   D. Küppers, in Proc. 7th Int. CVD Conf., T. H. Watson and H. Lydtin, (Eds.), Electrochem. Soc., Princeton, N. J., 1979, 159.
[214]   S. D. Allen and M. Bass, J. Vac. Soc. Technol., 16 (1979) 431.
[215]   T. F. Deutsch, D. J. Ehrlich and R. M. Osgood, J. Appl. Phys. Lett., 35 (1979) 175.
[216]   H. M. Kim, Sung-Shan Tar, S. L. Groves and K. K. Schuegraf in Proc. 8th Int. CVD Conf., J. M. Blocher, jr. G. E. Vuillard and G. Wahl, (Eds.), The Electrochem. Soc., Inc., 10 South Main Str., Pennington N. J. 08534, 1981, p. 258.
[217]   S. D. Allen, A. B. Trigubo and Y. C. Liu, in Proc. 8th Int. CVD Conf., J. M. Blocher, jr. G. E. Vuillard and G. Wahl, (Eds.), The Electrochem. Soc., Inc., 10 South Main Str., Pennington N. J. 08534, 1981, p. 267.
[218]   R. Bilenchi, M. Musci, in Proc. 8th Int. CVD Conf., J. M. Blocher, jr. G. E. Vuillard and G. Wahl, (Eds.), The Electrochem. Soc., Inc., 10 South Main Str., Pennington N. J. 08534, 1981, p. 275.
[219]   O. Tabata, S. Kimura, S. Tabata, R. Macabe, S. Matsuura in Proc. 8th Int. CVD Conf., J. M. Blocher, jr. G. E. Vuillard and G. Wahl, (Eds.), The Electrochem. Soc., Inc., 10 South Main Str., Pennington N. J. 08534, 1981, p. 272.
[220]   S. Dushman and J. M. Lafferty, Scientific Foundations of Vacuum Technique, 2nd edn., Wiley, New York, 1962.
[221]   M. Wutz, Therorie und Praxis der Vakuumtechnik, Vieweg, Braunschweig, 1965, revised and enlarged 2nd edn., 1982.
[222]   J. Yarwood, High Vacuum Technique, 4th edn., Chapman and Hall, London, 1967.
[223]   R. Glang, R. A. Holmwood and J. A. Kurtz, High-Vacuum Technology, Chapter 2 in L. I. Maissel and R. Glang, (Eds.), Handbook of Thin Film Technology, McGraw-Hill, New York, 1970.
[224]   L. Holland, W. Steckelmacher and J. Yarwood, Vacuum Manual, Spon, London, 1974.
[225]   A. Roth, Vacuum Technology, North-Holland, Amsterdam, 1976.
[226]   BALZERS Publication Nr. PK 800 046 PE 7910.
[227]   BALZERS Publication Nr. PP 800 014 PE.
[228]   BALZERS Publication Nr. BP 800 106 PE 7910.
[229]   BALZERS Publication Nr. PM 800 032 PE.
[230]   BALZERS Publication Nr. BP 800 108 PE 8001; BP 800 097 PE 7910; BP 800 102 PE 910; BP 800 094 PE 7910;
[231]   W. Gaede, German Patent Nr. 286 404 (1913); Ann. Physik, 46 (1915) 357.

[232] G. Rettinghaus, BALZERS Hochvakuum Fachbericht VKD 1,February 1974.
[233] G. Rettinghaus, W. K. Huber, Vacuum 24 (6) (1974) 249.
[234] W. Becker, Vakuum Techn., 7 (1958) 148.
[235] W. Becker, Proc. 1st Intern. Congr. Vacuum Techn., Namur, 1960, p. 173.
[236] W. Becker, Vakuum-Techn., 15 (9) (1966) 211; 15 (10) (1966) 254.
[237] C. Krüger, MIT Division DSR 7-8120.
[238] R. Buhl, Vakuum-Techn., 30 (6) (1981) 166.
[239] P. J. Gareis and G. F. Hagenbach, Ind. Eng. Chem., 57 (1965) 27.
[240] B. D. Power, High Vacuum Pumping Equipment, Chapman and Hall, London 1966.
[241] P. A. Redhead, J. P. Hobson and E. V. Kornelson, The Physical Basis of UHV, Chapman and Hall, London, 1968.
[242] A. J. Kidnay and M. J. Hiza, Cryogenics, 10 (1970) 271.
[243] J. P. Hobson, J. Vac. Sci. Technol., 10 (1973) 73.
[244] R. A. Haefer, Kryo-Vakuumtechnik, Springer, Berlin, 1981.
[245] R. W. Moore, Cryopumping in the free-molecular flow regime, Trans. 2nd Int. Vac. Congress, Pergamon, Oxford, 1962, p. 426.
[246] M. Faraday, Phil. Trans., 147 (1857) 145.
[247] R. Nahrwold, Ann. Physik, 31 (1887) 467.
[248] A. Kundt, Ann. Physik, 34 (1888) 473.
[249] R. Pohl and P. Pringsheim, Verhandl. Deut. Physik Ges., 14 (1912) 506.
[250] M. Auwärter, BALZERS, Austrian Pat. Nr. 192 650 (1952); US Pat. No. 2.920.002 (1960); FRG Pat. Nr. 1.104, 283 1968);
[251] D. S. Brinsmaid, G. J. Koch, W. J. Keenan and W. F. Parson, KODAK, US Pat. No. 2,784,115 (1957), Application made on May 4, 1953.
[252] L. I. Maissel, in L. I. Maissel and R. Glang, (Eds.), Handbook of Thin Film Technology, McGraw-Hill, New York, 1970, Chapter 1.
[253] L. Holland and W. Steckelmacher, Vacuum, 2 (1952) 346.
[254] H. A. MacLeod, Thin Film Optical Filters, Adam Hilger Ltd., London, 1969, Chapter 9.
[255] G. Deppisch, Vakuum-Techn., 30 (1981) 67.
[256] E. B. Graper, J. Vac. Sci. Technol., 10 (1973) 100.
[257] S. Schiller and U. Heisig, Bedampfungstechnik, VEB Verlag Technik, Berlin, 1976.
[258] L. Holland, Vacuum Deposition of Thin Films, Wiley, New York, 1956.
[259] K. H. Behrndt, in G. Hass and R. E. Thun, (Eds.), Physics of Thin Films, Vol. 3, Academic Press, New York, 1966.
[260] K. H. Behrndt, Trans. 10th Nat. Vac. Symp. AVS, 1963, Macmillan, New York, 1963, p. 379.
[261] H. Anders, Dr. H. Anders Company, D-8470 Nabburg, Germany.
[262] H. Dönz, Thesis, 1981, Inst. of Experimental Physics, University of Innsbruck, Austria.
[263] Th. Kraus, Vakuum Techn., 31 (1982), 130.
[264] B. O'Brien and T. A. Russel, J. Opt. Soc. Am., 24 (1934) 54.
[265] J. Strong, Procedures in Experimental Physics, Prentice Hall, New York, 1938.
[266] J. Strong and E. Gaviola, J. Opt. Soc. Am., 26 (1936) 153.
[267] L. G. Schulz, J. Opt. Soc. Am., 37 (1947) 349; 37 (1947) 509; 38 (1948) 432.
[268] J. Strong, Astrophys. J., 83 (1936) 401.
[269] R. A. Fischer and J. R. Platt, Rev. Sci. Instr., 8 (1937) 505.
[270] J. Strong, J. Phys. Radium, 11 (1950) 441.
[271] S. Schiller, H. Förster, G. Lenk, G. Kühn and W. Kunack, Neue Hütte, 13 (1968) 705.
[272] W. Reichelt, W. Dietrich and A. Hauff, Metalloberfläche, 20 (1966) 474.
[273] S. Schiller, G. Beister and G. Zeissig, Die Technik, 23 (1968) 775.
[274] M. Auwärter, in M. Auwärter, (Ed.), Ergebnisse der Hochvakuumtechnik und der Physik dünner Schichten, Bd. 1, Wiss. Verlagsges. mbH, Stuttgart, 1957, p. 14.
[275] H. K. Pulker, G. Paesold and E. Ritter, Appl. Opt., 15 (1976) 2986.
[276] N. N., Physical Vapour Deposition, Airco Temescal, Berkeley, California 94710 USA, 1976.
[277] H. M. Smith and A. F. Turner, Appl. Opt. 4, (1965) 147.
[278] G. Hass and J. B. Ramsey, Appl. Opt., 8 (1969) 1115.
[279] G.Groh, J. Appl. Phys., 39 (1968) 5804.
[280] W. P. Barr, J. Phys. E, 2 (1969) 1112.
[281] V. N. Rudenko, Teplofizika Vysokikh, Temperatur, 5, (5) (1967) 877.
[282] R. Vesper, Laser, 1 (1969) 15.
[283] P. D. Zavitasanos, Microel. Reliab., 8 (1969) 303.
[284] V. N. Ban and B. E. Knox, Int. J. Mass Spectrom. Ion. Phys., 3 (1969) 131.

[285] N. Mandani and K. G. Nichols, J. Phys. London, 3 D (1970) L7 – L9.
[286] F. P. Gagliano and U. C. Paek, Appl. Opt., 13 (1974) 274.
[287] G. Wulff, Vakuum Techn., 26 (1977) 39.
[288] K. Hayek, Diffusion und Keimbildung bei der Kondensation an festen Oberflächen, Publikationstelle der Universität Innsbruck, Austria, 1973.
[289] M. Krohn and H. Bethge, J. J. Métois and R. Kern, and Nucleation Chapter, 4[th]ITFC, Laughborough, 1978, in Thin Solid Films, 57 (1979) 227 – 292.
[290] H. Poppa, Kl. Heinemann and A. G. Elliot, J. Vac. Sci. Technol., 8 (1971) 471; J. F. Pócza, A. Barna and P. B. Barna, J. Vac. Sci. Technol., 6 (1968) 472.
[291] H. K. Pulker, Habilitation paper, Universitiy of Innsbruck, Austria, 1973.
[292] C. A Neugebauer and R. A. Ekvall, J. Appl. Phys., 35 (1964) 547.
[293] E. Ritter, Proc. Colloq. Thin Films, Budapest, 1965, p. 79.
[294] H. K. Pulker and E. Jung, Thin Solid Films, 4 (1969) 219.
[295] E. Bauer, in M. Auwärter, (Ed.), Ergebnisse der Hochvakuumtechnik und Physik dünner Schichten, Bd. I, Wiss. Verlagsges. mbH, Stuttgart, 1957, p. 39.
[296] H. K. Pulker and Ch. Zaminer, Thin Solid Films,5 (1970) 421.
[297] H. K. Pulker and E. Ritter in M. Auwärter, (Ed.), Ergebnisse der Hochvakuumtechnik und Physik dünner Schichten, Bd. II, Wiss. Verlagsges. mbH, Stuttgart, 1971, p. 244.
[298] E. Ritter, J. Vac. Sci. Technol., 3 (1966) 225.
[299] P. Goldfinger and M. Jeunehomme, in J. D. Waldron, (Ed.), Advances in Mass Spectrometry, Pergamon Press, London, 1959.
[300] A. Preisinger and H. K. Pulker, Jap. J. Appl. Phys. Suppl., 2 (1) (1974) 769.
[301] E. Ritter and R. Hoffmann, J. Vac. Sci. Technol., 6 (1969) 733.
[302] W. Heitmann, Z. Angew. Phys., 21 (1966) 503.
[303] A. Etstathion, D. M. Hoffmann and E. R. Levin, J. Vac. Sci. Technol., 6 (1969) 383.
[304] G. G. Summer and L. L. Reynolds, J. Vac. Sci.Technol., 6 (1969) 493.
[305] S. K. Bahl and K. L. Chopra, J. Vac. Sci. Technol., 6 (1969) 561.
[306] L. H. Gadgil and A. Goswami, J. Vac. Sci. Technol., 6 (1969) 591.
[307] J. L. Richards and P. B. Hart, L. M. Callone, J. Appl. Phys., 34 (1963) 3418.
[308] J. E. Johnson, J. Appl. Phys. 36 (1965) 3193.
[309] P. Bourgeois and P. Mock, Le Vide, 119 (1965) 376.
[310] E. B. Dale, J. Vac. Sci.Technol., 6 (1969) 568.
[311] W. Hänlein and K. G. Günther, Advances in Vacuum Science and Technology Vol. 2, E. Thomas (Ed.), Pergamon, London, 1960, p. 727.
[312] K. G. Günther, in J. C. Anderson (Ed.), The Use of Thin Films in Physical Investigations, Academic Press London, New York, 1966, p. 213.
[313] J. E. Davey and T. Pankey, J. Appl. Phys., 39 (1968) 1941.
[314] J. R. Arthur and J. J. Lepore, J. Vac. Sci. Technol., 6 (1969) 545.
[315] G. Zinsmeister, Vak.-Techn., 13 (1964) 233.
[316] T. M. Mills, Electrochem. Soc. Extended Abstr., 76–2 (1976) 785.
[317] W. Himes, B. F. Stout and R. E. Thun, in G. H. Bancroft (Ed.), Trans. 9th Nat. Vac. Symp., Macmillan, New York, 1962, p. 144.
[318] D. W. Pashley, Phil. Mag., 10 (1964) 127.
[319] D. W. Pashley, in Thin Films, Chapter 3, Amer. Soc. Met., Chapman and Hall, London, 1964.
[320] W. R. Grove, Phil. Trans Roy. Soc. London, 142 (1852) 87.
[321] J. Plücker, Pogg. Ann., 103 (1858) 88.
[322] P. Wright, Am. J. Sci., 13 (1877) 49.
[323] G. K. Wehner and G. S. Anderson, in L. I. Maissel and R. Glang, (Eds.), Handbook of Thin Film Technology, McGraw-Hill, 1970, Chapter 3; L. Maissel, ibid, Chapter 4.
[324] J. Stark, Z. Elektrochem., 14 (1908) 752; 15 (1909) 509.
[325] C. Lehmann and P. Sigmund, Phys Stat. Sol., 16 (1966) 507.
[326] H. Fetz, Z. Phys., 119 (1942) 590.
[327] G. K. Wehner, J. Appl. Phys., 25 (1954) 270.
[328] G. K. Wehner, J. Appl. Phys., 26 (1955) 1056.
[329] J. M. Fluit,Colloqu. Int. Centre Nat. Rech. Sci., Bellevue, 1961.
[330] O. Almen and G.Bruce, Nucl. Instr. Methods, 11 (1961) 257; 11 (1961) 279.
[331] G. K. Wehner, Sputtering by ion bombardment, in Advances in Electronics and Electron Physics, 7 (1955) 239.
[332] P. Davidse and L. Maissel, J. Appl. Phys. 37 (1966) 574.
[333] G. Anderson, W. Mayer and G. K. Wehner, J. Appl. Phys., 33 (1962) 2991.

[334] R. Stuart and G. K. Wehner, J. Appl. Phys., 33 (1962) 2345.

[335] W. M. Thompson, Phil. Mag., 18 (1968) 377.

[336] P. Sigmund, Phys. Rev., 184 (1969) 383.

[337] I. S. T. Tsong and D. J. Barber, M. Mater. Sci., 8 (1973) 123.

[338] L. Gurmin et al., Fiz. Tverd. Tela, 10 (1968) 411.

[339] B. Navinsek, J. Appl. Phys., 36 (1965) 1678.

[340] R. Glang, R. A. Holmwood, and P. C. Furois, Trans. 3rd Int. Vacuum Congr., Stuttgart, 1965, Pergamon Press, 1967, Vol. 2, p. 643.

[341] C. R. C. Priestland and S. D. Hersee, Vacuum, 22, (1973) 103.

[342] N. Laegreid and G. K. Wehner, J. Appl. Phys., 32 (1961) 365.

[343] R. M. Valletta, J. A. Perri, and J. Riseman, Electrochem. Technol., 4 (1966) 402.

[344] N. F. Jackson, E. J. Hollands, and D. S. Cambell, Proc. Joint IERE/IEEE Conf. Appl. Thin Films in Electronic Eng., Imperial College, July 1966, p. 13.

[345] D. M. Mattox and G. J. Kominiak, Physical properties of thick sputter-deposited glass films, Abs. 172, Vol. 72–2, Extended abstracts 142nd Nat. Meeting Electrochem. Soc., Miami Beach, Florida, October 1972, to be published in J. Electrochem. Soc.

[346] P. D. Davidse and L. I. Maissel, J.Vac. Sci. Technol, 4 (1967) 33.

[347] A. Southern, W. R. Willis, and M. T. Robinson, J. Appl. Physics, 34 (1963) 153.

[348] C. H. Weijsenfeld, A. Hoogendoorn, and M. Koedam, Physica, 27 (1961) 963.

[349] G. K. Wehner, Rep. No. 2309, General Mills, Minneapolis, 1962.

[350] J. L. Vossen, Unpublished observations, 1974.

[351] C. H. Weijsenfeld and A. Hoogendoorn, Proc. 5th Conf. Ion. Phenom. Gases, München, 1961, Vol. 1, p. 124.

[352] D. McKeown and Y Y. Cabezas, Ann. Rep. Space Sci. Lab., General Dynamics, July 1962.

[353] F. Keywell, Phys. Rev., 97 (1955) 1611.

[354] J. Comas and C. B. Cooper, J. Appl. Phys., 37 (1966) 2820.

[355] D. McKeown, A. Cabezas, and E. T. Machenzie, Ann. Rep. Low Energy Sputtering Stud., Space Sci. Lab., General Dynamics July 1961.

[356] R. C. Krutenat and C. Panzera, J. Appl. Phys., 41 (1970) 4953.

[357] R. Frericks, J. Appl. Phys., 33 (1962) 1898.

[358] J. Vossen, J. Vac. Sci. Techn., 8, 5 (1971) 12.

[359] C. Overbeck, J. Opt. Soc. Am., 23 (1933) 109.

[360] G. Veszi, J. Brit. Inst. Radio Eng., 13 (1971) 225.

[361] T. Lakshmanan and J. Mitchell, Trans. 10th Nat. Vac. Symp. AVS, 1963, p. 335.

[362] D. Gerstenberg and C. Calbick, J. Appl. Phys., 35 (1964) 402.

[363] G. Deppisch, Vak.Techn., 30 (1981) 106.

[364] B. Berghaus, German Patent Nr. 683414, Klasse 12i, Gruppe 37, 1939.

[365] D. M. Mattox, Sandia Corp. Rep. SC-DR-281-63 (1963), J. Vac. Sci. Technol., 10 (1973) 47; Japan J. Appl. Phys. Suppl., 2, (1) (1974) 443.

[366] D. G. Teer, in Proc. Ion Plating and Allied Techniques Conf., Edinburgh, 1977, CEP Consultants Ltd. Edinburgh, UK, p. 13.

[367] G. Carter and J. S. Colligan, The Ion Bombardment of Solids, Heinemann, London 1968.

[368] D. M. Mattox and G. J. Kominiak, J. Vac. Sci. Technol., 9 (1972) 528.

[369] R. D. Bland, G. J. Kominiak and D. M. Mattox, J. Vac. Sci. Technol., 11 (1974) 671.

[370] R. F. Bunshah and R. S. Juntz, J. Vac. Sci. Technol., 9 (1972) 1404.

[371] C. T. Wan, D. L. Chambers and D. C. Carmichael, J. Vac. Sci. Technol., 11 (1974) 379; Proc. Int. Conf. Vac. Metallurgy, Tokyo, 1973.

[372] J. L. Vossen, J. J. O'Neill, et al., R. C. A. Review, (1970) 293.

[373] G. W. White, A high rate source for ion plating, Vac. Metall. Conf., 1974, AVS-VAC. Metall. Div., Pittsburgh, PA, June, 1974.

[374] F. A. Maxfield and R. R. Benedict, Theory of Gaseous Conduction, McGraw-Hill, New York 1941, p. 343.

[375] T. J. Killian, Phys. Rev., 35 (1930) 1238.

[376] J. D. Cobine, Gaseous Conductors, Dover, New York 1958, p. 158.

[377] I. Langmuir and H. Mott-Smith, Gen. Electr. Rev., 27 (1924) 762.

[378] D. L. Chambers and D. C. Carmichael, Research/Development, May 1971, p. 32.

[379] J. N. Avaritsiotis and R. P. Howson, Thin Solid Films, 65 (1980) 101.

[380] C. Hayashi, H. Ohtsuka and S. Komiya, Proc. III. C. I. P., Nice, France 11–14 Sept. 1979.

[381] W. C. Herrmann Jr. and J. R. MacNeil, Proc. SPIE Technical Meeting, Los Angeles, USA, January 1982, paper 325–14.

284

[382] K. Suzuki, K. Matsumoto and T. Takatsuka, Thin Solid Films, 80 (1981) 67.
[383] M. I. Ridge, M. Stenlake, R. P. Howson and C. A. Biship, Thin Solid Films, 80 (1981) 31.
[384] M. I. Ridge, R. P. Howson, C. A. Bishop, Proc. SPIE Technical Meeting, Los Angeles, USA, January 1982, paper 325– 06.
[385] G. Carter and D. G. Armour, Thin Solid Films, 80 (1981) 13.
[386] T. Spalvins, Proc. 24th Ann. Technical Conf., 1981, Society Vac. Coaters, p. 42.
[387] D. Hoffman and D. Leibowitz, J. Vac. Sci. Techn., 8 (1971) 107.
[388] F. Soddy, Proc. Roy. Soc., 78 (1907) 429.
[389] I. Langmuir, J. Am. Chem. Soc., 35 (1931) 931.
[390] W. Heitmann, Appl. Optics, 10 (1971) 2414.
[391] R. F. Bunshah and A. C. Raghuram, J. Vac. Sci. Techn., 9 (1972) 1385.
[392] E. Ritter, J. Vac. Sci.Techn., 3 (1966) 225.
[393] K. Kerner, Trans. 3rd I. V. C., Vol. 2, Pergamon Press, 1966, p. 31.
[394] A. Itoh and S. Misawa, Proc. 6th I. V. C., Part 1, 1974; Jap. J. App. Phys. Suppl., 2 (1) (1974) 467.
[395] H. Küster and J. Ebert, Thin Solid Films, 70 (1980) 43.
[396] J. Ebert, Proc. SPIE Technical Meeting, Los Angeles, January 1982, paper 325 – 04.
[397] T. H. Allen, Proc. SPIE Technical Meeting, Los Angeles, January 1982, paper 325 – 13.
[398] N. Schwartz, Trans. Natl. Vac. Symp. 10th, 1963, Boston 1964, p. 325.
[399] G. Perny, Le Vide, 21 (1966) 106.
[400] J. Pompei, Proc. 2nd Symp. Deposition Thin Films Sputter., Univ. Rochester, 1967, p. 127.
[401] E. Hollands and D. S. Campbell, J. Mater. Sci. 3, (1968) 544.
[402] J. Pompei, Proc. 3rd Symp. Deposition Thin Films Sputter., Univ. Rochester, 1969, p. 165.
[403] J. Heller, Thin Solid Films, 17, (1973) 163.
[404] K. G. Geraghty and L. F. Donoghey, Proc. 5th Conf. Chem. Vap. Dep., Electrochem. Soc., Princeton, New Jersey, 1975, p. 219.
[405] F. Shinoki and A. Itoh, J. Appl. Phys., 46, 3381 (1975).
[406] T. Abe and T. Yamashina, Thin Solid Films, 30, (1975) 19.
[407] B. Goranchev, V. Orlinov, and V. Popova, Thin Solid Films, 33, (1976) 173.
[408] L. F. Donaghey and K. G. Geraghty, Thin Solid Films, 38, (1976) 271.
[409] P. Clarke, Proc. 24th Ann. Techn. Conf. 1981, Soc. of Vac. Coaters, p. 39.
[410] W. T. Pawlewicz, P. M. Martin, D. D. Hays and I. B. Mann, Proc. SPIE Technical Meeting, Los Angeles, January 1982, paper 325 – 15.
[411] S. Schiller, G. Beister, S. Schneider and W. Sieber, Thin Solid Films, 72 (1980) 475.
[412] W. T. Pawlewicz and R. Busch, Thin Solid Films, 63 (1979) 251.
[413] S. Schiller, G. Beister, S. Schneider and W. Sieber, Thin Solid Films, 72 (1980) 475.
[414] E. Leija, T. Pisarkiewicz and T. Stapinski, Thin Solid Films, 76 (1980) 283.
[415] K. Budzynska, T. Horodyski and A. Kolodziej, Thin Solid Films, 100 (1983) 203.
[416] R. P. Howson, J. N. Avaritsiotis, M. I. Ridge and C. A. Bishop, Thin Solid Films, 58 (1979) 379.
[417] H. K. Pulker, Reactive Ion Plating, 2nd Coll. Int. Pulverisation Cathodique et ses Applic., CIP 76, Nice, France, May 18–21, 1976.
[418] H. A. McLeod, Proc. IPAT 79, London, July 1979, p. 74, CEP Consultants Ltd, Edinburgh.
[419] M. I. Ridge, R. P. Howson, J. N. Avaritsiotis and C. A. Bishop, Proc. IPAT 79, London, July 1979, p. 21, CEP Consultants Ltd., Edinburgh.
[420] B. Zega, Proc. IPAT 79, London, July 1979, p. 74, CEP Consultants Ltd, Edinburgh, p. 225.
[421] A. Matthews, D. G. Teer, Proc. IPAT 79, London, July 1979, p. 74, CEP Consultants Ltd, Edinburgh, p. 11.
[422] J. Machet, J. Guille, P. Saulnier and S. Robert, IPAT 81, Amsterdam, July 1981, Thin Solid Films, 80 (1981) 149.
[423] D. O. Hayward and B. M. W. Trapnell, Chemisorption, Butterworth, London, 1964.
[424] E. Ritter, Thesis, University of Innsbruck, 1958, Austria.
[425] H. K. Pulker, Thesis, University of Innsbruck, 1961, Austria.
[426] K. Kerner and G. Mutschler, Bosch Techn. Ber., 3 (1970) 3.
[427] G. Wedler, Adsorption, Chem. Taschenbuch Nr. 9, Verlag Chemie, Weinheim, 1970.
[428] W. Heitmann, Appl. Optics, 10 (1971) 2685.
[429] H. K. Pulker, Thin Solid Films, 34 (1976) 343.
[430] H. K. Pulker, G. Paesold and E. Ritter, Appl. Optics, 12, (1976) 2986.
[431] BALZERS Ltd. US Pat. No. 3.980 044, Sept. 1976.
[432] G. Kienel and H. Walter, Sputtering Optical Thin Films on Large Surfaces, Research Development, Nov. 1973, 49 – 56.

[433]   J. L. Vossen, RF Sputtered Transparent Conductors. The System In_2O_3–SnO_2, RCA Review, 32 (1971) 289 – 296.

[434]   G. Kienel and G. Gallus, The production of conductive and transparent thin films, Proc. 6th Int. Vac. Congr. 1974; Japan. J. Appl. Phys. Suppl., 2 (1) (1974) 479 – 482.

[435]   G. Kienel, Herstellung von transparenten leitenden Schichten durch Kathodenzerstäubung, Vakuumtechnik, 26 (1977) 108 – 115.

[436]   G. Kienel and W. Stengel, Herstellung optischer Schichten durch Kathodenzerstäubung, Vakuumtechnik, 27 (1978) 7.

[437]   W. T. Pawlewicz, R. Busch, D. D. Hays, P. M. Martin and N. Laegreid Reactively Sputtered Optical Coatings for Use at 1064 nm, in:Laser-Induced Damage in Optical Materials: 1979, H. E. Bennett, A. J. Glass, A. H. Guenther and B. E. Newnam, (Eds.) NBS Spec. Publ. 508, 1980, pp. 359 – 374.

[438]   T. M. Donovan, J. O. Porteus, S. C. Seitel and P. Kratz, Multithreshold HF/DF Pulsed Laser Damage Measurements on Evaporated and Sputtered Silicon Films, in:Laser-Induced Damage in Optical Materials, 1980, H. E. Bennett, A. J. Glass, A. H. Guenther and B. E. Newnam, (Eds.), NBS Spec. Publ. 620, (1981) pp. 305 – 312.

[439]   W. T. Pawlewicz, I. B. Mann, W. H. Lowdermilk and D. Milam, Laser damage resistant transparent conductive indium tin oxide coatings, Appl.Phys. Lett., 34,(1979) 196.

[440]   W. T. Pawlewicz, D. D. Hays and P. M. Martin, High-band-gap oxide optical coatings for 0.25 and 1.06 μ fusion laser, Thin Solid Films, 73, (1) (1980) 169 – 175.

[441]   a) K. Suzuki and R. F. Howson, Proc. Internat. Ion Engineering Congr. – ISIAT 83 and IPAT 83, Kyoto, 1983, p. 889; publ. by Institute of Electrical Engineers of Japan, Tokyo, T. Takagi (Ed.).
        b) M. El-Sherbiny, F. Salem and A. Aboukhashaba, Proc. Internat. Ion Engineering Congr. – ISIAT 83 and IPAT 83, Kyoto, 1983, p. 901; publ. by Institute of Electrical Engineers of Japan, Tokyo, T. Takagi (Ed.).

[442]   G. Herzberg, Atomic Spectra and Atomic Structure, Dover, New York, (1944).

[443]   a) G. Herzberg, Spectra of Diatomic Molecules, Van Nostrand, New York, 1950.
        b) G. Herzberg, Electronic Spectra and Electronic Structure of Polyatomic Molecules, VAn Nostrand, New York, 1966.

[444]   K. K. Yee, in Proc. 5th Int. Conf. on CVD, J. M. Blocher jr., H. Hintermann and L. H. Hall, Eds., The Electrochem. Soc., Princeton, New Jersey, 1975.

[445]   B. Zega, Review of current PVD methods, IPAT Workshop 1982, LSRH Neuchâtel, Switzerland, June 2 – 4, 1982.

[446]   L. Holland, Thin Solid Films, 27 (1975) 185.

[447]   F. K. McTaggart, Plasma Chemistry in Electrical Discharges, Elsevier, New York, 1967.

[448]   The Application of Plasmas to Chemical Processing, R. F. Baddour and R. S. Timmins (Eds.), MIT Press, Cambridge, MA, 1967.

[449]   A. M. Mearns, Thin Solid Films, 3, (1969), 201.

[450]   Chemical Reactions in Electrical Discharges, B. D. Blaustein (Symposium Chairman), Am. Chem. Soc., Washington, DC, 1969.

[451]   P. M. Hay, in Chemical Reactions in Electrical Discharges, B. D. Blaustein (Chairman), Am. Chem. Soc., Washington, DC, 1969, p. 350.

[452]   D. D. Neiswender, in Chemical Reactions in Electrical Discharges, B. D. Blaustein (Chairman), Am. Chem. Soc., Washington, DC, 1969, p. 338.

[453]   Engineering, Chemistry and Use of Plasma Reactors, J. E. Flinn, Am. Inst. Chem. Eng., New York, 1971, Chem. Eng. Prog. Symposium Series No. 112.

[454]   A. T. Bell, Models for High Frequency Electric Discharge Reactors, American Institute of Chemical Engineers, New York, 1971, Chemical Eng. Prog. Symp. Series No. 112.

[455]   M. Venugopalan, Reactions Under Plasma Conditions, Wiley-Interscience, New York, 1971.

[456]   E. Nasser, Fundamentals of Gaseous Ionization of Plasma Electronics, Wiley-Interscience, New York, 1971.

[457]   M. R. Havens, M. E. Biolsi and K. G. Mayhan, J. Vac. Sci. Technol., 13 (1976) 575.

[458]   H. Yasuda, in J. L. Vossen and W. Kern, (Eds.), Thin Film Processes, Part. IV, Academic Press, New York, 1978, Chapter 2, p. 361.

[459]   Papers of Session 9 5th Int. Symp. on Plasma Chemistry, 1981, Symp. Proc., Vol. 1, B. Waldie and G. A. Farnell, (Eds.), Int. Union of Pure and Applied Chemistry.

[460]   M. M. Millard, J. J. Windle, and A. E. Pavlath, J. Appl. Polym. Sci., 17 (1973), 2501.

[461]   H. Kobayashi, A. T. Bell, and M. Shen, J. Appl. Polym. Sci. 17 (1973) 885.

[462]   H. Kobayashi, M. Shen, and A. T. Bell, J. Macromol. Sci.-Chem., A8, (1974) 1345.

286

[463]   A. R. Denaro, P. A. Owens, and A. Crawshaw, Eur. Polym. J., 4 (1968) 93.
[464]   A. R. Denaro, P. A. Owens, and A. Crawshaw, Eur. Polym. J., 5 (1969) 471.
[465]   A. R. Denaro, P. A. Owens, and A. Crawshaw, Eur. Polym. J., 6 (1970) 487.
[466]   H. Kobayashi, M. Shen, and A. T. Bell, J. Macromol. Sci.-Chem., A8 (1974) 373.
[467]   E. S. Lo, and S. W. Osborn, J. Org. Chem., 35 (1970) 935.
[468]   S. Morita, T. Mizutani, and M. Ieda, Jpn. J. Appl. Phys., 10 (1971) 1275.
[469]   K. C. Brown and M. J. Copsey, Eur. Polym. J., 8 (1972) 129.
[470]   A. R. Westwood, Eur. Polym. J., 7 (1971) 363.
[471]   L. F. Thompson and K. G. Mayhan, J. Appl. Polym. Sci., 16 (1972) 2317.
[472]   J. C. Lee, Appl. Optics, 17 (1978) 2645.
[473]   S. Morita, J. Tamano, S. Hattori and M. Ieda, J. Appl. Phys., 51 (1980) 3938.
[474]   R. Szeto and D. W. Hess, J. Appl. Phys., 52 (1981) 903.
[475]   P. J. Martin, H. A. Macleod, R. P. Netterfiled, C. G. Pacey, and W. G. Sainty, Appl. Opt. 22 (1983) 178.
[476]   J. R. Sites, P. Gilstrap and R. Rujkorakarn, Opt. Eng., 22 (1983) 447.
[477]   R. P. Howson, M. I. Ridge, C. A. Bishop and K. Suzuki, Proc. Internat. Ion Engineering Congr. – ISIAT 83 and IPAT 83, Kyoto, 1983, p. 933; publ. by Institute of Electrical Engineers of Japan, Tokyo, T. Takagi (Ed.).
[478]   K. Murata, K. Enjouji and S. Nishikawa, Proc. Internat. Ion Engineering Congr. – ISIAT 83 and IPAT 83, Kyoto, 1983, p. 963; publ. by Institute of Electrical Engineers of Japan, Tokyo, T. Takagi (Ed.).
[479]   M. I. Ridge and R. P. Howson, Proc. Internat. Ion Engineering Congr. ISIAT 83 and IPAT 83, Kyoto, 1983, p. 1011; publ. by Institute of Electrical Engineers of Japan, Tokyo, T. Takagi (Ed.).

CHAPTER 7

7. FILM THICKNESS

7.1 GENERAL CONSIDERATIONS

The thickness of an individual film or of a series of many films represents a dimension which figures in practically all equations used to characterize thin films. However, when discussing film thickness, definitions are required since there one has to distinguish between various types of thicknesses, such as geometrical thickness, mass thickness and optical thickness. The geometrical thickness, often called physical thickness, is defined as the step height between the substrate surface and the film surface. This step height multiplied by the refractive index of the film is termed the optical thickness and is expressed generally in integer multiples of fractional parts of a desired wavelength. Finally, the mass thickness is defined as the film mass per unit area obtained by weighing. Knowing the density and the optical data of a thin film, its mass thickness can be converted into the corresponding geometrical as well as optical thickness. However, with ultrathin films ranging between a few and several atomic or molecular "layers", the concept of a film thickness often becomes meaningless, since no closed film may exist as a consequence of the possible open microstructure of such minor deposits. Although film thickness is a length, the measurement of it, obviously cannot be accomplished with conventional methods for length determinations but requires special methods. The great efforts made to overcome this problem led to a remarkable number of different, often highly sophisticated, film-thickness measuring methods reviewed in some articles such as [1 – 5]. With some of the methods, it is possible to carry out measurement during and after film formation – other determinations can only be undertaken outside the deposition chamber after the film has been produced. Many of the methods cannot be employed for all film substances, and there are various limits as regards to the range of thickness and measuring accuracy. Furthermore, with these methods the film to be measured is often specially prepared or dissolved during measurement and therefore becomes useless for additional investigations or applications. If only those methods which can be employed during film deposition are considered, then the very large number of methods is considerably reduced. In-situ measurements, however, are not only desirable, but many basic investigations and most industrial applications require precise knowledge of the film thickness at any instant to enable termination of the deposition process at the desired optimum moment. Apart from a few exceptions in film deposition by PVD methods, only optical measuring units and mass determination monitors are used.

It is also very important to measure and control the rate of deposition since

TABLE 1

FILM THICKNESS AND RATE MEASURING METHODS

Method	Film materials	Thickness Minimum	Maximum nm	Measurement during (+) or after (Δ) deposition possible	Rate measurement possible	Calibration necessary	Comments	References
optical intensity measurement methods	metals dielectrics	few monolayers		+, Δ	−	yes*	*with indirect measuring technique	[3, 6–11, 99, 100, 106]
wide-band optical monitoring	metals dielectrics	fractions of quarterwave thickness		+, Δ	−	no		[101 – 105]
interference filter method	dielectrics	0,3		Δ	−	no		[12]
optical interference methods	metals dielectrics	multiple beam: 3 two beam: 50	1000 1000	Δ Δ		no		[13–17, 18–22]
X-ray interference method	metals	15	200	Δ	−	no		[23–25]
polarisation methods ellipsometry	metals* dielectric	monolayers	50	+	−	no	*depending on the special method used	[26, 27]
micro weighings gravimetric micro-balances	metals* dielectrics	monolayers	none	+ Δ	yes	yes	*very sensitive vibrations	[1, 28–34]

Method	Film materials	Thickness Minimum nm	Thickness Maximum	Measurement during (+) or after (Δ) deposition possible	Rate measurement possible	Calibration necessary	Comments	References
oscillating quartz microbalance	metals dielectrics	monolayers*	several μm*	+ Δ	yes	yes	*depending on the resonance frequency of the quartz used	[35–42]
chemical analysis	metals	monolayers	non	Δ	–	yes	many useful reactions can be found in textbooks of chemical microanalysis	[1, 43–45]
quantitative evaporation	metals* dielectrics		none	+	–	yes	*only substances which do not decompose or form an alloy with the crucible material	[2, 4]
resistance measurement	metals			+ Δ	–	yes		[4]
capacity measurement	dielectrics	few 0.1		Δ	–	yes		[4]
electron emission measurement	alkali alkaline earth	fraction of an atomic layer	an atomic layer	+ Δ	–		only important for basic studies	[46–48]

Method	Film materials	Thickness nm		Measurement during(+) or after (Δ) deposition possible	Rate measurement possible	Calibration necessary	Comments	References
		Minimum	Maximum					
X-ray fluorescence analysis electron induced X-ray emission	elements as from atomic number 12 and their compounds	few monolayers	100–400	+ Δ	yes	yes	depending on the atomic number of the element	[49–52] [73]
vapour ionisation methods	metals dielectrics	few 0.1	none	+	yes	yes		[53–61]
optical emission spectroscopy method	metals	few monolayers	unlimited	+	yes	yes		[5, 62]
atomic absorption method	metals	few monolayers	unlimited	+	yes	yes		[63–65]
stylus methods	metals dielectrics*	15	none	Δ	–		*soft easily scratched films are unsuitable	[66, 67]
β-ray absorption	metals dielectrics	5	2000	+ Δ	–			[68–71]
radio activity measurement*	metals dielectrics	monolayers		+ Δ	–		*special precautions required! only for scientific research	[72]

film structure and properties are affected by the rate to some extent. Ratemeters can usually be used for all film materials. This is a distinct advantage in multilayer deposition of metals and dielectrics. The simultaneous measurement of thickness and rate with one and the same measuring principle is certainly of advantage, as is having a thickness and rate monitoring system which can be actively integrated in an automatic process control.

Comparison of the various film thickness and deposition rate measuring methods available is given in Table 1. In the following, some of the most important methods are discussed.

7.2 METHODS APPLICABLE TO ALL TYPES OF FILMS

7.2.1 INTERFERENCE METHODS

Multiple beam interferometry has a high accuracy and also represents the most reliable measuring method for the calibration of other techniques. If light falls onto an arrangement consisting of a semi-reflecting surface and a fully reflecting surface which form an air-wedge, a set of interference lines will be seen. Using a parallel beam of monochromatic light at normal incidence, a set of parallel interference lines is obtained, whose separation is equivalent to a thickness change of the wedge of $\lambda/2$. These lines are termed fringes of equal thickness. In practice, the sample with the film stepheight to be measured is fully silvered over the whole surface area. The upper semi-transparent mirror is a semi-silvered glass optical flat. Since the interference fringes follow lines of equal wedge thickness, the effect produced is that of contouring the surface. The pattern obtained consists of parallel dark lines on a bright background, showing a displacement at the position of the film step as can be seen in Fig. 1.

The film thickness t is obtained by measuring the distance L and the displacement ΔL of the lines and by multiplication of the formed quotient with the half wavelength of the monochromatic light according to:

$$t = \frac{\Delta L}{L} \frac{\lambda}{2} \tag{1}$$

The method is performed using a microscope equipped with a micro-photographic camera. To produce the monochromatic light, sodium or mercury spectral lamps or a He/Ne laser are used. A detailed description of the method is given by Tolansky [13,14]. Many technical variants for rapid and reliable measurements have been published, see [16 and 17].

Two-beam interference microscopes operating according to the principle of the Michelson interferometer and accessory devices converting an ordinary microscope into a two-beam interferometer are commercially available. In such microscopes, collimated monochromatic light is half reflected onto the sample surface and half transmitted to an adjustable flat reference mirror by a beam splitter. The two reflected beams recombine in the microscope and the resulting variation in the optical path difference of the beams produces parallel interference lines of equal thickness which are also displaced at the position of the film step. The lines obtained are, however, relatively broad limiting the resolution and the accuracy of such measurements by the uncertainty in selecting the line centre.

An elegant two-beam interference method, based on the action of a special double prism, is used in the interference accessory device according to Nomarski [18–20]. This device can be inserted in polarizing microscopes in vertical illumination. It enables a very rapid thickness determination especially of thicker films.

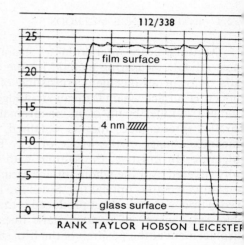

Fig. 1

Fringe pattern of a 154 nm thick TiO₂ film on glass.
Multiple beam interferometry, $\lambda = 546$ nm.

Fig. 2

93 nm thick MgF₂ film on glass.

7.2.2 STYLUS METHODS

In stylus instruments, a diamond needle of very small dimensions (e. g. 13 μm in tip radius) is used as the arm of an electromechanical pick-up. The position of the pick-up is determined by a slider which is pressed on the area surrounding the needle to give the mean zero position. The thickness t of a film is measured by traversing the blunt stylus over the edge of the deposit, the thickness being recorded on a pen recorder as the difference in level between the film and the substrate. The substrate should generally be smooth compared with the thickness of the deposit. To ensure this, measurements are often made on a sample specimen deposited on an optically flat substrate, a portion of which is masked off. Build-up of deposit at the edge of the step is sometimes revealed. The resolution obtainable is slightly better than that of a two-beam interferometer. A typical thickness measurement trace is shown in Fig. 2. A well known instrument of this type is the Talystep constructed by Rank Taylor–Hobson, England [67a]. The stylus force is adjustable from 1 to 30 mg and vertical magnification from 5000 × to 1 000 000 × is possible. Magnifications along the surface are determined by the speed of traverse of the pick-up. Although the measuring unit is mounted on an anti-vibration platform, the high sensitivity of the device demands its use on a rigid table standing on a firm floor.

A newer instrument of this type containing a micro computer is the Dektak II from Sloan Technology Corp. [67b]. With this device the measuring and leveling parameters can be preprogrammed. A single switch then initiates the measurements. Vertical movement of the stylus is digitized in real-time and the resulting scan is displayed and can be printed by a printer. The computer automatically levels the scan and autoranges the profile to best fit a screen. It print the scans as displayed or zoom magnify an area of particular interest. A complete graphic editor is obtained. Further data manipulation or mass storage of measurement results can be performed. The Dektak II includes a calibration standard that may be used for periodic verification of calibration. The measurement range is between 50 nm and 65000 nm. The vertical resolution is 1 nm; and the tolerable maximum sample thickness is about 20 mm.

7.3   METHODS APPLICABLE TO PVD FILMS

7.3.1   OPTICAL REFLECTANCE AND TRANSMITTANCE MEASUREMENTS

In evacuated coating plants, it is usual to mount the parts to be coated on rotating work holders permitting the uniform subsequent or simultaneous deposition of evaporated sputtered or ion-plated deposits of any combination of metals or dielectric materials. The film thickness on the sample itself or on test glass can be monitored optically by transmitted and/or reflected light using a photometer. Such instruments are commercially available, for example, the BALZERS modulated beam optical film thickness measuring instrument GSM 210 shown in Fig. 3e. With this instrument, the optical intensity variations on a test glass coated simultaneously with the substrate is measured.

The effective transmission or reflection may be measured, for example, in the production of semi-transparent films. In most cases, however, the optical film thickness of transparent dielectric films is determined, exploiting the fact that the reflection and transmission change periodically with increasing film thickness (provided that the refractive indices of the film and the substrate are different). The following connection between geometrical and optical film thickness is applicable:

$$t = x \frac{\lambda}{4n}$$

(2

λ = wavelength of the light used for measurement;
n = refractive index of the film;
x = the number of reflection and transmission peaks.

Typical measuring arrangements of the GSM 210 , namely the two-beam principle and the mono-beam principle, are shown in Fig. 3a. A specia arrangement for measuring the thickness of highly reflective coatings on optica elements during deposition while coating all substrates equally which was developed by HONEYWELL is shown in Fig. 3b.

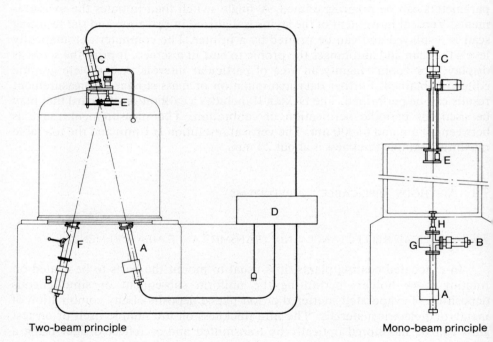

Two-beam principle   Mono-beam principle

Fig. 3 a
Measuring arrangement of the optical film thickness monitor BALZERS GSM 210.
A Modulated light source   E Test glass changer
B Receiver Refl.   F Light beam deflection
C Receiver Transm.   G Deflector
D Indicating instruments   H Intermediate-piece

The use of modulated light eliminates interference from daylight and from light emitted by evaporation sources or a glow discharge plasma. The instrument can be operated in the spectral range between 350 nm and 2500 nm Monochromatic light is produced from a tungsten filament lamp by narrow band

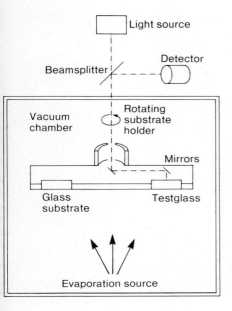

Fig. 3b
Measuring arrangement of the optical film thickness monitor of HONEYWELL.

interference filters. For the intensity detection, a blue and red sensitive photodetector is used. The sensitivity of this instrument covers the range of minimum light levels specified for anti-reflection coatings on glass surfaces measured in reflection to maximum light levels as in the measurement of nearly 100 percent transmittance. The use of two photodetectors arranged to detect reflected and transmitted light respectively, enables measurement of both intensities in rapid succession.

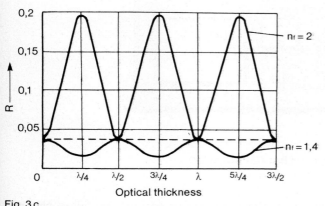

Fig. 3 c
Reflectance as function of the optical thicknesses $n_f \cdot t_f$ of non-absorbing dielectric films. The refractive index of the testglass is $n_s = 1.5$. The refractive indices of the films are $n_f = 1.4$ and $n_f = 2$.

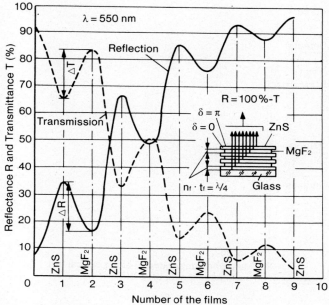

Fig. 3 d
Reflectance and transmittance of a dielectric multilayer system
as function of the number of the alternating deposited ZnS and
MgF_2 quarterwave layers [8b].
$\triangle R$ and $\triangle T$ are the differences of R and T between neighbouring
extreme values, δ = phasejumps at the various interfaces.

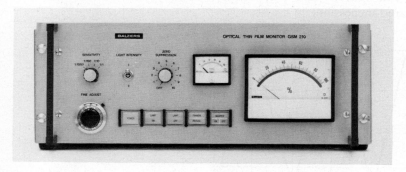

Fig. 3 e
Optical thin film measuring instrument BALZERS GSM 210.

Depending on the plant and the application, various test glass changers
holding up to 150 test glasses are available for transmission and/or reflection
measurement. The test glass change is accomplished by either manual or remote
control according to which variation of the test glass changer is fitted.

The single layers, for example of a dielectric multilayer system, can be measured individually each on a separate test glass or all together on a single test glass. Typical photometer curves are shown in Figs. 3c and 3d.

For control of the rate of deposition, a quartz crystal monitor can additionally be used. Because of the importance of this method, it will be treated in more details in the next section.

7.3.2 OSCILLATING QUARTZ-CRYSTAL MICROBALANCE

The deposition of noble metals onto oscillating quartz crystals of the thickness shear type, for fine adjustment of their frequency, has already been carried out for many years by frequency standard manufacturers. The idea of using the frequency decrease by mass deposition to determine the weight of the coating is comparatively new. Sauerbrey [35] and Lostis [36] were the first to propose the quartz-crystal microbalance. The AT-cut crystal oscillating in a thickness shear mode was found to be best suited for this purpose. The thickness x_q of an infinite quartz plate is directly related to the wavelength λ of the continuous elastic transverse wave, the phase velocity v_q of that wave and the frequency ν_q (i. e. the period τ_q) of the oscillating crystal, as shown in Fig. 4:

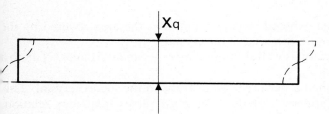

Fig.4
Schematic representation of a quartz crystal oscillating in the thickness shear mode (AT- or BT-cut).

$$x_q = \frac{1}{2}\lambda = \frac{1}{2}(v_q / \nu_q) = \frac{1}{2}v_q\tau_q \tag{3}$$

By introducing the area density m_q of the quartz mass and the quartz density ρ_q $(m_q = \rho_q x_q)$, one obtains the exact relation between the relative frequency and the relative area mass density of the infinite thickness shear resonator:

$$\frac{d\nu_q}{\nu_q} = - \frac{dm_q}{m_q} \tag{4}$$

Changing to differences Δ results in

$$\frac{\Delta v_q}{v_q} \simeq -\frac{\Delta m_q}{m_q} \tag{5}$$

The fact that eqn. (5) is already an approximation formula was not stated in the literature for a long time.

Sauerbrey substituted the area mass density Δm_q of an additional quartz wafer by the area mass density m_f of the deposited foreign material. He made the assumption that for small mass changes, the addition of foreign mass can be treated as an equivalent mass change of the quartz crystal itself. His relation for the frequency change of the loaded crystal is

$$\frac{\Delta v}{v} \simeq -\frac{m_f}{m_q} \tag{6}$$

The restrictions of this formula are as follows:
    it is not mathematically rigorous;
    the elastic properties of the deposited material are different from those of the quartz crystal;
    the size of technical quartz crystals is finite;

Furthermore, the area exposed to deposition, i. e. the area covered by the electrodes, is usually different from that of the crystal.

Nevertheless, this relation is well supported by experimental data [37] up to a mass load of $m_f / m_q \leqslant 2 \%$.

Stockbridge [38] applied a perturbation analysis to the loaded crystal. This mathematically rigorous approach predicts the mass versus frequency relation, eqn. (6), as the first-order term of the series expansion. However, the method did not find practical use, since the higher-order terms require experimentally

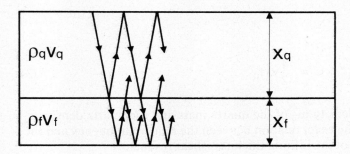

Fig. 5

Compound resonator model. x_q is the quartz wafer thickness, x_f the film thickness, $\rho_q v_q$ is the acoustic impedance for unit cross-sectional area of the crystal and $\rho_f v_f$ that of the film.

measured constants. Miller and Bolef [74] applied a continuous acoustic wave analysis to the compound resonator formed by the quartz crystal and the deposited film. As shown in Fig. 5, they considered a continuous acoustic wave of frequency v propagating in the form of a damped plane wave from the quartz into the deposited film. At the interface, the wave is partially reflected and partially transmitted. Provided that the acoustic losses in the quartz and thin film are small, one can use the following relation to determine the complete set of resonance frequencies v (fundamental and harmonics) for the compound resonator:

$$\tan \frac{\pi v}{v_q} = - \frac{\rho_f v_f}{\rho_q v_q} \tan \frac{\pi v}{v_f} \tag{7}$$

v_f is the mechanical resonance frequency of the deposited film determined by $v_f = v_f / 2x_f$, ρ_f is the density of the film and v_f the shear wave velocity in the film. Equation (7) shows that the compound resonator frequency v is dependent on the factor $\rho_f v_f / \rho_q v_q$, which is the ratio between the shear-mode acoustic impedance of the deposit material and that of the quartz.

Behrndt [75] pointed out that the multiple-period measurement technique may have some advantages over the frequency-meter technique. While, for example, the frequency shift is non-linear in Sauerbrey's eqn. (6), the period τ of the oscillation increases linearly with mass loading.

$$\Delta \tau = \frac{2}{v_q \rho_q} m_f = \frac{2 \rho_f}{v_q \rho_q} \tag{8}$$

With modern digital electronics there is no problem in indicating multiple period time instead of frequency. By proper selection of the multiple period measurement time, the dependence of thickness on material density can be considered and the film thickness can be displayed directly. Equation (8) is valid in the range up to $m_f / m_q < 15\%$.

Lu and Lewis [76] proved the excellent performance of Miller and Bolef's continuous acoustic wave analysis for mass loads up to $m_f / m_q < 25\%$. It is easy to show that Behrndt's eqn. (8) of the period measurement method and Sauerbrey's eqn. (6) of the frequency measurement method are approximations of Miller and Bolef's eqn. (7) for small m_f / m_q values.

However, with such high mass loads, the mechanical stability of the system quartz-crystal deposited thick films decreases. Thus the fact that materials with different elastic properties will obey different mass-frequency relations requiring correction by the acoustic impedance ratio $Z = Z_q / Z_f$, eqn. (5), is of less importance practically.

All the approaches mentioned above have in common the fact that the mass sensitivity of the crystal is based on an infinite resonator model. Today monitors based on frequency as well as on period measurement technique are commercially available. Following the frequency measurement concept, insertion of the frequency constant $N = v_q x_q$ ($N_{AT} = 1656$ kHz mm) into eqn. (6) yields the well known equation (9):

$$\Delta v \;\simeq\; -\,\frac{v_q^2}{N\,\rho_q}\,\Delta m_f \;\simeq\; -\,C_f\,\Delta m_f \tag{9}$$

From eqn. (9), it follows that the mass sensitivity C_f of a quartz crystal micro-balance can be calculated from the general properties of the quartz plate. Calibration was performed with various film materials using electromechanical microbalances and chemical microanalysis. As can be seen in Table 2, the experimentally found sensitivity values $C_{f\,exp}$ are in very good agreement with calculated values.

TABLE 2

INITIAL INTEGRAL MASS SENSITIVITY C_f OF OSCILLATING QUARTZ CRYSTALS OF DIFFERENT RESONANCE FREQUENCY AND CRYSTAL SHAPE, EXPERIMENTALLY DETERMINED AND CALCULATED BY VARIOUS AUTHORS.

v_q		X_q (mm)	$C_{f\,calc.}$ (Hz cm^2 g^{-1})	$C_{f\,exp}/C_{f\,calc}$	References
2.5 MHz, AT-cut 5th overtone,	p	0,662	0,141	0,98	1966 [38]
5 MHz, AT-cut fundamental mode	p	0,331	0,565	1,003	1966 [86]
	p			0,985	1967 [87]
	p, pc			0,967	1971 [75]
	pc			0,979	1975 [88]
5.5 MHz, AT-cut fundamental mode	p	0,301	0,684	0,991	1965 [89]
6 MHz, AT, cut fundamental mode	p	0,267	0,815	1,035	1971 [90]
8 MHz, AT-cut fundamental mode	p	0,207	1,450	–	–
10 MHz, AT-cut fundamental mode	p	0,165	2,265	0,991	1968 [91]
				1,010	1969 [92]
14 MHz, AT-cut fundamental mode	p	0,118	4,437	–	–

p = plane crystal, pc = plano-convex crystal; crystal diameters are between 12 and 17 mm, electrode diameters are between 6 and 8 mm.

Depositing more material onto the same quartz plate, its sensitivity should decrease, since C_f is proportional to the square of the crystal frequency. To overcome this difficulty, Behrndt [3,77] suggested a correction by replacing v_q in eq. (9) with v_c, where the index c indicates the resonant frequency of the quartz crystal with the deposited film mass:

$$\Delta v \;\simeq\; -\,\frac{v_c^2}{N\,\rho_q}\,\Delta m_f \;\simeq\; -\,C_f'\,\Delta m_f \tag{10}$$

The difference between C_f and C_f' is easy to calculate. For a mass corresponding to a 2 % shift in starting frequency, the difference in areal density using C_f' instead of C_f is about 4 %. For accurate mass determinations, this variable sensitivity has to be considered. The quartz crystal monitor can also record the rate of deposition, as already mentioned above. Rate measurements are very important especially in reactive deposition. The rate is obtained by electronic differentiation of the mass-dependent frequency change with respect to time. The slightly varying mass sensitivity with increasing mass load need not be considered for rate measurements.

As regards the instrumentation required to operate a quartz crystal monitor, it is common practice to use, in addition to the sensor oscillator, a reference oscillator and to generate a difference frequency between the two oscillators. These difference frequencies can be measured accurately to within $\pm$ 1 Hz with digital or analog instruments. The effective accuracy of the monitor depends on the stability of the electronic circuit, on the temperature coefficient of the crystal used, on the proper construction of the quartz sensor head and on a reasonable positioning with respect to the hot vapour sources. If a frequency change of 1 Hz – caused by a deposited mass alone – could be detected in a 5 MHz quartz crystal monitor, the corresponding sensitivity obtained under these ideal conditions would be in the 10^{-8} g cm^{-2} range. In practice, however, the attained sensitivity is in the 10^{-7} g cm^{-2} range. This accuracy of quartz-crystal microbalances for mass determination is by far sufficient for most requirements in thin-film deposition. Commercially available thickness and rate monitors use quartz crystals with a frequency of 5 or 6 MHz. Fig. 6a shows the BALZERS QSR 101 thickness and rate monitor and Fig. 6b shows the various types of quartz crystal holders. The high sensitivity, excellent mechanical stability and the controllable thermal

Fig. 6a
Oscillating quartz crystal film thickness and rate monitor BALZERS QSR 101.

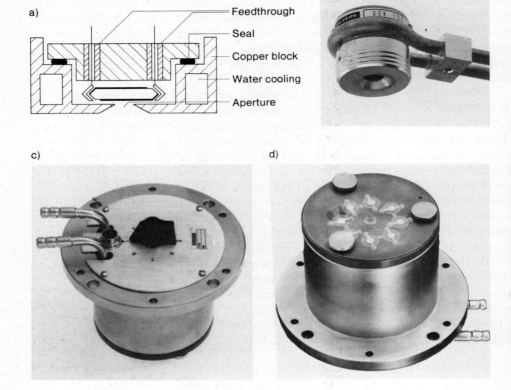

a)

Feedthrough

Seal

Copper block

Water cooling

Aperture

b)

c)

d)

e)

Fig. 6 b

a: Schematic representation of a quartz crystal holder with freely suspended crystal.
The contact via the facette ensures low-loss suspension. The holder has two additional feed-throughs to lead the thermocouple wires out which are not marked.
b: Quartz crystal holder for one crystals; BALZERS QSK 300.
c and d: Manually operated quartz crystal holder for 8 crystals; BALZERS QSK 810.
e: Motor driven quartz crystal holder for 6 crystals; BALZERS QSK 610.

influences make the quartz-crystal microbalance a valuable tool. Determinations of the physical film thickness, however, may be more of a problem since the density ρ_f of a thin film, depending on deposition method, chosen parameters and film thickness, is generally different from the density of the bulk material being considered. At film thicknesses below 100 nm, the density discrepancy is greater than with thicker films. This seems to be valid for metals as well as for dielectrics. It is, however, usually possible to reproduce the film density. For the deposition process, this demonstrates that the chosen vacuum and deposition parameters must be carefully controlled and maintained exactly to always attain the same film density. In its manner and after calibration, the geometrical film thickness can be exactly reproduced by quartz-crystal monitoring.

Similar problems exist during the reproducible production of optical thickness and electrical characteristics of thin films by quartz-crystal monitoring.

Investigations have shown that the optical constants of thin films are determined by chemical composition, structure, microstructure and incorporated or adsorbed gases and vapours. The prior conditions required to render the optical constants are unchanged geometrical configuration and exact reproduction of the vacuum and coating parameters. These parameters are primarily the substrate temperature, the residual gas pressure and composition, and the deposition rate.

Thin-film optical interference systems of transparent materials are usually designed with the aid of a computer. Optimization of such systems consisting of optically homogeneous films is often effected by varying the layer thickness while keeping the refractive indices unchanged at experimentally obtained values. In the design process, multilayer stacks are obtained with layers where the film thickness shows no simple relationship to quarter waves of one reference wavelength. For monitoring the deposition of such multilayers, the quartz-crystal film thickness and rate monitor is best suited [78,42].

Since the quartz-crystal oscillator is detecting the mass, its sensitivity remains practically unchanged, a situation which is of great advance for exact measurements [79]. Film systems for application in the ultraviolet, visible and infrared region can be monitored with the same quartz-crystal device. Furthermore, it is ideally suited for automatic deposition control using additional special process controllers [80]. A frequent argument against quartz-crystal monitoring is that the reproducibility of the optical thickness is worse than with photometric monitoring. This has been shown not to be the case [78,42, 81-85]. To obtain proper results in fabricating an optimized optical multilayer system, that means, to obtain experimentally a transmission versus wavelength curve in accordance with the calculated values, it is necessary to reproduce exactly all the previously chosen process parameters and geometrical configurations. Otherwise, the refractive indices of the films and/or the thickness distribution may change to some extent and, in an undefined way, modify the expected results.

As can be seen, for example in Fig. 7, the experimentally obtained transmission curve is in good accordance with the calculation. When a quartz-crystal oscillator is used in sputtering equipment, the main problem is excessive heating by charged particle bombardment, which causes frequency

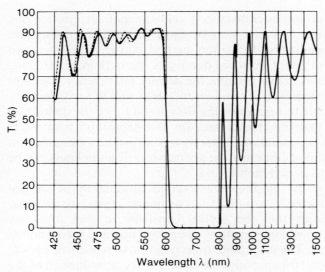

Fig. 7
Transmittance curve of a special low pass multilayer filter.
$n_L = 1.46$, $n_H = 2.40$.
The stack consists of 23 layers, 6 of them are correcting layers.
To control thickness a quartz crystal monitor was used.
Full line = measured values, dotted line = calculated values.

instabilities. A grounded front electrode of the sensing crystal prevents electron bombardment, and better still contact with the whole plasma is avoided by a grounded aluminium grid in front of the aperture of the quartz holder [93], although this slightly decreases sensitivity. A more convenient way to prevent heating problems due to charged particle bombardment is to introduce in front of the crystal holder aperture a deflecting magnetic field of about 1000 gauss [94].

7.3.3 VAPOUR-DENSITY MEASUREMENT BY MASS SPECTROMETRY

A quadrupole mass spectrometer with a cross beam ion source is a highly sensitive instrument for detecting atomic as well as molecular vapour species. The ion current is a relative indication of the number density of a particular species in the vapour stream and can be used to measure the deposition rate.

For the deposition of mixed films or of complex alloys from two or more evaporation sources, it is necessary to control the evaporation rate of each material independently with sufficient accuracy. A single quadrupole mass spectrometer can measure the evaporation rates of the different materials in a time-multiplex process.

To prevent disturbances of the film composition all components of the control system – measurement and evaporation source control – must be fast ones.

The block diagram of a closed-loop rate controller is shown in Fig. 8. Here only two channels are considered, but more channels are possible.

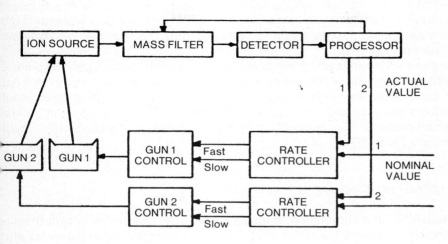

ig. 8
Mass spectrometer rate controller [95].

Part of the vapour beam of the evaporated materials passes through the ionization region of the mass spectrometer ion source. The different types of ions are separated according to their mass-to-charge ratios and measured. A control logic (BALZERS QMG 511: autocontrol/demultiplexer) allows up to 12 ion species to be measured in fast sequence (5 ms/channel) and produces steadily updated analog output signals for the different rates.

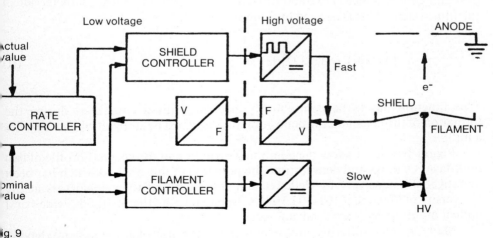

ig. 9
Block diagram of dual loop control [95].

The actual values of the rates are processed by independent analog controllers. The evaporation rates of electron beam evaporation guns are controlled by a new dual loop control mechanism. Not only is the filament temperature of a BALZERS electron beam gun varied (meaning slow response/high dynamic range), but also the shield voltages (meaning fast response/less dynamic range) is controlled [95].

In cases where an interference of the rate signal with residual gas background is a limiting factor, the beams can be modulated by a mechanical chopper in front of the quadrupole mass spectrometer. The difficulties arising with conventional size mass spectrometers, to locate the chopper at a suitable position in the deposition chamber have been solved by the development of well shielded small instruments [59,96,97].

7.4 TRENDS IN MONITORING TECHNOLOGY

Reliability and automation in thickness and rate monitoring of thin films is increasingly demanded [78,42,98,107], but new measuring technologies are also being developed. One of these is wide-band optical monitoring. For that purpose, a rapid scan spectrometer is used in transmission T or reflection R directly measuring the substrate. The method consists of comparing continuously, over the whole useful spectral range, the actual spectral profile of the assembly during the formation with the desired spectral profile which the system should possess when the thickness of the last layer reaches its correct value. The transmittance corresponding to the thickness t can be written as $T(\lambda, n_1, t_1, \dots n_{i-1}, t_{i-1}, n_i, t_i)$ which can be written as $T_i(\lambda, t)$ and that corresponding to the theoretically correct thickness t_i as $T_i(\lambda, t_i)$.

The operation of a method based on this principle requires the possibility of evaluating the distance between the actual and the desired optical characteristic of the film system. This is defined by a merit function:

$$f_i = \int_{\lambda_1}^{\lambda_2} [T_i(\lambda, t_i) - T_i(\lambda, t)] d\lambda \tag{11}$$

This function has to be continuously calculated (by a computer) during the formation of the layer and the deposition process must be stopped when it reaches a null.

Figure 10 and 11 show the computed evolution of the spectral profile of two multilayer coatings and demonstrate the potential of this method which has been investigated extensively by Pelletier and Bousquet and their coworkers at the University of Marseille [101-105,108]. Compared with other methods, wide-band optical monitors have some advantages.

Many spectral filtering problems require the use of layer systems having thicknesses which bear no obvious relationship to each other. With wide-band

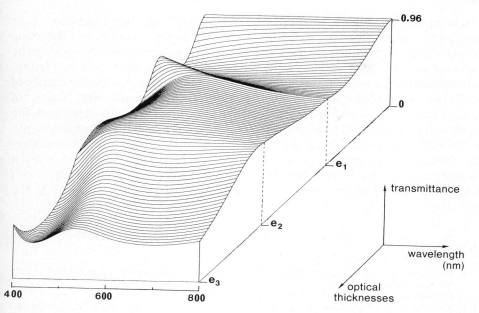

0.96

0

e_1

e_2

e_3

transmittance

wavelength
(nm)

optical
thicknesses

400   **600**   **800**

Fig. 10

Computed evaluation of the spectral profile during the deposition of an all dielectric beam splitter (according to E. Pelletier, 1981).

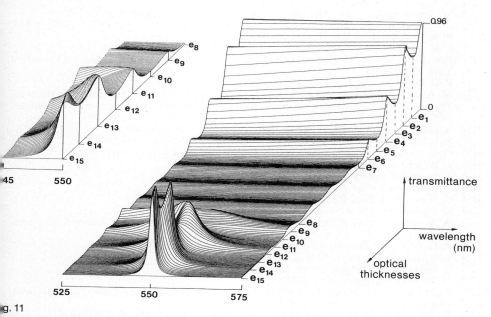

e_8
e_9
e_{10}
e_{11}
e_{12}
e_{13}
e_{14}
e_{15}

45   550

0.96

0

e_1
e_2
e_3
e_4
e_5
e_6
e_7

e_8
e_9
e_{10}
e_{11}
e_{12}
e_{13}
e_{14}
e_{15}

transmittance

wavelength
(nm)

optical
thicknesses

525   550   575

g. 11

omputed evaluation of the spectral profile during the deposition of a Fabry-Perot interference ter (according to E. Pelletier, 1981). e = optical thickness of the individual layers.

optical monitoring, there is no problem of controlling film deposition with non-quarter-wave thickness. A further advantage is the possibility of detecting coating errors in the early stages. The errors can be corrected by varying the thickness of subsequent films through real-time computation and correction. With commercially available devices e. g. [109] the reflectance or transmittance of thin films is determined through real-time measurements. By means of microprocessors it is possible to precalculate layer systems, to determine the equipment parameters to be preset and to automatically read the information into the device before the beginning of each individual film. The microprocessor offers the possibility to record the transmission or reflection spectrum with a servomotor driven monochromator at the end of the deposition of each layer and to set such spectra in comparison to the theoretical curves computed in the beginning.

REFERENCES

[1]   H. Mayer, Physik dünner Schichten, Vol. 1, Wissenschaftliche Verlagsgesellschaft mbH., Stuttgart, 1950.
[2]   H. K. Pulker and E.Ritter, Vak. Techn., 4 (1965) 91.
[3]   K. H. Behrndt, in G. Hass and R. E. Thun (Eds.), Physics of Thin Films, Vol. 3, Academic Press, New York, 1966, p. 1.
[4]   R. Glang, in L. I. Maissel and R. Glang (Eds.), Handbook of Thin Film Technology, McGraw-Hill, New York, 1970, Chapter 1.
[5]   C. Lu, Monitoring and Controlling Techniques for Thin Film Deposition Processes, 1981.
[6]   A. Ross, Vak. Techn., 8 (1959) 1.
[7]   W. Steckelmacher, J. M. Parisot, L. Holland and T. Putner, Vacuum, 9 (1959) 171.
[8]   a) W. Steckelmacher and J. English, Trans. 8th AVS Symp. 1961, Pergamon N.Y., 1961, p. 852.
      b) W. Steckelmacher, Vak.-Techn. 20 (1971) 139.
[9]   P. M. Schaible and C. L. Standley, IBM Tech. Disclosure Bull., 6 (1) (1963) 112.
[10] P. D. Davidse and L. I. Maissel, Trans. 3rd Int. Vac. Congr., 1965, Vol. 2, Pergamon Press, New York, 1965, p. 651.
[11] P. D. Davidse and L. I. Maissel, J. Appl. Phys., 37 (1966) 574.
[12] M. Auwärter, R. Haefer and P. Rheinberger in M. Auwärter (Ed.), Ergebnisse der Hochvak. Technik u. d. Physik dünner Schichten, Vol. 1, Wissenschaftl. Verlagsges. m.b.H., Stuttgart, 1957, p. 22.
[13] S. Tolansky, Multiple Beam Interferometry, Clarendon, Oxford, 1948.
[14] S. Tolansky, Surface Microtopography, Interscience Publ., New York, 1960.
[15] G. D. Scott, T. A. McLauchlan and R. S. Sennett, J. Appl. Phys., 21 (9) (1950) 843.
[16] L. Bachmann, Mikroskopie, 13 (1958) 250.
[17] H. K. Pulker, Naturwiss., 53 (1966) 224.
[18] G. Nomarski, J. Phys. Rad., 16 (Mars) (1955) 9.
[19] G. Nomarski, J. Phys. Rad., (1955) 1 – 3.
[20] G. Nomarski and A. R. Weill, Rev. Métall., LII (2) (1955) 121.
[21] W. Weintraub and H. J. Degenhardt, Armed Services Technical Inform. Agency (USA), Sept. 1961.
[22] R. Lignon, Mémoires Scientifiques, Rev. Métall., LXI (2) (1964) 85.
[23] H. Kiessig, Ann. d. Phys., 5 (10) (1931) 715.
[24] Y. Fujiki and T. Yoshida, J. Phys. Soc. Jpn., 14 (1959) 1828.
[25] N. Wainfan, Electronics, 36 (51) (1963) 17.
[26] R. M. A. Azzam and N. M. Bashara, Ellipsometry and Polarized Light, North Holland, Amsterdam, 1977.
[27] K. Kinosita, M. Yamamoto and and S. Hirasawa, Thin Solid Films, 90 (1982) 19.
[28] K. H. Behrndt, Z. Angew. Phys., 8 (1956) 453.
[29] R. E. Hayes and A. R. V. Roberts, J. Sci. Instr., 39 (1962) 428.

309

[30]   J. A. Poulis, P. J. Meeusen, W. Dekker and J. P. de Mey, in Vacuum Microbalance Techniques, Vol. 6, Plenum Press, New York, 1967, p. 27.
[31]   Th. Gast in Vacuum Microbalance Techniques, Vol. 6, Plenum Press, New York, 1967, p. 59.
[32]   S. P. Wolsky, E. J. Zdanuk, C. H. Massen and J. A. Poulis in Vacuum Microbalance Techniques, Vol. 6, Plenum Press, New York, 1967, p. 37.
[33]   L. Cahn and H. R. Schultz, in Vacuum Microbalance Techniques, Vol 3, Plenum Press, New York, 1963, p. 29.
[34]   D. Hacman, 7th Meeting Evaporat. Panel, Malvern (England), Sept. 1962.
[35]   G. Sauerbrey, Physikal. Verhandl., 8 (1957) 113; Z. Phys., 155 (1955) 206 – 222.
[36]   P. Lostis, Rev. Optique, (1959) 38.
[37]   H. K. Pulker, Z. Angew. Phys., 20 (1966) 537.
[38]   C. D. Stockbridge, in K. H. Behrndt (Ed.), Vacuum Microbalance Techniques, Plenum Press, New York, 1966, p. 193.
[39]   H. K. Pulker, E. Benes, D. Hammer and E. Söllner, Thin Solid Films, 32 (1976) 27.
[40]   H. K. Pulker, Vacuum, 29 (1979) 309.
[41]   C. Lu, J. Vac. Sci. Technol., 12 (1975) 578.
[42]   H. K. Pulker and J. P. Decosterd, in A. W. Czanderna and C. S. Lu (Eds.), Application of Piezoelectric Quartz Crystal Microbalances, Chapter 3, Elsevier, Amsterdam, 1983.
[43]   J. Terwellen, Z. Phys. Chem., A 153 (1931) 52.
[44]   R. Suhrmann and H. Schnackenberg, Z. Phys., 119 (1942) 287.
[45]   N. Wainfan, N. J. Scott and L. G. Parrat, J. Appl. Phys., 30 (10) (1959) 1607.
[46]   I. Langmuir, Phys. Rev., 22 (1923) 357.
[47]   M. C. Johnson and A. F. Henson, Proc. Roy. Soc. (London), A 165 (1938) 148.
[48]   R. Haefer, Z. Phys., 116 (1940) 604.
[49]   R. Weyl, Z. Angew. Phys., 13 (6) (1961) 283 – 288.
[50]   J. J. Finnegan and P. R. Gould, Trans. 9th Nat. Vac. Symp. 1962, Pergamon Press, New York, 1962.
[51]   K. Hirokawa and T. Shimanuki, Z. analyt. Chem., 190 (1962) 309.
[52]   F. F. Cilyo, US Patent No. 3,373,278, (1968).
[53]   H. Schwarz, Rev. Sci. Instrum., 32 (1961) 104.
[54]   R. H. Hammond, C. H. Meyer, Jr., B. L. Gehman and G. M. Kelly, US Patent No. 3,419,718, (1968).
[55]   C. Dufour and B. Zega, Vide, 104 (1963) 180.
[56]   Sloan Technology Corp., Iotron Deposition Control System, 1973.
[57]   Nortec Corp., Automatic Deposition Control Systems Featuring the Nortec Sweeping Beam Sensor, 1973.
[58]   L. L. Chang, L. Esaki, W. E. Howard and R. Ludeke, J. Vac. Sci. Technol., 10 (1973) 11.
[59]   G. B. Bayard and K. Lin, Uthe Technol. Int. J., 1 (6) 1974.
[60]   J. L. Terry and J. D. Michaelsen, Proc. 6th Int. Vac. Congr., 1974, p. 483.
[61]   H. Lutz, D. G. Dimmler, J. Strozier, H. Wiesmann, M. Strongin and D. H. Dougless, J. Vac. Sci. Technol., 13 (1976) 389; H. Lutz, J. Vac. Sci. Technol., 15 (1978) 309.
[62]   H. R. Smith, Jr., US Patent No. 3,609,378, (1971).
[63]   C. Lu, Ph. D. Dissertation, Syracuse University, 1975.
[64]   C. Lu, M. J. Lightner and C. Gogol, J. Vac. Sci. Technol., 14 (1977) 103.
[65]   C. A. Gogol, J. Vac. Sci. Technol., 16 (1979) 884.
[66]   X. Schwartz, X. Brown, Trans. Vac. Symp., 1961.
[67]   a) Talystep, Rank-Taylor Hobson Report.
        b) Dektak II, Sloan Technology Corp. 535 East Montecito Street, Santa Barbara, CA 93103.
[68]   R. Berthold, Z. d. VDI, 95 (7) (1953) 207 – 210.
[69]   B. W. Schuhmacher and S. S. Mitra, Electr. Reliab. Minin., 1 (1962) 321 – 333.
[70]   H. Hart, Elektrie, 16 (3) (1962) 82 – 88.
[71]   D. Kantelhardt and O. Schott, Z. Angew. Physik, 15 (4) (1963) 307 – 309.
[72]   L. E. Preuss and C. Bugenis, Am. Vac. Soc. Proc., 7th Nat. Symp., (1961) 260 – 272.
[73]   W. Weisweiler and R. Neff in G. Pfefferkorn (Ed.), Beiträge zur Elektronenmikroskopischen Direktabbildung von Oberflächen, Band 12/1, R. A. Remy, Münster, FRG, 1979, p. 229.
[74]   J. G. Miller and D. I. Bolef, J. Appl. Phys., 39 (1968) 5815.
[75]   K. H. Behrndt, J. Vac. Sci. Technol., 8 (1971) 622.
[76]   C. Lu and O. Lewis, J. Appl. Phys., 43 (1972) 4385.
[77]   K. H. Behrndt, Proc. 4th Int. Vacuum Congr., 1969, p. 579.
[78]   H. K. Pulker and E. Girardet, J. Vac. Sci. Technol., 6 (1969) 131.

310

[79]   H. K. Pulker, Thin Solid Films, 32 (1976) 27.
[80]   BALZERS Reports BG 800 118 PE and BB 800 108 PE.
[81]   P. B. Clapham, Thin Solid Films, 4 (1969) R39.
[82]   E. L. Church, Appl. Opt., 13 (1974) 1274.
[83]   H. A. MacLeod, Vacuum, 27 (1977) 383.
[84]   C. J. van der Laan and H. J. Frankena, Vacuum 27 (1977) 391.
[85]   H. A. MacLeod, Appl. Opt., 20 (1981) 82.
[86]   H. L. Eschbach and E. W. Kruidhof in K. H. Behrndt (Ed.), Vac. Microbalance Techniques,
       Vol. 5, Plenum Press, New York, 1966, p. 207.
[87]   H. K. Pulker, Thin Solid Films, 1 (1967/68) 400.
[88]   H. K. Pulker and E. Söllner, Technical University, Wien, Austria, unpublished results, 1975.
[89]   D. Hillecke and R. Niedermayer, Vak.-Techn., 3 (1965) 69.
[90]   D. King and G. R. Hoffman, J. Phys. E, 4 (1971) 993.
[91]   R. M. Mueller and W. White, Rev. Sci. Instr., 39 (1968) 291.
[92]   R. M. Mueller and W. White, Rev. Sci. Instr., 40 (1969) 1646.
[93]   R. Chabicovsky, H. K. Pulker and W. Schädler, Vak.Techn., 19 (1970) 101.
[94]   C. Lu, US Patent No. 3.732.778.
[95]   W. K. Huber, U. Wegmann and K. Wellerdieck, Proc. 8th Int. Vacuum Congr. 1980, Cannes,
       France, Vol. II Vac. Technol. and Vac. Metallurg., p. 542, Suppl. Revue Le Vide, les Couches
       Minces, No. 201.
[96]   P. Genequand, J. Vac. Sci. Technol., 11 (1974) 357.
[97]   BALZERS Product Information BG 800 151 PD 8205.
[98]   W. P. Thöni, Thin Solid Films, 88 (1982) 385.
[99]   H. A. Macleod and E. Pelletier, Optica Acta, 24 (1977) 907.
[100] H. A. Macleod, Appl. Opt., 20 (1981) 82.
[101] B. Vidal, A. Fornier and E. Pelletier, Appl. Opt., 17 (1978) 1038.
[102] B. Vidal, A. Fornier and E. Pelletier, Appl. Opt., 18 (1979) 3851.
[103] B. Vidal and E. Pelletier, Appl. Opt., 18 (1979) 3857.
[104] B. Vidal, Opt. Comm., 31, (1979) 259.
[105] J. P. Borgogno, P. Bousquet, F. Flory, B. Lazarides, E. Pelletier and P. Roche, Appl. Opt., 20
       (1981) 90.
[106] D. A. Walsh, G. N. Pukite and J. R. Fenter, Rev. Sci. Instrum., 47 (1976) 932.
[107] R. Herrmann and A. Zöller, in J. R. Jacobsson (Ed.), SPIE Proceedings Vol. 401 Thin Film
       Technologies, (1983) paper No. 401 – 09.
[108] F. Flory, E. Pelletier, B. Schmitt and H. A. MacLeod, in J. R. Jacobsson (Ed.), SPIE Proceedings
       Vol. 401 Thin Film Technologies, (1983) paper No. 401 – 12.
[109] Leybold-Heraeus, Leaflet 15-140.1/2, OMS 2000 Process Photometer.

CHAPTER 8

8. PROPERTIES OF THIN FILMS

The properties of thin films are primarily determined by the type of chemical element or compound they comprise and by the film thickness. Their optical, electro-optical, electrical and mechanical behaviour is also determined by structure, microstructure, surface and interface morphology, chemical composition and homogeneity. These are strongly influenced by the film preparation method, the chosen parameters, and by the post-deposition treatment.

The properties and characteristics of films in the optical and electro-optical categories are discussed because these types constitute the most important applications of coatings on glass and plastics. Finally, some tables containing pertinent film data are appended.

8.1 STRUCTURE

The arrangement of atoms in a material is determined chiefly by the strength and directionality of the interatomic bonds. Thus one needs to distinguish between strong or weak and directional or non-directional bonds. Typical for strong bonds such as covalent, ionic and metallic bonding is a pronounced lowering of electron energies. Weak bonds are less easily defined, but can be viewed in terms of weak dipole attractions and quantum mechanically determined interatomic forces, the so-called van der Waals forces.

The local arrangement of atoms in a solid may be either regular (crystalline) or irregular (polymeric or glassy); however, paracrystalline states are also known. The arrangement depends partly on whether the bonding is directional, represented by a bonding polyhedron determined by the bonding angles, or non-directional, depending on the relative sizes of the atoms or ions represented by a coordination polyhedron. The regular three-dimensional arrangement of atoms or ions in space constitutes the crystalline structure. These highly ordered structures of atoms or ions in a space lattice arise from geometrical conditions which are imposed by directional bonding and close packing with the aim of minimizing the energy of the solid. However, many solids have non-crystalline structures. Their subunits are packed together randomly. They lack the long-range order of crystals because they have only limited mobility or are sterically hindered at the equilibrium solidification temperature.

Non-crystalline materials are generally classified according to whether they are composed of individual long-chain molecules, three-dimensional networks or arrangements between these two limiting cases.

Most non-crystalline three-dimensional rigid networks are formed by the bulk inorganic glasses and some special organic polymers both are discussed in Chapter 2. However, there are also many non-crystalline films which are either true polymers or are formed because of insufficient energy supply during their deposition.

A solid may also contain more than one phase. There are many ways in which crystalline and non-crystalline aggregates of atoms and molecules can be arranged in a solid. The shapes and distributions of phases are important features of bulk materials and thin films.

Determination of the crystal structure of solids is carried out by means of X-ray diffraction (von Laue 1912) or electron diffraction (Davidson and Germer 1927) e. g. [1]. An X-ray incident on a group of atoms transmits its electric field to the electrons of the atoms so that these start to oscillate. However, a vibrating electric charge is always a source of electromagnetic waves, the wavelength of which is identical with that of the impinging ray. Hence every atom acts as an emitter, the waves radiated from the various atomic layers (crystall planes) causing interference effects. This means that the wave train constructively interferes in certain directions, and in others destructively, depending on whether or not the waves are in phase. These events cause the well known diffraction phenomena which can be described with the classical Bragg equation

$$n \lambda = 2d \sin \theta. \tag{1}$$

Here, the number n determines the order of the reflection, that is the number of wavelengths through which the phases of the waves are shiftet. λ = wavelength, d = crystal interplanar spacing, θ = angle of incidence.

The intensity of the diffracted beams on the other hand is determined by the distribution of the atoms, i. e. their relative geometric arrangement within a unit cell. As already mentioned, the periodic arrangement of the electron density $_e\rho$ in the crystal is essential for the diffraction of X-rays. Also, according to Bragg, the latter can be described in a Fourier series as a function of the structure factors F_{hkl}, which determine the intensity of the reflections:

$$_e\rho_{(xyz)} = \frac{1}{V} \sum_h \sum_k \sum_l F_{hkl} \exp\left[-2\pi i(hx+ky+lz)\right] \tag{2}$$

where V represents the volume of the elementary cell, h, k, and l the indices of the relevant reflections and x, y and z the atomic space coordinates. Here indices are understood as integral factors which indicate the relative orientation of the various planes in the crystal or the reflections caused by them.

With the discovery of diffraction of electrons in crystals, it was clear that electron diffraction, like X-ray diffraction, should in principle enable the determination of crystal structure.

The diffraction of electrons can also be regarded as a special case of the electro magnetic wave equation, where the electrostatic potential of the atom replaces electron density $_e\rho$ (X-ray diffraction). This potential is composed of the effect of the atomic nucleus and that of the electron cloud. The spatial distribution of the potential corresponds approximately to that of the electron density, but falls off less rapidly. The maxima of the Fourier series, however, correspond in both cases to the position of the atomic nucleus.

The intensensity of the diffracted beams is proportional to the square of the structure factor Φ for electrons; hence an equation results analogous to that of X-ray diffraction:

$$\varphi_{(xyz)} = \frac{2\pi h^2}{me} \frac{1}{V} \sum_h \sum_k \sum_l \Phi_{hkl} \exp\left[-2\pi i(hx+ky+lz)\right] \tag{3}$$

Here the structure factor Φ signifies the vectorial sum of the waves scattered by the single atoms which show amplitude f and phase γ. Every atom contributes a scattered wave to the whole diffraction effect, the amplitude of which is proportional to the so-called form factor. The phase is thus defined by the position of the atom in the elementary cell, whilst the form factor is a characteristic constant for every sort of atom which represents a measure of its scattering power. Hence no special differences exist in the positions of the diffracted beams, which in both X-ray and electron diffraction cases satisfy the geometric relations between lattice constant and X-ray or material wavelengths, according to the Bragg equation. However, there are definitely differences in their intensities.

The reason lies in the non-applicability of the kinematical approximation (theory of single scattering) in the case of diffraction of electrons by single crystals. This, in comparison with X-ray quanta, is due to the much stronger scattering of the electrons at the lattice atoms and the consequent very high intensity of the singly and multiply diffracted electrons.

When investigating polycrystalline materials with disordered or partially ordered orientation, ring diagrams or texture diagrams respectively, are obtained. In these cases, there is sufficient conformity with the prediction of the kinematical theory. Therefore, the texture diagram is the most important type for structure analysis with electrons. Vainshtein, Pinsker and co-workers [2–7], as well as Cowley [8, 9] evaluated the basis, about 30 years ago, of the determination of crystal structure by means of electron diffraction.

The main advantages of the diffraction of fast electrons compared with X-ray diffraction, as regards structure determination, lie in (i) the better determinability of hydrogen atoms and ions, which is not possible by X-ray diffraction because of the lack of an electron cloud, and (ii) the low amount of substance required (~ 0.01 mg). Furthermore, many substances occur only in an extremely finely distributed state which is ideal for electron diffraction but almost unsuitable for X-ray diffraction.

Therefore, with electrons, one can also investigate very thin layers, even on the surface of solid substrates. The classical case of the determination of an unknown crystal structure from diffraction data is, however, extremely rare when investigating deposited thin films. Usually, only a limited number of possible crystal structures in a film to be investigated is under consideration, since a strong limitation is imposed by the knowledge of the chemical composition and the film-production conditions. The crystal structure in a vapour-deposited film can often be determined quickly after measuring the diffraction pattern and by comparison of these measured data with those tabulated, say, in the ASTM index [10]. One often has to determine whether a condensate is amorphous, polycrystalline, with or without texture, or a single crystal. Other interesting questions concern particle size and crystal perfection. Therefore it should be stated here that the degree of lattice disorder in thin films is usually far greater than in compact materials [11].

The range of possible crystal structures can, however, for a given material in the form of a thin film also be larger than for the bulk material because of the many variations possible in the film production. An example of this is the evaporation of Si films on Si single-crystal substrates under ultra high vacuum [12]. Silicon evaporation onto 800°C hot substrates resulted in single-crystalline films of such high perfection that there was practically no distinction betweeen the new material and the single-crystal substrate. Evaporation at 550°C still resulted in single-crystal films but there were many stacking faults. With a substrate temperature of 320°C, polycrystalline films were obtained and finally if the substrate was not heated up, amorphous films were produced. With the exception of amorphous structures which are more frequently present in evaporated films than in bulk material, structures can also occur in films which are not a stable modification of the relevant material in the compact state. Reports are given of face-centered cubic structures in sputtered Ta, Mo, W, Re, Hf and Zr films, although these materials otherwise have a body-centered cubic or body-centered hexagonal structure [13].

The structures of some vacuum-deposited films discussed here were investigated with an electron diffractograph [14] in reflection and transmission. For this purpose, the films were evaporated onto either glass substrates (reflection electron diffraction) or onto thin carbon or silica films (transmission electron diffraction). Evaporation of the films for the reflection diffraction was usually carried out directly in the diffractograph. Using an acceleration voltage of 80 kV the wavelength of the electrons applied for diffraction is 0.042 Å. In order to exclude artefactive changes of the structure by the electron beam as much as possible, the beam current should usually be kept below 10 μA. Since dielectric films generally show charging effects from the electron beam, a discharger, operated, say, with very low energy argon ions must be employed. The diffractograms obtained were photographed and then measured carefully.

Ag and Au films

Evaporated silver films have often been investigated with respect to the production of epitaxial condensates. For such experiments, evaporation was usually carried out on heated single-crystal surfaces of various materials under extremely clean vacuum conditions. See, for instance, [15] which also contains many references.

For optical applications, however, the structure of Ag condensates on unheated and amorphous substrates is of interest. For this purpose, glass substrates are coated with silver at $\sim 10^{-6}$ mbar residual gas pressure. The condensation rates are between 1 and 10 nm s^{-1}. The electron-optical investigations show that the films obtained are polycrystalline and consist of crystallites of cubic-densest-packed structure. Thus the same lattice type exists in the films which is characteristic of bulk silver. Very thin condensates (mass thickness $\leqslant 5$ nm) are built up of small, isolated, three-dimensional micro-crystals. The presence of H_2O would appear to favour the formation of three-dimensional crystals [16]. These crystallites are large compared with average film thickness. When adding more material, the crystallites gradually grow together and from 15 nm upwards form a closed film. Films 2nm thick already show a relatively sharp Debye-Scherrer ring diagram. With progressive evaporation, because of better formed larger Ag crystals, the sharpness of the diffraction rings increases and remains unchanged from ~ 30 nm film thickness upwards. Gold films behave in a similar way. A typical transmission diffraction pattern of polycrystalline Au is shown in Fig. 1.

Fig. 1
Electron diffraction pattern of a 30 nm thick polycrystalline Au film.

MgF$_2$ films

The structure of 10 to 200 nm thick MgF$_2$ films evaporated at normal incidence onto glass substrates of 30°C and 350°C is crystalline in all cases according to diffraction investigations [17a]. This confirms the evidence given by Heavens [18] and Bauer [19] of a crystalline structure of magnesium fluoride even in very thin deposits. The reflection diffraction of thin films evaporated onto cold glass substrates showed ring diagrams with broadening of lines, i.e. these films are polycrystalline and consist of very small micro crystals with many lattice defects. Thicker films show diffraction fringes of better crystallized magnesium fluoride. From about 100 nm thickness, in agreement with [20], a <110> growth texture is found. Films evaporated onto substrates of higher temperature are better crystallized even when having small thickness, and hence sharper diffraction fringes are obtained.

As a function of the substrate temperature T_s, various textures were also observed:

$T_s = 185°C$ .... $<111>$ orientation [21]
$T_s = 400°C$ and $540°C$ ..... $<110>$ orientation [20, 21].

Other textures arise when vertical deposition is not used.

Cryolite films

Structure investigations of cryolite films [22] show, in agreement with Bauer [19], that films evaporated onto unheated glass substrates are polycrystalline and consist of crystallites with high lattice defects. This occurrence is also partly caused by the phase separation of NaF and AlF_3 [22] observed during evaporation. In the case of short-period tempering between $250°C$ and $300°C$, sharp diffraction fringes are obtained. Thick films also have a fiber texture. Evaluation of the patterns showed that in addition to cryolite, the films also often still contain NaF and $Na(AlF_4)$ as crystalline components.

ZnS films on glass substrates

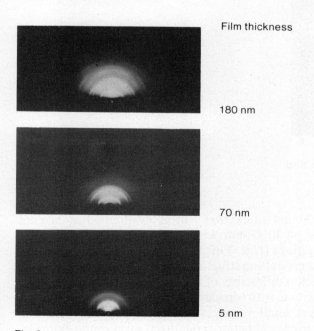

Film thickness

180 nm

70 nm

5 nm

Fig. 2
Reflection electron diffraction patterns of evaporated ZnS films of varying thicknesses.

ZnS films

Zinc sulphide exists in two forms, as cubic zinc blende and as hexagonal wurtzite. Of the two modifications, the latter is stable at higher temperatures. The ZnS films obtained by simultaneous vapour condensation of sulphur and zinc onto unheated substrates are crystalline even from 5 nm mass thickness. However, the diffraction pattern shows that these small, isolated, three-dimensional ZnS microcrystals contain a large number of stacking faults in all three spatial coordinates. In thicker films (from about 20 nm), the diffraction pattern shows better-ordered crystallites of the zinc blende type. ZnS films deposited at normal incidence have a clearly distinct <111> growth texture becoming noticeable from about 100 nm, as can be seen in Fig. 2 [17b].

TiO_2 films

Titanium dioxide exists in nature in three crystalline forms: rutile, anatase and brookite. Brookite is an alkali stabilized modification and has never been observed in evaporated films but was found in dip-coated ones. In the case of reactive evaporation of TiO (vapour phase : $Ti + TiO$) and condensation of the vapour on hot substrates (glass, or SiO and C films) depending on the temperature of the substrate T_s, various TiO_2 phases are obtained:

$T_s \leqslant 280°C$	diffuse electron diffraction pattern, i.e. amorphous TiO_2 films
$T_s \geqslant 310°C$	sharp diffraction pattern, polycrystalline TiO_2 films, contain rutile and anatase
$T_s = 380°C$ to $470°C$	sharp diffraction fringes, polycrystalline TiO_2 films consisting purely of rutile [23].

These results were confirmed by the author [17b].

Amorphous TiO_2 films (starting material TiO) which were obtained by condensation on glass substrates of 30°C, crystallized under strong electron radiation (60 kV, current density = 15 A cm^{-2}, measured with a Faraday cup) into rutile and anatase, as determined with selected area diffraction. Crystallization was found to be due to a temperature effect.

Compared with rutile, the anatase crystals were found to be very large (10 μm). Anatase twin formation was also observed [23]. When evaporating TiO_2 (anatase) onto sodium chloride cleavage surfaces, amorphous films are obtained which also crystallize into rutile and anatase under strong electron radiation [24, 25].

TiO_2 films which are produced at 150°C by chemical vapour deposition (hydrolysis of tetra-isopropyltitanate) are amorphous. Tempering in air at 350°C leads to anatase formation, and from 700°C, to rutile [26].

Pure titanium films (i.e. mostly free of TiO and TiN) have a hexagonal lattice as indicated by diffraction (a = 2.95 Å, c = 4.67 Å). By tempering in air at 400°C, the metal films can be completely oxidized to TiO_2. The TiO_2 films obtained show a diffraction pattern of the rutile lattice (a = 4.58 Å, c = 2.98 Å) [27].

TiO$_2$ films produced by oxidizing Ti films, which contain the cubic phases TiO and TiN, (a$_{TiO}$ = 4.23 Å, a$_{TiN}$ = 4.166 Å), by tempering in air between 400° C and 450° C show solely tetragonal anatase (a = 3.73 Å, c = 9.37 Å). Only after heat treatment at 600° C are rutile diffraction rings also obtained in addition to anatase rings from the TiO$_2$ films thus produced [27].

In addition to electron diffraction, the measurement of the film refractive index also enables distinction between anatase and rutile. According to Hass [27], polycrystalline rutile films have a refractive index value of $_R$n$_{546}$ = 2.70 and that of the anatase films is $_A$n$_{546}$ = 2.39.

Measurements carried out [28] on vapour-deposited TiO$_2$ films, which in all cases were produced with the method of reactive evaporation but using different starting materials, gave the following results: When using the starting material TiO (vapour-phase species : Ti + TiO), at least for the first $\lambda/4$ thick TiO$_2$ film which was condensed on glass at a substrate temperature of 450° C, a refractive index n = 2.63 was obtained. By evaporating Ti metal (vapour-phase species : Ti) on hot glass substrates, high refracting TiO$_2$ condensates (n$_{546}$ = 2.6 to 2.65) can also be obtained.

TiO$_2$ films which are produced by evaporation of Ti$_3$O$_5$ (vapour-phase species : TiO) onto 450° C hot glass substrates under otherwise similar conditions, however, showed a refractive index value of n$_{546}$ = 2.33. Slight variations in the TiO$_2$ refractive index can be produced by MoO$_X$ and WO$_X$ incorporation, which stems from chemical reactions with the boat walls. This undesirable contamination can be avoided by electron-beam evaporation or by using Ta boats. As a result, a higher film refractive index is also obtained [28].

TiO$_2$ films produced by reactive evaporation of the starting materials TiO or Ti contain at least a predominant amount rutile; whereas TiO$_2$ films produced by the same method but from the starting material Ti$_3$O$_5$ consist of anatase. Tempering at higher temperatures and irradiation with electrons produce even from amorphous TiO$_2$ films frequently rutile films [29]. Structure investigations of TiO$_2$ films produced in different ways give the following results.

During oxidation of pure Ti metal films, the hexagonal titanium lattice apparently transforms preferentially into the tetragonal rutile lattice with a = 4.58 Å, c = 2.95 Å and space group D $_{4h}^{14}$. In the rutile lattice, every Ti is surrounded by six O in form of a slightly deformed octahedron, whilst every O is located at the center of mass of a nearly equilateral triangle, the corners of which are occupied by Ti ions. Furthermore, it is remarkable that in the case of reactive evaporation, always when all or a certain quantity of the particles contained in the vapour are Ti atoms, at least part of the TiO$_2$ crystallites in the film are evident as the rutile phase even at relatively low temperature. In reactive sputtering of TiO$_2$ films from a Ti target, predominantly rutile films are also obtained, the amount depending on substrate temperature and oxygen gas pressure, as can be seen in Fig. 3 [30]. In both cases this seems to be caused by easier and higher mobility of the Ti atoms. On the other hand, it is evident that in the case of oxidation of TiO, the transformation of the cubic NaCl lattice of TiO with an applied temperature of 450° C always takes place in the direction of the tetragonal anatase lattice with

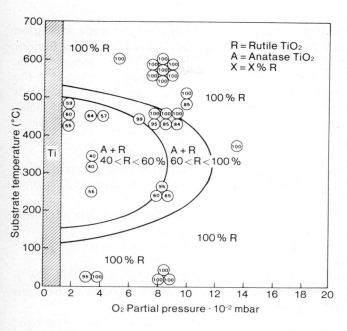

Fig. 3
Diagram of the structural composition of TiO₂ coatings produced
by reactive sputtering as function of substrate temperature and
oxygen partial pressure according to Pawlewicz et al. [30].

$a = 3.73$ Å, $c = 9.37$ Å and space group D_{4h}^{19}. However, the anatase lattice can be understood as an NaCl lattice deformed along the c axis, where half the cations are removed and the anions are slightly displaced in the direction of the c axis. Hence, in the case of oxidation, the transfer of the cubic TiO lattice to tetragonal anatase is apparently preferred for spatial and energetic reasons. Anatase is in fact not as stable as rutile, but when the phase has once been formed high temperatures are necessary in order to bring about a transformation. The transformation of the bulk material from anatase to rutile takes place between 800°C and 1100°C [31, 32].

SiO₂ films

Although SiO₂ occurs in several crystalline forms in nature, evaporated films which are condensed onto glass substrates or carbon films between 30°C and 400°C are always vitreous amorphous according to electron diffraction. The refractive index of the compact quartz glass is $n_{546} = 1.46$. However, evaporated quartz glass films frequently show a somewhat higher value, $_{film} n_{546} = 1.472$ which indicates a different degree of polymerization compared with solid quartz glass. Whilst in the quartz glass the polymerization always takes place via $\equiv$ Si-O-Si $\equiv$ bonding, in the films, because of small oxygen deficit, $\equiv$ Si-Si $\equiv$ bonds may also occur. The properties of glass like SiO₂ films depend on the preparation conditions such as type of deposition technique, substrate temperature and

320

formation rate. Thus, for example, independent of reaction type, a fast deposition rate generally produces quite porous films with a good deal of bond strain. Similar behaviour can be observed by deposition at lower substrate temperatures. Chemical and structural differences in the glassy SiO_2 films can be detected by the use of infrared spectrometry [33-38, 40]. The observed absorption bands correspond to vibration modes due to the deformation of the SiO_2 basic tetrahedra. Four classes can be distinguished, including vibration modes of reaction products with water or hydrogen:

– Si-O stretching modes involving displacements associated primarily with the oxygen atoms. For glassy SiO_2 films, their spectral range is between 1250 cm^{-1} (8μm) and 934 cm^{-1} (10.7μm).
– Si stretching modes with displacements primarily of the silicon atoms occurring between 909 cm^{-1} (11μm) and 588 cm^{-1} (17 μm).
– Si-O bending modes having a spectral range location between 588 cm^{-1} (17 μm) and 384 cm^{-1} (25 μm).
– O-H stretching regions and other modes connected with Si-H (2260 cm^{-1}), Si-OH (3640 cm^{-1}) and H_2O (3400 cm^{-1}) (1620 cm^{-1}), vibrations are found between 3700 cm^{-1} (2.7 μm) and 1600 cm^{-1} (6.3μm).

The most important stretching band is located at about 1000 cm^{-1}. This band is used to determine the stoichiometry of silica films. The band shifts from 960 cm^{-1} to 1085 cm^{-1} in going from SiO to SiO_2 [39, 40].

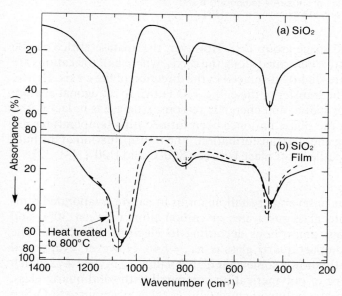

Fig. 4

Infrared absorption spectra of vitreous quartz (a) and a SiO_2 film prepared by low temperature CVD (b) [38].
Both samples have comparable thicknesses.

Compared with vitreous quartz, porosity and strain in films cause the 1085 cm^{-1} and 460 cm^{-1} absorption bands to be broadened and shifted to lower wavenumbers, whereas the 800 cm^{-1} band is shifted to higher wavenumbers. As can be seen in Fig. 4, after proper densification (i.e. heat treatment between 800 and 1000°C), a pure SiO$_2$ film deposited with no oxygen deficiency by low-temperature CVD becomes essentially indistinguishable from high-temperature thermally grown silicon dioxide [38].

Broader and shifted absorption bands are also found in SiO$_2$ films deposited under vacuum with electron-beam evaporators. Here porosity, strain and oxygen deficiency shift the bands. As can be seen in Fig. 5 obtained from attenuated total

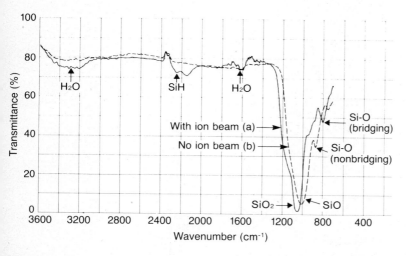

Fig. 5

Infrared absorption spectra of SiO$_2$ films deposited with (a) and without (b) activation (oxygen ion beam of 200 mA). The absorption spectrum of the latter film exhibit several absorption bands indicating an oxygen deficient film [40].

reflectance measurements [40], such films contain many non-bridging, dangling bonds $\equiv$ Si-O $\cdot$ $\cdot$ Si $\equiv$ which may be partially hydrogenated if hydrogen (possibly also water vapour) is in the residual gas. Reactive evaporation of SiO$_2$ in activated oxygen yields narrower bands located at greater wavenumbers and shows only bridging bonds $\equiv$ Si-O-Si $\equiv$ and also incorporated water. The appearance of silanol, Si-OH (3640 cm^{-1} and 935 cm^{-1}) and of water, H$_2$O (3330 cm^{-1}), absorption bands as a consequence of reactions with water vapour by exposing freshly prepared films to humid atmosphere is shown in Fig. 6. Water and the silanol groups are removed under densification by short tempering in dry atmosphere [38].

More frequently than with inorganic materials, infrared spectrometry is used to characterize organic polymer films [41].

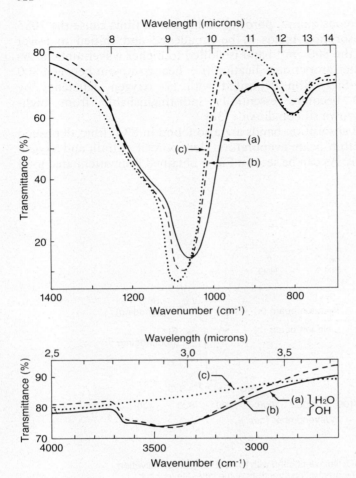

Fig. 6

Influence of humidity on an electron beam evaporated SiO₂ film deposited on a substrate of 400°C [38].
(a) Initial spectrum
(b) 24 days exposure at 85°C and 85% rel. humidity
(c) densified 10 min. in N₂ at 938°C

8.2 MICROSTRUCTURE

Condensation, nucleation and growth phenomena have been investigated both theoretically and experimentally for films made under conditions typical for physical vapour deposition in high vacuum. A number of papers concerning this topic are listed in ref. [42].

The nucleation and growth steps during the formation of a continuous film are usually as follows:

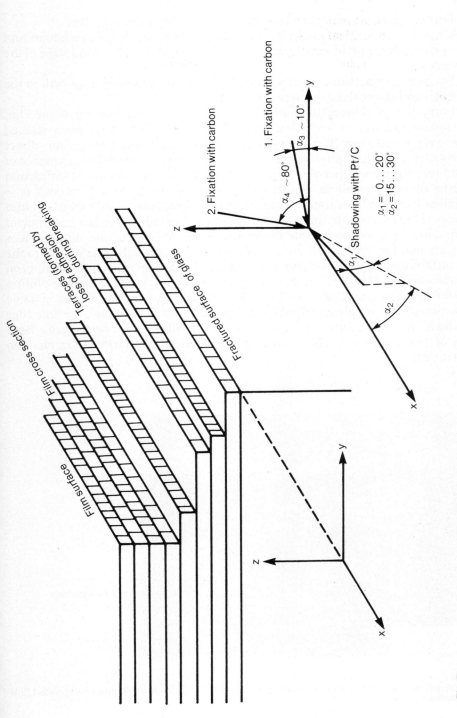

Fig. 7a
Graph showing the vapour beam directions required to obtain suitable preshadowed surface replicas for cross sectional investigation of film microstructures in TEM [50].

Formation of subcritical and critical nuclei in the nucleation stage.

Growth of the critical nuclei by the addition of surface diffusing adatoms and by coalescence of the small grains to larger aggregates in the island stage of the deposit.

Formation of networks with secondary nucleation and nucleus growth in the holes and channels of the network.

Formation of a closed film habitus and preferred thickness growth of the film. The microstructures of very thin metal films particularly have been studied extensively employing electron microscopy. However, for a long time direct microstructure investigations of thicker films were rarely undertaken. Evidence of an indirect nature was usually obtained from the line broadening of diffraction patterns of polycrystalline condensates according to $\Delta(2\theta) = (\lambda/c)\cos\theta$ (c = crystallite size) and from transmission electron micrographs from replicas of film surfaces. However, both methods supply only limited and often insufficient information. In order to be able to determine the microstructure directly, it is necessary that the whole cross-section of the film from the substrate surface to the film surface can be observed and also measured. Such investigations have been undertaken since 1965 [43, 44]. For this purpose, use was made of high-resolution evaporated surface replicas (Pt/C) of the exposed film profile, or else suitable deposition of thin supporting films having micro-holes (0.1 – 1 μm), the hole edge of which was the substrate for the growing film, was performed. Both types of specimen were subsequently investigated in the transmission electron microscope.

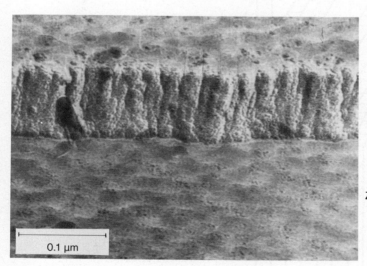

MgF$_2$-film

ZnS-film surface

0.1 µm

Fig. 7b
Electron micrograph showing the microstructure of a MgF$_2$ film deposited onto an unheated ZnS film substrate.

Generation of the film profile is best accomplished via fracturing. During the fracture process, the material always breaks along the lines of weakest cohesion, i. e. along the grain boundaries or, with amorphous materials, along points of weak polymerization. The film cross sections thus obtained are largely free of artefacts. Electron micrographs of single films of the following materials investigated in the described manner have been published:

Al [43], SiO-Cr [44], MgF_2, $Na_3(AlF_6)$, ZnS [45], CdSe [46], Au [47], Al, Ag, Au, Be, Cr, Sn, Bi, Pb, Te, Ge, Si, CdTe, GeTe [48].

Only few investigations [49, 50] are known regarding the microstructure of evaporated multilayer systems. In [49], micrographs of the cross-sections of narrow-band interference filters (Ag, ZnS, $Na_3(AlF_6)$) and in [50] those of dielectric interference mirror coatings (ZnS/MgF_2, ZnS/ThF_4, TiO_2/SiO_2, Al_2O_3/SiO_2) were published.

Figure 7 shows the microstructure of a MgF_2 film of 115 nm thickness, which was evaporated in vacuum at normal incidence onto a 70 nm thick ZnS film of 30°C. The diagram in Fig. 8 shows the correlation between film thickness and crystallite diameter of ZnS films. Also, micrographs of the microstructure of ZnS/MgF_2 and TiO_2/SiO_2 multilayer systems can be seen in Figs. 9 and 10.

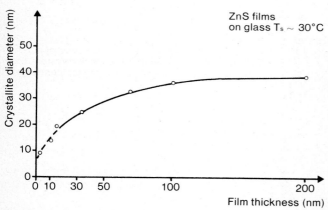

Fig. 8
Correlation between crystallite diameter and film thickness of ZnS films deposited on glass substrates.

Interesting information has been obtained from these investigations: Most evaporated films have a microstructure consisting of more or less close-packed columns or columnar crystals. Evaporated SiO_2 films, however, showing homogeneous vitreous microstructure, number among the few exceptions. For films developing columnar microstructure, it is remarkable that with oblique evaporation direction the columns grow inclined, sometimes almost parallel to

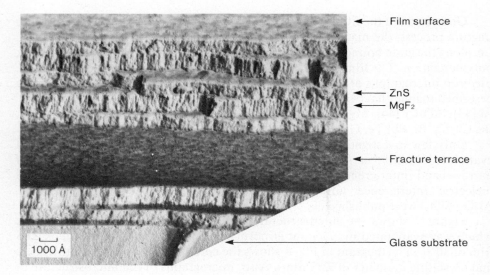

← Film surface

← ZnS
← MgF₂

← Fracture terrace

← Glass substrate

1000 Å

Fig. 9
Electron micrograph showing the microstructure of an alternating film system of a total of 7 λ/4 layers from ZnS/MgF₂ with fracture terraces. All films are evaporated at normal incidence, the glass substrate was at about 30°C.

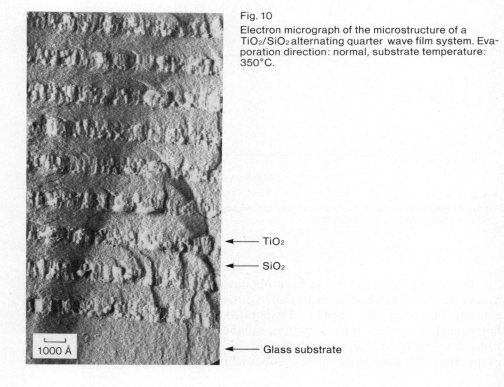

Fig. 10
Electron micrograph of the microstructure of a TiO₂/SiO₂ alternating quarter wave film system. Evaporation direction: normal, substrate temperature: 350°C.

← TiO₂

← SiO₂

1000 Å

← Glass substrate

the incident vapour beam [43, 48, 50, 51]. A careful evaluation of experimental work yielded the tangent rule: $\tan\alpha = 2\tan\beta$ [51]. The relation is shown in Fig. 11 [52].

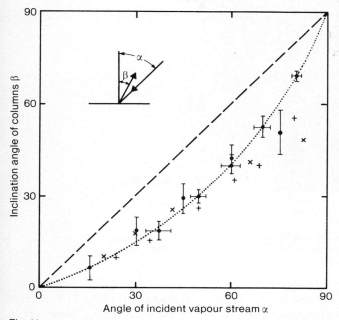

Fig. 11

Column orientation β versus angle of vapour incidence α [51].
Tangent rule (dotted curve):
$\tan \alpha = 2 \tan \beta$
(after Nieuwenhuizen and Haanstra)

However, the columnar microstructure does not always signify well formed crystallites. Electron diffraction of ThF_4 films and Al_2O_3 films condensed onto 300°C hot substrates, which both grow in columns, showed in the case of ThF_4 an extremely high number of growth defects and the Al_2O_3 films were even amorphous.

Changes in the substrate temperature can have a great influence on the microstructure of the film. This is demonstrated by the example of the TiO_2/SiO_2 multilayers shown in Fig. 12. When evaporated onto hot glass substrates (350°C), TiO_2 grows in crystalline columns, whilst on cold substrates ($\sim 30°$C) columnar growth hardly occurs [50]. However, SiO_2 evaporated films showed a smooth homogeneous microstructure at every substrate temperature. These vitreous amorphous condensates can be influenced, as regards the degree of densification, by the substrate temperature and the residual gas atmosphere. Particularly those films condensed onto hot substrates show a microstructure very similar to that of the molten quartz or borosilicate glass. Films evaporated onto cold substrates have a somewhat rougher homogeneous microstructure due to the lower degree of

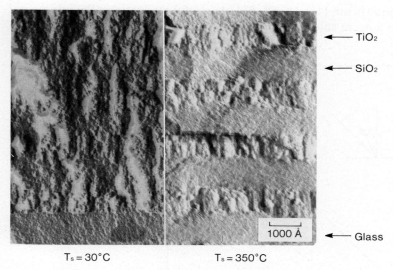

Fig. 12

Transmission electron micrograph of Pt/C surface replicas of the cross section of TiO₂/SiO₂ multilayers deposited by reactive evaporation onto glass substrates of 30°C and 350°C [50].

polymerization of the SiO_2 components. The lateral resolution of the Pt/C replica method ($\leqslant$ 50 Å) used was not sufficient to determine small pores. However, a loose, sponge-like consistency can be assumed from water-vapour sorption measurements. The vapour-deposited films of almost any materials are porous. The degree differs but can be influenced by the production conditions. The porosity which produces considerably larger inner surfaces and causes a reduction of density [45] can be recognized from the profile pictures of films which consist of discrete columns. The column diameters of some evaporated films measured from the electron micrographs are given in Table 1.

TABLE 1

COLUMN DIAMETER OF FILMS EVAPORATED AT NORMAL INCIDENCE ONTO GLASS SUBSTRATES AT ROOM TEMPERATURE

Film material	Ag	MgF_2	$Na_3(AlF_6)$	ZnS	TiO_2*
Geometrical thickness (nm)	52	115	90	70	68
Column diameter (Å)	280 – 350	130 – 200	160 – 230	240 – 280	150 – 200

*T_S = 350°C

Contrary to the findings of Pearson [49], which refer to the observation of micrographs of the fracture cross-sections of ZnS/Cryolite multilayers, the cross-sections of ZnS/MgF_2 and TiO_2/SiO_2 multilayer systems, investigated by Pulker and Günther [50], showed no recognizable influence of the microstructure on the subsequent layers. According to measured data, there is also no large deviation in the average column diameter with increasing number of layers, although nucleation no longer takes place on a fairly plane surface even from the second layer. Most films show rough surfaces . The roughness of a film surface may be statistical but it can also arise in the case of films formed from discrete crystalline columns as a result of the various speeds of growth (caused by the various crystallographic directions) of the single columns. Because of the roughness of the film surfaces in the multilayer systems, there are no sharp transitions between the single layers. There are instead boundary zones of varying thickness, where a gradual change in material properties takes place in the interfaces.

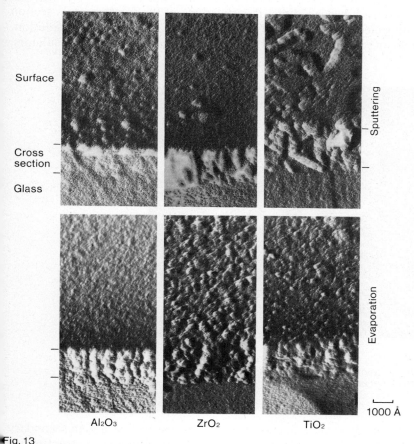

Surface

Cross section

Glass

Sputtering

Evaporation

1000 Å

Al_2O_3 ZrO_2 TiO_2

Fig. 13

Transmission electron micrographs of various oxide films deposited onto glass substrates at Ts = 350°C by reactive evaporation and by reactive sputtering (BALZERS Sputron). Differences in film microstructure and surface topography can be seen.

As can be seen in Fig. 13, dielectric films of about 100 nm thickness deposited reactively in a triode sputtering system show quite different microstructures compared with those prepared by high vacuum evaporation. Thus, the columns in sputtered TiO_2 films are irregularly packed and the film surfaces of TiO_2 and ZrO_2 films show areas which are rather smooth. In sputter-deposited Al_2O_3 films, no columnar microstructure was detected. Furthermore, it seems that in general ion-assisted physical-vapour-deposition technologies, especially with biased substrates, yield films of high density. Oxide films deposited using one of the low-temperature chemical-vapour-deposition technologies, however, are generally less dense and show often amorphous microstructures.

As regards the observation techniques for microstructure determination, it was found that transmission electron microscopy of cross-section surface replicas is generally required for thin and fine-grained films. With thicker and coarse-grained films, however, direct investigations of the fracture cross-section with the scanning electron microscope can also be used with sufficient resolution. Figure 14 shows an example. To avoid disturbing charging effects with insulators,

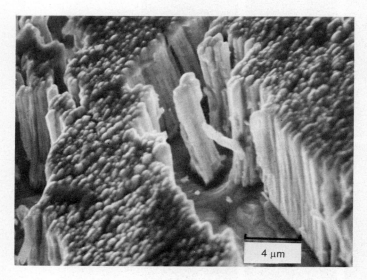

4 µm

Fig. 14
Scanning electron micrograph of a 6 micron thick rhodium film showing large columnar crystals.

and to obtain a high electron reflection coefficient η, as well as a high secondary emission coefficient δ, the samples are coated with thin and almost structureless heavy metals (e.g. Pt, Pd, Ir) [53]. Movchan and Demchishin [54] showed that the microstructure of metals and dielectric films is determined mainly by the ratio of substrate temperature to the melting point of the film material, T_s/T_m. In the case of a ratio below about 0.45, pronounced columnar microstructure will always be obtained with columns formed in the direction of growth. This important criterion is almost invariably satisfied with coatings on glass [55]. As regards the question of the origin of the frequently observed columnar film microstructures, it seems that the most promising theory is one in which the significant factors are limited mobility of the condensing atoms or molecules, together with a shadowing of uncoated portions of the substrate by atoms or molecules already condensed [52, 56, 57]. Very interesting hypothetical studies on film deposition with zero or limited particle mobility simulated by a computer resulted in loosely packed chain like structures and in dendritic chains of several molecules diameter [52, 57]. Recently made computer-simulated deposition investigations have been extended, also taking into account substrate surface defects, in the search for an explanation of the sometimes observed growth of nodular film defects [58]. However, the dendrite diameters are much smaller than the column diameters actually observed.

To overcome this difficulty, results of experimental film-growth investigations by Messier and co-workers [59-61] led Messier to assert the following model: In the first stages of film deposition, small structural basic units acting as the basis for later growing dendrites are formed and the voids in the films between these basic units are of appropriate size. During further film thickness growth, the structural units tend to cluster into larger groups. The voids in the group are closed, such that larger voids now appear between the groups rather than between the dendrites, and this groups of dendrites form the observed columns. The agglomeration to larger columns, following that mechanism, proceeds with increasing film thickness. This is now in accordance with experimental observation. The void content is often higher in dielectric films than in metal films. This becomes obvious when comparing the packing density p of the deposits, which is defined as:

$$ p = \frac{\text{volume of the solid part of the film}}{\text{total volume of the film (solid + voids)}} \tag{4} $$

Columnar or dendritic microstructure is found also in most metal and many compound films formed by chemical vapour deposition in certain ranges of condition. As a result of the same mechanism of uninterrupted crystal growth towards the direction of material supply it is further found in films obtained by some deposition techniques from solution, particularly in electroplated films.

It can be stated that the microstructure of most thin films is quite different from that of the bulk material. It is therefore not surprising that the properties of thin films differ from that of the bulk. This is essentially true for the density of the films, for roughness, sorption and diffusion behaviour (water-vapour uptake, diffusion along inter-columnar space), optical and electrical behaviour (increased light scattering, (form) birefringence and reduced electrical conductivity), and for mechanical properties (such as relatively high film stresses due to grain boundary interactions).

8.3 CHEMICAL COMPOSITION

8.3.1 SURFACE ANALYSIS

There are many different methods which can be used to determine the chem. composition of thin films, ranging from classical chemical microanalysis to the more powerful modern physical analysis techniques. The material available to perform the analyses is generally only a few milligrams, and in such small quantities an impurity content of 0.1 % or less should be found qualitatively and often even be determined quantitatively. Therefore, generally, detecting sensitivities ranging from 10^{-6} (ppm) to 10^{-9} (ppb) are required. Impurities and other chemical elements in low concentrations can be detected in thin films by (listed with increasing detecting sensitivity) ultraviolet spectrometry, neutron-induced

TABLE 2

EXCITATION AND EMISSION PROCESSES WHICH CAN BE USED FOR THE ANALYSIS OF SOLID SURFACES AND THIN FILMS [88].

Excitation	Emission analysis		
	Photons	Electrons	Ions
Photons	XRF IRS	ESCA (UPS) XAES	
Electrons	EPMA (APS) (CL)	AES HEED (LEED)	
Ions	IIR (IIX)	(IAES)	SIMS ISS (SSMS) (RBS)

In Table 2, the following abbreviations for the various investigative techniques are used:

XRF	=	X-ray fluorescence	SSMS	=	spark-source mass spectrometry
ESCA	=	electron spectrometry for chemical analysis	IRS	=	infrared spectrometry
			UPS	=	ultraviolet photon spectrometry
XAES	=	X-ray-induced Auger electron spectrometry	EPMA	=	electron probe microanalysis
APS	=	appearance potential spectrometry	CL	=	cathode luminescence
			IIX	=	ion-induced X-ray
IIR	=	ion-induced radiation	HEED	=	high-energy electron diffraction
AES	=	Auger electron spectrometry	IAES	=	ion-induced Auger electron spectrometry
LEED	=	low-energy electron diffraction	ISS	=	ion surface spectrometry
SIMS	=	secondary ion mass spectrometry	RBS	=	Rutherford back scattering

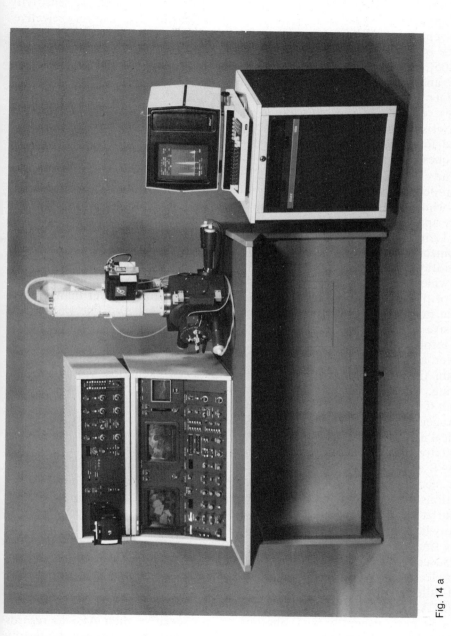

Fig. 14 a

Scanning Electron Microscope PHILIPS SEM 505 equipped with LaB$_6$-Cathode, Specimen Current Detector, Back-scattered Electrons Detector, Signals Mixing Unit, Computer-Interface, SDR-Unit and EDAX PV 9100 Energy Dispersive X-Ray Analyzer.

activation analysis and the various techniques of mass spectrometry [62-65]. The first method can be applied with about 70 elements and has a detecting sensitivity in the ppm range. The activation analysis is very sensitive but also very selective. For all chemical elements, however, mass spectrometry can be applied. With samples of a few micrograms, the detecting sensitivity lies in the ppb range [103-105]. Very high detecting sensitivities can be achieved by the use of modern surface analysis techniques, some of which were discussed in Chapter 3 Section 3.1.2. Of numerous interaction processes of photons, electrons and ions with solid surfaces resulting in the emission of excited particles and radiation, or in a characteristic change of the incident species properties, those which can be applied for surface analysis are listed in Table 2. The most frequently used techniques are boxed in the table.

The electron probe microanalysis, EPMA [66-68], can be successfully used for non-destructive chemical investigations of films and film systems in all cases where whole-sample information (qualitative as well as quantitative) is sufficient. Surface analysis is not possible because of the large depth yielding information of about 1 μm. The most pronounced advantage of the microprobe is the possibility of connecting it with a simple scanning electron microscope. Of the two possibilities of analyzing the emitted X-rays, the energy dispersive analysis is in much wider use than the wavelength dispersive technique. However, the detection range of the first technique is usually limited to elements heavier than sodium (atomic number 11) and the resolution is only $120 - 150$ eV. The wavelength dispersive analysis detects lighter elements down to boron (atomic number 5) and offers resolutions of 1 eV. A quantitative analysis of thin films on a substrate is, however, a little complicated because of the relatively small excitation volume in the thin films, compared to the widely broadened excitation volume in the substrate, which causes high background radiation. The situation becomes less difficult if unsupported thin films can be prepared for the analyses.

Auger electron spectrometry, AES [69-74], and electron spectrometry for chemical analysis, ESCA [75, 76], are well established techniques of performing quantitative analysis of the outermost atomic layers of solid surfaces [77].

AES enables elemental identification by means of energy analysis of the Auger electrons which are emitted as a consequence of the radiationless Auger transitions when an atom is ionized at a core level, e. g. KL_1L_2 transitions. AES is a true surface-sensitive technique, because the mean free path of Auger electrons lies only between 0.5 and 3 nm [72]. However, as the number of Auger electrons is very small compared to that of the secondary electrons, they can be detected more successfully using the first derivative dN/dE (N = number, E = energy of the Auger electrons). Variations in chemical binding can be detected as peak shifts, the so-called chemical shifts.

ESCA, using the emission of electrons from a solid surface by excitation with X-rays or more generally with photons, is less sensitive than AES as regards impurity detection when excitation is performed with X-rays, since the deeply penetrating X-rays do not generate as many core holes in the surface region. However, an advantage of ESCA is that X-rays produce less severe damage than electrons, so that ESCA is especially useful in the analysis of organic materials. Furthermore, chemical shift of the core levels enables conclusions on the binding to be more easily drawn for this one-electron process than for the complicated three-electron process of AES. When excitation is performed with ultraviolet light in the range of 4 to 40 eV (UPS) instead of X-rays of 1200 to 1500 eV, only weakly bound valence-shell electrons react. Such electron spectra are sensitive to molecular orbital effects and are used for their investigation, but they do not possess elemental significance.

Secondary ion mass spectrometry, SIMS [78-82], involves the collection and analysis of positive and negative secondary ions, which are ejected from a surface which is bombarded with an ion beam. This technique is very sensitive but can be destructive for chemical compounds. Nevertheless, it can yield excellent qualitative information. However, quantitative results are difficult to obtain.

8.3.2 DEPTH PROFILING

Unless non-destructive Rutherford backscattering spectrometry is used [83-85], only composition information of a single film or of a multilayer system on a substrate can be obtained by destructive profiling techniques. Depth profiling can be performed with various techniques and in various ways depending on the total thickness profile which has to be analyzed [86, 87, 88].

Cross-sections of thick coatings, such as $\geqslant 2\ \mu$m, on flat surfaces can be obtained advantageously by taper-sectioning. A taper is polished in the surface and the angle-lapped region exposes the entire depth to be analyzed [89-92]. It is generally necessary to use a grazing angle α in the range of 0.1° to 1°. Thus, the real film thickness t appears in the cross-section c enlarged by $c = t/\sin\alpha$. The technique is useful but the provision of an angle of less than 1° is very difficult.

A similar enlargement results from ball-cratering, a technique developed recently to overcome many of the difficulties associated with taper-polishing [93, 94]. In ball-cratering, a rotating steel sphere covered with fine diamond paste is indented into the coating surface. The radius of the ball R is known, so that the depth d of the crater produced is given by $d = D_2^2/8R$, where D_2 denotes the upper (film/atmosphere) crater diameter. With D_1 as the lower (film/substrate) crater diameter, the film thickness t can be obtained according to $t = (D_2^2 - D_1^2)/8R$.

These two techniques demand fine spatial resolution which is achievable with AES but not with ESCA. In both cases, the compositional depth profile is then conveniently obtained using a narrow electron beam in a line scan instrument such as a scanning Auger microprobe, SAM [95].

The maximum vertical depth Δz analyzed using an electron beam of diameter b is

for taper-polishing: $\qquad \Delta z = b \tan\alpha \qquad\qquad$ (5) and

for ball-cratering: $\qquad \Delta z = (b/R)\,[2R(d-y)]^{1/2} \qquad$ (6)

y is the depth at the point of analysis.

Further techniques to investigate compositional depth profiles of thicker films or film systems are stepwise-erosion by spark [96] or by laser pulses [97]. One shot erodes less than 0.1 μm film material, however, a proper uniformity in erosion is hard to obtain.

Sputter-depth profiling is the most frequently used technique and is ideal especially for thin-film analysis. In a review paper by Günther [88] on «Non optical characterization of optical coatings», one of the topics is depth profiling. The following descriptions on sputter profiling are taken partly from this review.

Sputter-depth profiling means sputter etching. This technique is convenient as ion beams of only a few keV energy are used. The incident ions (Ar+) transfer some of their energy to target atoms or molecules (i.e. those of the films to be analyzed) by multiple collisions. As these target atoms collide with others, a collision cascade is generated, and atoms which gain more than the surface binding energy will be sputtered [98, 99]. Thus, atoms and/or molecules are removed in succession from the solid, and a new surface is continuously created which can be analyzed by common surface analysis techniques (ESCA, AES). If the identification of sputtered particles (by SIMS or IIR) is used for compositional depth profiling, the influence of particles emitted from the crater edge must be taken into consideration [100-102].

Certain precautions must be taken when interpreting depth profiles obtained by sputter etching, since artefacts can arise. The collision cascade will cause an intermixing of constituents from sections closer to the surface together with those in deeper sections, and vice versa. This collisional mixing results in a broadening of a given compositional profile [103].

The depth resolution Δz in sputter profiling can be derived from a simple sequential layer sputtering (SLS) model, see [104], as

$$\frac{\Delta z}{z} \propto z^{-1/2} \tag{7}$$

z being the sputtered depth [105, 106]. Although this relationship has been found to be in good agreement with a number of experimental results, other measurements show a dependence

$$\frac{\Delta z}{z} \propto z^{-1} \tag{8}$$

This dependence can be described by an extended SLS model including surface transport [107]. In addition to sputter removal from the surface layer, surface atoms can be transferred into deeper layers, also known as the knock-on effect in ion sputtering [108].

Preferential sputtering [109, 110], caused in multicomponent targets by different sputtering yields of the surface constituents, may cause systematic errors in sputter profiling but will not necessarily limit depth resolution [103]. However, the depth resolution is sensitive to effects of surface roughness, which will be enhanced or developed during ion etching [111]. Sputter-induced surface roughness can be reduced using two ion guns, each inclined to the surface normal symmetrically [112].

Because of the different and mostly unknown sputtering coefficients of the various coatings, it is necessary to calibrate sputter profiles – which are usually plotted as a function of sputter time – in order to obtain true compositional profiles as a function of depth. The profiles can be calibrated in various ways [113]. For optical interference coatings, it is best to take the layer sequence from the raw profile and to adjust the thicknesses in a manner such that the computed spectral characteristic of the overall layer system fits the measured one [114, 115]. Another possibility is to measure the depth of the etching crater interferometrically or with a stylus instrument after a defined period of time. Often it is not necessary to know the exact thickness of the layers contained in an optical coating, but knowledge of the chemical composition and comparison with standard designs can give important clues about the cause of a faulty coating. Obviously, ion etching and electron-beam analysis of dielectrics is more difficult than that of metals. Problems arise with the distortion of thin films and glass surface layers which may result from ion-beam [116, 117] and electron-beam [118-120] irradiation. Alkaline and earth-alkaline compounds, such as MgF_2, dissociate strongly even under relatively weak electron irradiation [121, 122]. Alkali ions (sodium) in the surface region of glass substrates or in dielectric thin films may move during ion bombardment or because of local heating effects [118, 123].

Very peculiar to dielectrics are charging effects, which may be caused either by immission or by emission of electrons or ions. Charging can seriously influence the resolution and the sensitivity of almost every analysis technique based on excitation by and detection of charged particles. Hence, special precautions to prevent or minimize charging of dielectric samples must be taken. Numerous experimental tricks have been developed for this purpose: embedding of the sample in conductive material like silver paste [124], shielding with a thin foil or a diaphragm (such as aluminium, tantalum) [121, 125], or depositing a conducting grid [126]. Conductive coatings (silver or gold) may also be used, not only with scanning electron microscopy, the usual case, but also with depth profiling by ion etching. The intimate contact of a silver coating with the dielectric surface and a ground contact enables the excess charge to diminish. The silver coating can easily be removed inside the ion spot because of its high sputter rate [127]. The use of an auxiliary electron gun or a filament emitting low-energy electrons which flood the dielectric surface for charge compensation is not necessarily successful. Sometimes, only proper adjustment of the etching ion beam together with the primary analyzing electron beam, backed by some experience, will yield useful analysis results [121].

To perform depth profiling of any desired sample, as well as AES, ESCA or SIMS equipment, the universal ion-etching device BALZERS IEU 100, shown in Fig. 15a, can be used [128]. In order to be able to carry out many of the possible preparations and analytical techniques, the IEU 100 must be equipped with an ion density profile probe for controlled adjustment of the focused and the parallel ion

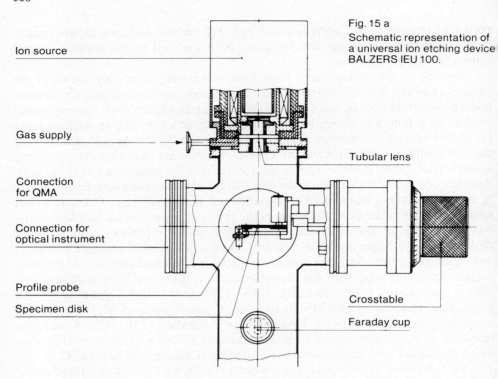

Ion source

Gas supply

Connection
for QMA

Connection for
optical instrument

Profile probe

Specimen disk

Tubular lens

Crosstable

Faraday cup

Fig. 15 a
Schematic representation of
a universal ion etching device
BALZERS IEU 100.

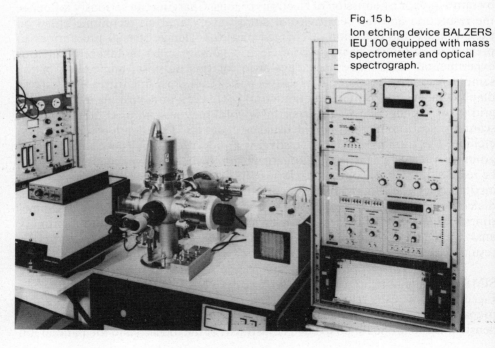

Fig. 15 b
Ion etching device BALZERS
IEU 100 equipped with mass
spectrometer and optical
spectrograph.

beams and the device should be supplied with a quadrupole mass spectrometer for element analysis or for analysis of the concentration profile. Furthermore, one can attach a spectrograph and where appropriate a combination of line filters and light-guiding fiber optics with photomultiplier equipment for spectrochemical analysis. A fully equipped IEU 100 is shown in Fig. 15b. By recording the current of the secondary ions and/or the intensity of the emitted photons, both of which originate at the bottom of the etch pits during ion bombardment, the material can be analyzed. Beside depth profiling with film systems and bulk materials, sample areas of a few microns to several millimeters in diameter can also be thinned to a thickness that makes the material transparent to light. The thinned samples are best suited for various electron microscopical investigations and their light transmittance enables in addition the performance of ultraviolet or infrared absorption spectra measurements.

To illustrate the application of such combinations we present results of an investigation obtained from the application of profile analysis in the IEU 100 and reflection electron diffraction, as well as the energy dispersive analysis of a surface layer.

The layer to be investigated consists of three oxide films: TiO_2, SiO_2 and TiO_2 each with an optical thickness of $\lambda/4$. The layers were taken from an organic solution and transferred to a soda lime glass substrate through pyrolysis. Reference to [28] of Chapter 6 should be made, for the production of such layers. In addition, the bottom layer contained sodium in a concentration of several mol %. This information could be taken from the intensity/time profiles obtained from spectral analysis in the visible region with continuous ion beam etching, compare Fig. 16a. It is now also of interest to determine whether or not the Na contained especially in the bottom layer is part of a compound, and whether any compound present can be identified. Earlier investigations showed that at least a part of the Na occurs

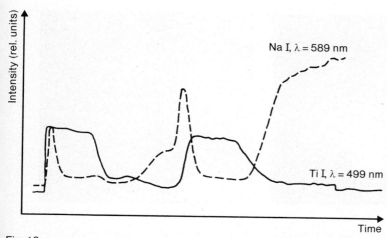

Fig. 16 a

Concentration profile of Na and Ti in a triple coating of $TiO_2/SiO_2/TiO_2/Glass$.

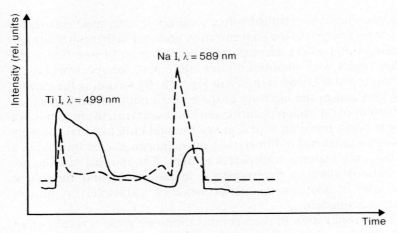

Fig. 16 b
Concentration profile to establish relative etching depths.

in the compound Na_xTiO_2 [27] of Chapter 6. By etching areas on the surface with a beam of parallel ions, it is possible to expose a plane of the bottom layer.

The depth of the exposed bottom of the etch pit, i. e. the position of the etched areas under the surface, can be adjusted using the concentration profile. Figure 16 b shows the Ti concentration profile obtained when the bottom TiO_2 layer is etched. A comparison of the entire profile in Fig. 16b with that in Fig. 16a shows clearly that a bit more than a third of the bottom layer was etched away. If the film thickness is not known, the depth distribution can be calibrated by interferometric measurement of the etch pit depths. Figure 16c shows the reflection electron

Fig. 16 c
Reflection electron diffraction pattern of the etched surface shown in Fig. 16 d (Na_2O enrichment).

Fig. 16 d
Scanning electron micrograph of the etched film in a depth corresponding to the concentration profile of Fig. 16 b.

diffraction pattern of this plane of the specimen. Surprisingly, the reflection for Na_2O (according to ASTM chart 23-528 A) appears in this pattern. In a simultaneously conducted transmission investigation of the same layer, in contrast, the diffraction rings for brookite or of Na_2TiO_3 were almost exclusively found and Na_2O only occurred sporadically. This result was obviously caused by the different preparation methods. Here, however, the diffraction rings of the TiO_2 modification recede completely in the reflection diffraction picture because Na_2O has a much lower sputtering rate and thus rises out of the etched plane shielding the TiO_2 crystals during the taking of the reflection diffraction picture. Also illustrating this is Fig. 16d, which is a SEM magnification showing, in addition, that the yield of secondary electrons from the Na_2O is much higher than that of the environment.

The investigation of the lighter areas in the picture by energy dispersive analysis of the excited characteristic radiation confirms the validity of this interpretation of the picture. Furthermore, the same considerations are valid here as for bulk specimens. The intensity/time profile shown in Fig. 16a shows the concentration profile with a distortion caused by the different resistances of sputter rates TiO_2 and Na_2O. Meanwhile, the convolution function taken from the intensity/time profile is relatively easy to determine [129].

8.4 MECHANICAL PROPERTIES

The mechanical properties of thin films produced by various deposition techniques are strongly dependent on structure, microstructure, chemical composition and in particular incorporated impurities. They can thus be influenced by the production technology used and the parameters chosen. The published data concerning mechanical properties are not as extensive as those concerning optical, electrical and magnetic properties, and although there are some excellent reviews on this topic, see for example [130] and [131], new insights have been gained in the last decade.

At this time, we are beginning to understand, at least in principle, the mechanical properties of metallic and some pure and stable compound films. It has become possible to calculate the tensile stress in polycrystalline films using elastic constants as well as structural and microstructural data of the films. The agreement between measured and calculated stress values of pure films is rather better than expected in such cases. Our knowledge concerning energetic interactions on grain boundaries has qualitatively and quantitatively improved. Although practically there are only very few measured data on surface free energy available, it is possible to influence in a semiquantitative way the stress behaviour of the films by changing the surface free energy of the crystallites composing the films. Unfortunately, knowledge on sometimes less stable compound films is much more rudimentary. Their data seem to depend very sensitively on the preparation conditions; also stoichiometry, structure and microstructure are not so well characterized. Such films often show compressive stresses. Although there are some model conceptions for the origin of compressive film stress, no quantitative treatment of this problem exists.

The importance of adhesion in preventing failure is well known and generally accepted, but also here quantitative treatments are difficult to find, as becomes obvious in studying Chapter 5, Section 5.2. The same is found with hardness and density of thin films.

8.4.1 STRESS

The stress behaviour of films is very important in all applications of thin films with respect to durability and hence use. This fact was recognized very early on. The first investigations regarding mechanical stresses in thin films were carried out in 1877 by Mills [132] on chemically deposited films. Thirty years later, more quantitative studies were undertaken by Stoney [133]. During further investigation, it was soon found that almost all films, whatever the production method, have mechanical stresses. Tensile and compressive stresses should be distinguished; tensile stresses arise if the film tends to contract parallel to its surface. Analogously, compressive stress is present if the film expands parallel to the surface.

For films produced by PVD techniques, it is approximately true that with metal films deposited on a substrate at room temperature and even higher, tensile stresses are predominant. The magnitude of the tensile stresses is between 10^8 and 10^{10} dyn cm^{-2}. Films of high-melting metals have the larger values, those of lower melting metals the smaller values in this range. Reactively deposited dielectric films and metal films modified by chemical and/or physical gas incorporation frequently have compressive stresses. Compared with metal films, dielectric films often have smaller stress values.

The presence of mechanical stresses in the films is usually undesirable. With good adhesion of the film-substrate system, strong deformation of the substrate often takes place. If the stresses are very high, depending on sign, then because of transient effects in the case of overstressing of the elasticity, undesired crack formation or blistering can occur in the films. In the worst cases, the films can even become completely detached from the substrate.

The observed total mechanical stress can consist of three components according to eqn. (9)

$$\sigma = \sigma_{ext.} + \sigma_{therm.} + \sigma_{intr.} \qquad (9)$$

The external stresses (σ ext.) arise from attack by external forces, for example in the case of poor assembly of a component supporting the film. Thermal stresses (σ therm.) always arise if large differences exist in the thermal expansion coefficients between film material and substrate material. Strong effects can occur even during film production, or the later warming up or cooling down. The portion of thermally caused stresses of the total stress (σ) can be quite considerable in the most unfavourable cases.

The last and most important part is the intrinsic stress (σ intr.) in the film itself. Intrinsic stress is a structure- and microstructure-sensitive property which is caused by the mode of film growth and microstructural interactions also influenced by some contamination. Usually, the intrinsic stresses in the films are the dominant part of the total stress so that many investigations have specifically dealt with them. The occurrence of mechanical stresses even in island films is most remarkable. During further growth, bridges between the islands are observed in the network stage of film growth and finally a closed film is formed. Distinct changes in stress accompanies the agglomeration of the single crystallites in particular. In addition to lattice disorder effects, changes in density and electrostatic charges have been considered [134-143]. Depending on material and deposition conditions closed films can be vitreous or amorphous, but most are crystalline, particularly polycrystalline. Considerations concerning the interactions between grain boundaries and boundary surfaces were regarded for explaining the causes, and especially the occurrence, of tensile stresses [144]. Frequently, pure films have tensile stresses, whose values depend on the kind of material, the deposition technique and the film thickness. Often the stresses reach a maximum when the last holes in the network film are filled. Additional thickness growth causes, along with recrystallization, a large decrease of the stresses, as can be seen in Fig. 17. The inclusion of foreign materials in the films can cause not only a reduction of tensile stresses but even a shift to compressive stresses, as can be seen in Fig. 18. Most especially, gas incorporation into a metal lattice often leads to compressive stresses resulting from a constraint metal lattice expansion.

A film deposited on a thin substrate deforms this substrate because of the mechanical film stress. The deformation can be measured and, knowing the dimensions and the elastic constants (E = Young's modulus, v = Poisson's ratio) of the substrate, it is possible to calculate the film stresses.

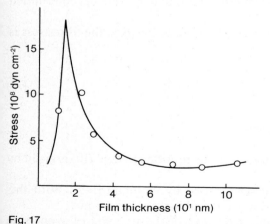

Fig. 17

Intrinsic mechanical stress of an Ag-film on glass as function of film thickness [145].

344

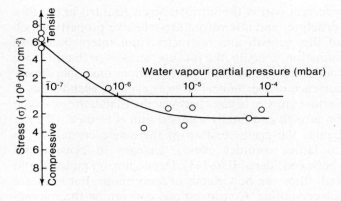

Fig. 18

Intrinsic mechanical stress of SiO$_x$ films as function of the residual water vapour pressure during deposition [146].

In the case of tensile stresses, a concave film surface forms; in the case of compressive stresses it is convex. The mathematical treatment of this problem was started by Brenner and Senderoff [147] and continued among others by Hoffman [144]. Various measuring methods [148a] have been derived for film stress measurement as regards substrate deformation; most commonly used are the bending of a disk and the bending-beam methods. The measurements can be carried out during or after film deposition. With the disk method, the deflection of the centre and the warping of the brim of a circular disk of glass, quartz or metal (initially plane) are measured under the influence of the stress produced by a deposited film. An optical plane test glass serves as reference. The Newton's interference rings arising in monochromatic light between the test glass and the disk are measured by a known method and the radius of curvature r is determined from this.

With the radius of curvature and the elastic data of the disk, the film stress is calculated according to:

$$\sigma = \frac{1}{6} \frac{E_s}{r(1-v)} \frac{t_s^2}{t_f} \tag{10}$$

t_s = substrate thickness, t_f = film thickness.

A review of the mathematical considerations resulting in eqn. (10) is given by Seidel [148b].

Figure 19 shows a measurement set-up used by Hoffman [149]. With this method, anisotropies can also be detected in the film-stress behaviour.

With bending-beam methods, the deflection of the free end of a one-side clamped substrate strip or the central deflection of a two-sided clamped strip is measured, which is produced under the influence of the film stress. The strip or

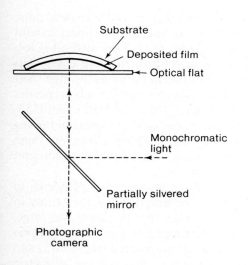

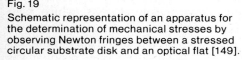

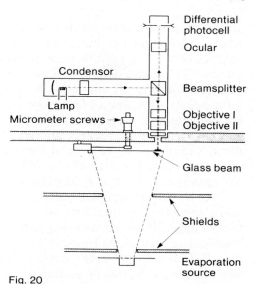

Fig. 19

Schematic representation of an apparatus for the determination of mechanical stresses by observing Newton fringes between a stressed circular substrate disk and an optical flat [149].

Fig. 20

Schematic representation of an apparatus for the determination of mechanical stresses using the bending beam technique with optical detection [150].

beam length is often ten times its width. The deflection is detected either optically, mechanically or electrically. The sensitivity achievable depends to a large extent on the system of detection. Highest sensitivities are achieved with optical and electrical detectors.

The radius of curvature of the beam is calculated from:

$$r = \frac{l^2}{2\delta} \tag{11}$$

l = substrate length, δ = deflection of the free end.

The stress is calculated according to eqn. (10). By inserting (11) into (10), eqn. (12) results:

$$\sigma = \frac{E_s t_s^2 \, \delta}{(1-\nu)3\, l^2 t_f} \tag{12}$$

A measuring apparatus with optical detections is shown schematically in Fig. 20 [150].

X-ray and electron diffraction methods are applied in order to measure atomic distances in the crystal lattice and their changes. Hence, the diffraction methods are also basically suitable for measuring the strain/stress behaviour in thin films. However, since the film thickness and the crystallite size in thin films are small, some line broadening already arises from this. In order to determine what contribution the mechanical stresses have on the diffuse lines, careful analysis of the line profiles must be undertaken [148, 151]. This method is less suitable for routine determination of stresses in thin films. In some cases, it is possible though rarely applied to determine the stresses in the films through their influence on other, known film properties, at least approximately. Such properties are, for example, the position of an absorption edge [152], the Hall effect [153, 154], electron spin resonance spectra [155] and in the case of superconducting films, variations in the critical transition temperature [156]. However, these effects can, unfortunately, also arise for other reasons, and thus these techniques can usually only be used as supplementary experiments.

As already mentioned, the stresses in thin films may present problems in many industrial applications of the films. Early observations of thin films in optical applications showed that particularly when the film thickness was large, cracks arose, which had cloudy marks and sometimes the films even became detached from the substrate. Exact measurements of evaporated single films and film systems indicated a partial stress compensation, especially in dielectric multilayers, since low refractive films often have tensile stresses, whilst high refractive films have compressive stresses [157, 158].

Furthermore, possibilities of influencing through manufacturing parameters were found. These are substrate temperature, condensation rate, residual gas composition and pressure and angle of incidence of the vapour beam.

For the theoretical interpretation of film stresses of crystalline films, interaction across grain boundaries often appears to be the dominating effect. Hoffman et al. [144] tried to explain the tensile stress behaviour of Ni films through a grain boundary model, and Pulker et al. were successful in interpreting the tensile stress behaviour of pure and doped MgF_2 films [159 a-c] and of pure Cr films [166] by analog mathematical treatments.

Because of its significance, the grain boundary model is explained briefly in the following.

Lattice relaxation forced by the energetic interaction between grain boundaries presents the elastic deformation which expresses itself macroscopically as observable mechanical stress. The close approach of two free surfaces (surface energy γ_s) in the formation of a grain boundary (grain boundary energy γ_{gb}) results in a change in energy as follows:

$$\Delta\gamma = 2\gamma_s - \gamma_{gb} \tag{13}$$

In the case of large-angle grain boundaries, $< \gamma_{gb} >$ equals $1/3$ γ_s and therefore the change in energy can be calculated according to $\Delta\gamma = 5/3$ γ_s. Part of the energy produces a constrained relaxation of the lattice atoms in the grain boundary, yielding a few percent of the unstrained lattice constant a. The elastic deformation is responsible for the macroscopically observed tensile stress (σ), according to Hooke's law

$$\sigma = E \, \varepsilon \tag{14}$$

$$\varepsilon = (x-a) / a = \Delta/d \tag{15}$$

The average value of Δ was calculated using the method of the interaction potential between two atoms [144].

The next possible separation of two particles, r_i, is generally limited by their ionic radii. The separation at which they can enter into interaction with each other is double the lattice constant. The energetic minimum, as can be seen in Fig. 21, is exactly at the lattice constant and its size is given by the above mentioned $5/3\gamma_s$. Now, only particles which arrive between r and 2a are considered, and a uniform probability is assumed for every position between these two limits. Then, it only rarely arises that a particle occupies the energetically favourable position at a. However, in an effort to nevertheless achieve this energetically favourable position, an expansion of the film takes place if a particle occupies a place between r and a, and a contraction occurs if a position between a and 2a is occupied. However, in both cases the energy difference arises in the form of deformation energy.

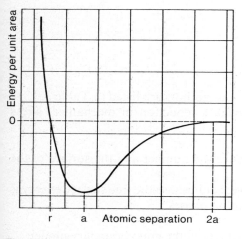

Fig. 21
Grain boundary potential after Doljack, Springer and Hoffman [144].

Since the potential is asymmetric, the predominant occurrence of tensile stresses can be explained.

The deformation energy per surface unit of the film is calculated, added to the potential energy per surface unit, and the new energy minimum ascertained. The difference in the two minima now gives a dimension for the magnitude of the deformation. The result is:

$$< \Delta > = \frac{\bar{a}\, r_i - r_i^2/2}{2\bar{a} - r_i} \tag{16}$$

$\bar{a}$ = average bulk lattice constant
r_i = the distance of closest approach (sum of the ionic radii).

The average force per unit area of boundary after relaxation must just equal the average elastic stress produced throughout the volume of the crystallite. Considering also Poisson's ratio v and the packing density of the film p, the intrinsic stress is given by

$$\sigma = \frac{E}{1-v} \frac{\Delta}{d}\, p \tag{17}$$

Multiplication by packing density p takes into account the presence of grain boundary separations exceeding $2\bar{a}$. For tensile stresses in MgF_2 films, good agreement was obtained between calculation and experiment [159 a-c]:

$\sigma = 6.7 - 7.3 \times 10^9$ dyn cm^{-2}
MgF_2 calculated

$\sigma = 6.9 - 7.1 \times 10^9$ dyn cm^{-2}
MgF_2 experimental.

The experimentally obtained results of various authors are listed in Table 3.
According to the grain boundary model, a reduction in mechanical stress arises if:

1. the crystallite size increases and hence the number and the surface of the grain boundaries reduces, and/or
2. the energetic interaction becomes weaker because the free surface energy is decreased owing to sorption or segregation phases.

Case 2 follows from a purely energetic treatment of the grain boundary model:

$$\sigma = (\frac{E_f}{1-v} \frac{4\gamma_s - 2\gamma_{gb}}{d_f}) \tag{18}$$

As presented in Fig. 22, a more or less fast evaporation at various pressures and varying amounts of water vapour in the residual gas by sorptive water incorporation into the film grain boundaries results in strong changes in stresses. However, the water adsorption depends on the partial pressure. For technical

TABLE 3

INTRINSIC TENSILE STRESS OF 100 nm THICK MgF$_2$ FILMS DEPOSITED ON UNHEATED GLASS SUBSTRATES

Authors		Residual pressure during deposition (mbar)	Deposition rate (nm s^{-1})	Intrinsic stress values measured under vacuum $(10^9 \text{ dyn cm}^{-2})$ $(1/1-\nu_s)$		Decrease by H$_2$O sorption during air inlet (%)
				published values	corrected values	
Schröder Schmidt	[170]	4×10^{-5}	5 10	3,3 3,8	4,2 4,9	60 – 110
Kinosita	[171]	2×10^{-5} to 4×10^{-5}	0,1 – 0,5 1 – 1,5	2,9 3,3	3,6 4,2	– –
Ennos	[158]	5×10^{-5} to 5×10^{-6}	4	4,8 to 5,6	5,7 to 6,7	30 – 40 –
Mäser Pulker	[159b]	5×10^{-6}	0,5	3,0	3,0	15 – 60
Kinbara	[169]	2×10^{-6} to 8×10^{-6}	0,5 – 1,5	2,4	3,1	–
Seidel Pulker	[172] [159c]	2×10^{-6} to 8×10^{-6} 2×10^{-5} to 5×10^{-5}	0,5 1,5 3,0 – 6,0 0,5 1,5 3,0 6,0	– – – – –	3,2 5,8 6,9 3,1 5,3 6,1	50 – 90 45 – 85 30 – 80 60 – 115 50 – 100 40 – 90

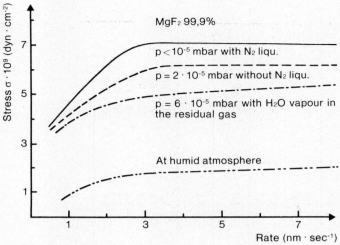

Fig. 22

Intrinsic stress values of pure MgF₂ films on unheated glass
substrates as function of the deposition rate and of the water
vapour partial pressure.

applications, it is therefore advisable to aim at foreign material coverage of
substances of low vapour pressure and low free surface energy. In the order: MgF_2,
CaF_2, BaF_2, H_2O, the free surface energy decreases.

Monomolecular segregation phases of CaF_2 and BaF_2 at the grain boundaries
of a MgF_2 film also distinctly reduce the mechanical stresses, as shown in Fig. 23.

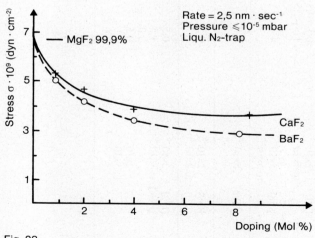

Fig. 23

Intrinsic stress values of MgF₂ films doped with CaF₂ and BaF₂
deposited on unheated glass substrates as function of the
doping.

TABLE 4

INTRINSIC TENSILE STRESS OF METAL FILMS

Thickness (nm) Material	10	15	20	40	60	80	100	200	Substrate material	Deposition rate (nm s^{-1})	Residual pressure (mbar)	T_s (°C)	Ref.
	Stress $\sigma \times 10^8$ (dyn cm^{-2})												
Ag	1		9	5	3,5	2,6	2	1	mica	0,4	10^{-8}	25	[161]
	1.5		5,5	3	2	1,5	1,1	1	mica	0,2	10^{-8}	25	[161]
	8	8	6,5	2.9	1,8		1,1		glass	0,8	10^{-6}	25	[141]
	6		4,3	0			1		MgF$_2$-film	0,25	8×10^{-8}	25	[162]
	14	8,3	5						SiO-film	0,25	8×10^{-8}	25	[162]
Al	20		20	23			8		aluminium	0,2 - 0,5	1,4×10^{-7}	−195	[163]
	25	1,8	5	3,7					aluminium	0,2 - 0,5	1,4×10^{-7}	− 54	[163]
			2	−5					fused silica	1,8	6 ×10^{-6}	25	[158]
		−0,9	−8,5	−8,1	8		−7,4		glass	0,5	7 ×10^{-5}	25	[164]
			3,2	(thickness = 24 nm)					fused silica	7	4 ×10^{-6}	25	[165]
Cr	26		28	55	83	90	95	110	fused silica	1	6×10^{-6}	25	[158]
	230		220	170	150	140	130		glass	0,2	7×10^{-7}	25	[166]
	160		180	160			130		MgF$_2$-film (with H$_2$O-film)	0,2	7×10^{-7}	25	[166]
	90		140			145	135		MgF$_2$-film (without H$_2$O-film)	0,2	7×10^{-7}	25	[166]
	120								MgF$_2$-film (without H$_2$O-film)	0,22	*1,3×10^{-8}	25	[167]
	42	46	45						H$_2$O-film	0,22	*3 ×10^{-6}	25	[167]

*O$_2$pressure

Monomolecular coverage is achieved at about 4 mol %. Larger doping results in only an insignificant stress reduction [159 a-c]. Here, we show possibilities of reducing the mechanical stresses which have barely been used yet. Of the many polycrystalline films employed in practice, however, little is known on stress behaviour as regards how dominant the grain boundary influence is, compared with other possible influences.

Evaporation, sputtering and ion plating have various parameters in the film production so that films of the same material but produced by different technology can have various stress values. Hence, the possibility exists of influencing stresses in a desired way. Systematic investigation of this is still in its initial stages.

Table 4 shows the intrinsic stress of various metal films made by evaporation and condensation in high vacuum [160]. Further stress measurements can be found in the literature, e. g. [158, 159, 148].

In practice, two aspects are important for the consideration of mechanical stresses:

    a) the reduction of the adhesion between film and substrate and
    b) the bending or warping of precision components bearing films.

If, in the process of film production, the deposited atoms are mobile and if only a low localized adhesion of the film atoms exists to the substrate, then the film atoms (molecules) slip in the boundary surface and only low (or no) stresses are formed. However, increasing localized adhesion often causes increasing film stresses.

The contribution of the physisorption to the adhesion is about 0.4 eV. The appropriate forces are between 10^4 and 10^{10} dyn cm^{-2}. The energetic portion of the chemical bonding ranges from 0.5 to about 10 eV. The forces here are larger than or equal to 10^{11} dyn cm^{-2}. It can be seen from this that, in the case of formation of a chemical compound boundary layer, the adhesion is not endangered by the mechanical film stresses. Only the weaker physisorption can be so strongly impaired by larger film stresses that detachment of the film sometimes occurs.

A numerical estimation of the surface mobility of atoms along a $<111>$ surface from one potential minimum to another in a cubic face-centered structure by the energetically easiest route is due to Carpenter and Campbell [168]. In order to overcome an adsorption energy of 0.2 eV (this corresponds to a physisorption), film stress values of 5×10^9 dyn cm^{-2} are thus required. These are relatively high stress values.

However, if the inhomogeneity usually present in the adhesion over a surface is considered, then, for example, crack formation in areas of reduced adhesion resulting from mechanical film stresses in the sense of a stress reduction process is easily to understand [169].

In order to reduce deformations of precision components, such as precision optics, low-stress film materials and low-stress-producing deposition technologies are vital. Increased effort is required here in order to determine materials and material combinations, and to develop suitable methods in the deposition processes.

8.4.2 HARDNESS AND ABRASION

The hardness of a material is a very important property and results primarily from interatomic forces and complicated interactions caused by deformation mechanisms in the material. It affects such properties as wear resistance [173 – 175], lubrication [176], and deformation bonding [177]. Of the typical film materials used for coatings on glass or plastics, it is mainly the oxides which show high hardness values, as can be seen in Table 5.

TABLE 5

BULK HARDNESS OF VARIOUS METALS, OXIDES AND FLUORIDES

Material	Knoop hardness (kg mm^{-2})	Ref.
Ag	60 – 90	[178]
Al	100 – 140	[178]
Cr	650 – 940	[181]
Al_2O_3	2100	[178]
Cr_2O_3	~ 2000	[181]
MgF_2	430	[178]
CaF_2	163	[179]
SrF_2	140	[179]
BaF_2	82	[179, 180]

Hardness of thin films is measured generally by indentation methods. The relation of the various hardness values obtained by various microhardness measuring techniques is shown in Fig. 24.

In principal, microhardness determination is rather simple. The indenter may be a square based pyramid of diamond or sapphire with a face angle of $\alpha = 136°$ for the Vickers test, or a rhombic-based pyramid with angles between the edges at the top of $\beta = 130°$ and $\gamma = 172°$ 30' for the Knoop test. The load P between 2 and 200 pond is chosen to produce an indentation that can be conveniently measured with a light or scanning electron microscope, and then the area is calculated from the size of the indentation. For the Vickers hardness it is found according to:

$$H_v = 2P \sin(\alpha/2)/d^2 = 1.8544 \, P/d^2 \tag{19}$$

d is the diagonal of the impression in mm and P is the load in kgf.

From a hardness test, the property measured is a plastic strength of the material; it is the amount of plastic deformation produced mainly in compression by a known force, the deformation varying from point to point in the region under the indenter. The depth of penetration of the indenter crystal should be no more

354

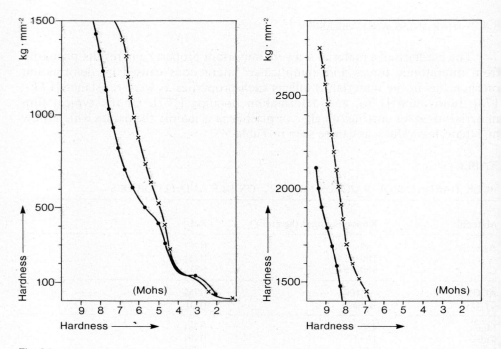

Fig. 24

Approximate correlation between various hardness values for minerals, ● = Knoop hardness, × = Vickers hardness [196].

than 10 % of the film thickness [182, 183]. There are often problems in measuring the hardness of many films because of their relative small thicknesses. For such films, ultralight indenter loads, or better still special measuring devices [184, 184a], and technical tricks must be used [184b]. Pure metal films deposited quickly on cold substrates have a disorder in their crystal structure and as a consequence a relatively high hardness value is observed. Tempering or deposition on heated substrates yields better ordered film structures and results in a decrease in hardness [185, 186]. With alloy films, an increase in hardness with decreasing grain size was observed, see [194, 195]. The variation of the microhardness with substrate temperature for oxide films is quite different from that of metals. The film hardness obtained at low temperatures is generally appreciably lower than the bulk hardness. The reason is a low packing density of the film. With incrasing substrate temperature it may increase slightly, remain constant or at first even decrease, as a consequence of structural changes from the tapered crystallites of zone 1, to columnar crystals of zone 2, of the zone model of Movchan and Demchishin [186] shown in Fig. 25 resulting from higher atom and molecule mobility.

   In contrast with metals, the hardness of oxides increases greatly as deposition temperature continues to rise [186, 187]. However, the required T_s is often higher

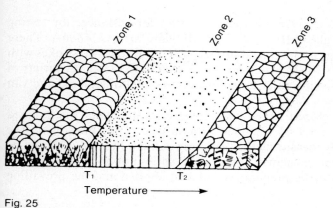

Fig. 25

Schematic representation of microstructural zones of condensates in dependence on temperature [54].

than 800° C, for example for Al_2O_3 [186], so that final increase in hardness can not be used in many optical film applications, because the optical film quality (light scattering) and the thermal stability of the substrate (optical glass) are limiting factors. With films made of the alkaline earth fluorides, an increase in substrate temperature to 350° C improves the film hardness appreciably. The reason seems to be also connected with an increase in packing density. Unfortunately, there are few quantitative results.

Abrasion, or more precisely abrasive wear, occurs when a hard rough surface, or a soft surface containing hard particles, slides against a softer surface, digs into it and ploughs a series of grooves. The material from the grooves is usually displaced in the form of loose wear particles. The abrasion rate of a surface, using any abrading medium, is inversely proportional to the hardness of the surface. In practice, the most common abrasive contaminant is SiO_2 in form of quartz sand, with a hardness of about 800 kp mm^{-2}. When abrasive wear should be prevented, it can be achieved by making this surface harder than the abrasive.

Considering thin films on glass, it is found, that coatings on front lenses or on other exposed surfaces require periodic cleaning. Usually cleaning is performed by gentle rubbing with a cloth or lens tissue; this may produce abrasive wear if inorganic dust particles such as the above mentioned SiO_2 sand is present. It is therefore an important requirement for many thin-film applications to have a film abrasion resistance as high as possible. The high abrasion resistance of a coating should protect the coating itself and the substrate surface beneath it against scratches. Ideally, it should improve the robustness of the whole system. Unfortunately, there is no test to determine the degree of thin-film abrasion quantitatively. However, some standard test procedures have been developed. One of these standard test involves a rubber pad loaded with a particular grade of emery. This special eraser can be drawn over the surface of the film to give a reasonably controlled amount of abrasion when the load and number of strokes specified.

Such specifications are given by the US Army for instance, for testing antireflection coatings on glass MIL-C-00675B(MU) and MIL-C-14806A. There should be no visible damage to the coated surface when rubbed for 40 strokes with an eraser conforming to MIL-E-12397 under a force of 0.91 to 1.13 kg. A variety of apparatus is in use to perform similar tests, for example that of Holland and van Dam [188].

Another method developed by Kraus [189] uses a stream of fine-grained quartz sand falling freely on a 45° inclined film surface. The abrasion on the surface can be inspected by measuring the scattering of light. Finally, it should be mentioned that the abrasion resistance of thin films varies not only with the film material used but also with the preparation method applied and the prevailing conditions.

8.4.3 DENSITY

In most thin metal and dielectric films, a gradient in film density can be found, see [190a, 190b, 191], which is 2 to 10 % (or even more) of the bulk, depending on the preparation method, experimental conditions and film thickness. Very thin films made by evaporation in vacuum generally show stronger deviations from bulk density than thicker films. The approach to bulk values occurs asymptotically. Physical and chemical affects are responsible for this phenomenon. The physical reasons may be a higher crystalline disorder, in general more vacancies in the crystals of the deposit, than in the bulk state; and holes and pores in the films produced by gas incorporation and special growth modes occurring more strongly in the first stages of film formation. Although, however, polycrystalline films may also be composed of relatively low-disordered individual crystallites – this case can be shown by electron diffraction – a lower film density is observed because of a loose arrangement - the packing density - of these film forming crystallites on the substrate. In many cases, a low packing density is even the dominating effect in causing a reduced film density. Table 6 shows the packing densities of some metal and fluoride films.

TABLE 6

PACKING DENSITY OF SOME METALS AND FLUORIDES

Film material	Packing density		Ref.
	$T_s = 30°C$	$T_s = 300°C$	
Ag	0,95	–	
Al	0,95		
Cr	0,96		
MgF_2	0,72	0,98	[45, 193]
CaF_2	0,65	–	[172, 192]
BaF_2	0,69	–	[172]

Besides higher substrate temperature also ion bombardment, especially during film growth, increases film density close to the value of the bulk material as was for instance shown recently for ZrO_2 films [272].

The chemical reasons for density variations are, for metal films, chemical compound formation during deposition of the first few hundred monolayers by unwanted reactions with the residual atmosphere. The density of the corresponding oxides, hydroxides or nitrides may be quite different from that of the metal, as can be seen in Table 7.

TABLE 7

DENSITY OF SOME METALS, OXIDES, HYDROXIDES AND NITRIDES [178]

Material	Density of				
	Me	Me_2O	$Me_2O_3\ 3H_2O$	$Me(OH)_3$	Me N
Ag	10.50	7.143	–	–	–
Al	2.702	–	2.43 – 2.53	2.42	–
Cr	7.20	–	5.21	–	5.90

With compound films, changes in stoichiometry by dissociation, fractionating and incomplete recombination or formation of other compounds may also contribute to density deviations.

8.5 CHEMICAL AND ENVIRONMENTAL STABILITY

Films are sometimes used at elevated temperatures or at very low temperatures, or they may be exposed to sudden changes in temperature. Variations in temperature, however, cause both reversible and irreversible changes in the films. Examples of reversible effects are the temperature dependence of the absorption coefficients of semiconductor films and the temperature dependence of the refractive index or of the electrical conductivity of dielectric and metals films. Another reversible effect is the refractive index variation caused by the temperature-dependent sorption and desorption of water vapour by a porous film structure. Irreversible effects include the possible increase in packing density, crystallinity, refractive index and electrical conductivity of films after intended or unintended annealing. The chemical resistance of the films is of great importance, particularly in connection with the influences of atmospheric moisture. In this respect, the solubility of the bulk material is a useful guide, although it must always be remembered that with a thin film, the ratio of surface to volume is extremely large and magnifies any tendency towards solubility present in the bulk material. The effect of humidity and salt water is of special importance in a tropical climate or at sea. There exist special simulators and precise test specifications to control film quality. Hygroscopic film materials may deteriorate rapidly under humid conditions. Absorption of moisture by some materials may result in swelling, which destroys their function and causes loss of physical strength and changes in other important mechanical properties. Insulating materials which absorb moisture may suffer degradation of their electrical and thermal properties. Therefore a humidity test should be applied to all films and film systems in contact with the atmosphere to determine the resistance to the effects of exposure to a warm, highly humid atmosphere such as is encountered in tropical areas. The test according to MIL-STD-810C (1975) is an accelerated environmental test, accomplished by the continuous exposure of the films to high relative humidity at an elevated temperature. Films should also be resistant to gases such as SO_2 or H_2S, to liquid cleaning agents, to acids, and sometimes also to fungus. Chemical compatibility between the film and substrate is also important. For example, the glass constituent PbO and film materials such as La_2O_3 may react and form optically absorbing metallic lead. At higher temperatures, reactions can also occur between adjacent thin-film materials, and decomposition can lead to optical absorption. A proper choice of materials must therefore also take chemical aspects into account.

A great deal of work has been done and is still carried out on the development of environmental tests especially for coatings on glass and plastics; for example, US Military specifications, United Kingdom specifications for aircraft equipment,

TABLE 8a

VARIOUS THIN FILM TEST SPECIFICATIONS

Requirements		MIL-C-675 A	MIL-C-00675 B	MIL-M-13508 C	MIL-C-14806 A	MIL-C-48497	MIL-F-48616	MIL-SID-810 B	DIN 58196/1	DIN 58196/2	DIN 58196/3	DIN 58390/2	DIN 58390/4	DIN 58390/6	DIN 58752	DIN 58197
Humidity		•	•	•	•	•	•	•				•				•
Salt spray (fog)		•	•	•	•		•	•					•			
Solubility	water (DI)					•	•		•							
	salt water	•	•		•	•	•									•
	solvents					•	•								•	•
Abrasion	eraser	•	•		•	•	•									
	cheesecloth		•	•		•	•									
Adhesion	tape		•	•		•	•									
	boiling water											•				•
Dust						•		•						•		
Temperature				•	•	•	•	•					•			
Thermal shock						•		•			•					
Radiation												•				•
Fungus						•		•								

British Standard Institute specifications, German DIN specification, etc. This has already resulted in a number of specifications which are equivalent to the most severe conditions ever likely to be met in both tropical and polar climates. The most often used test specifications are summarized in Table 8a.

Structure, hardness and density of some dielectric films deposited by various methods and under various conditions are listed in Table 8b. As can be seen in this table, film properties can be varied within extensive limits and with some deposition methods, values approaching those of the bulk materials are achieved.

8.6. OPTICAL PROPERTIES OF THIN FILMS

Materials are characterized optically by their optical constants, i.e. the extinction coefficient and refractive index. The extinction coefficient k' is the imaginary part of the complex refractive index $N = n - ik'$. It assumes the role of an index of attenuation. If this attenuation is caused by true absorption alone, it is termed the absorption coefficient k. The absorption (or extinction) constant is defined as $\alpha^{()} = 4\,\pi\,k^{()}\,\lambda^{-1}$. Finally, $1\ \mathrm{dB\ cm^{-1}} = 4.34\ \alpha^{()}$.

Thin films generally have properties which are somewhat different from the bulk starting materials. As regards optical properties, the observed values of the refractive indices and the absorption coefficients are often lower and higher, respectively, compared with the optical constants of the same bulk materials. Very thin films (t ⩽ 10 nm) can have such strong variations that the term «optical constants» may become problematic. This behaviour demonstrates, however, the

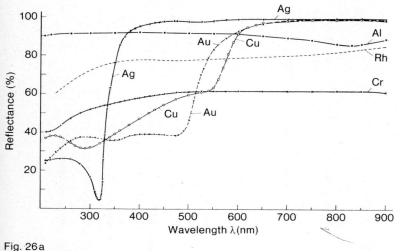

Fig. 26a

Spectral reflectance of various opaque metal films obtained at nearly normal incidence of the light beam. The films were deposited by evaporation in high vacuum. (according to Carl Zeiss, FRG).

TABLE 8b

STRUCTURE, HARDNESS AND DENSITY OF DIELECTRIC FILMS DEPOSITED BY VARIOUS METHODS

Deposition process	Residual pressure (mbar)	Starting material	Typical films formed	Typical deposition rate	Typical substrate temperature	Structure of the deposit	Film density		Hardness, abrasion resistance		Sources of impurities
							LT_s	HT_s	LT_s	HT_s	
Evaporation	10^{-5}–10^{-8}	ZnS, MgF$_2$	ZnS, MgF$_2$	0.5–1μm min^{-1}	25°–300°C	Poly-crystalline	−	+	low	medium	Crucible walls
Reactive Evaporation	10^{-4} (e.g. O$_2$)	SiO, TiO Al	SiO$_2$, TiO$_2$ Al$_2$O$_3$	<0.5μm min^{-1}	25°–400°C	Amorphous Poly-crystalline	−	+	low	high	Filament Crucible Walls
Sputtering	10^{-2}–10^{-3}(Ar)	SiO$_2$	SiO$_2$	<0.01μm min^{-1}	25°–300°C	Amorphous	+	++	high	high	Walls Gases used Sources
Reactive Sputtering	10^{-2}–10^{-3} (e.g. O$_2$/Ar)	Al, Zr	Al$_2$O$_3$ ZrO$_2$	<0.01μm min^{-1}	25°–300°C	Amorphous Poly-crystalline	+	++	high	high	Walls Gases used Sources
Reactive Ion-Plating	10^{-2}–10^{-3} (e.g. O$_2$/Ar)	ZnS SiO, TiO	ZnS SiO$_2$, TiO$_2$	<0.5μm min^{-1}	25°–250°C	Amorphous Poly-crystalline	++	++	high	high	Filament Crucible Gases used Walls
Low Pressure CVD	1–10^{-3} (O$_2$, NH$_3$/Ar)	SiH$_4$ TiCl$_4$	SiO$_2$, TiO$_2$	0.5–10μm min^{-1}	250°–550°C	Amorphous	−	++	medium	high	Walls Gases used
Plasma CVD	10^{-2} (O$_2$, N$_2$, NH$_3$/Ar)	SiH$_4$ TiCl$_4$	SiO$_2$, TiO$_2$	<0.5μm min^{-1}	25°–400°C	Amorphous	−	+ (++)	low	high	Walls Gases used

LT_s = Low substrate temperature

HT_s = High substrate temperature

− = low

+ = high

++ = near bulk

Deposition process	Residual pressure (mbar)	Starting material	Typical films formed	Typical deposition rate	Typical substrate temperature	Structure of the deposit	Film density		Hardness, abrasion resistance		Sources of impurities
							LT_s	HT_s	LT_s	HT_s	
Atmospheric Pressure CVD	10^3 (O_2, NH_3/Ar)	SiH_4	SiO_2	0.1–1μm min^{-1}	250°–500°C	Amorphous	–	+ (++)	medium	high	Walls Gases used
Spin- and Dip-Coating	10^3 Dry air or Ar	Proprietary liquids	SiO_2, TiO_2 Glasses	Non-linear ~1–5μm min^{-1}	25°C After annealing at about 400°C	Amorphous	–	+ (++)	low	high	Solvents Reagents
Spray Coating	10^3 Dry air or Ar	Proprietary liquids	SnO_2	0.1–1μm min^{-1}	200°–500°C	Amorphous	+	++	medium	high	Source materials Ambient

LT_s = Low substrate temperature

HT_s = High substrate temperature

– = low
+ = high
++ = near bulk

dependence of n and k on film thickness and to some extent on the substrat surface condition. For thicker films (t ⩾ 40 nm), the thickness dependenc decreases rapidly. The resulting film properties are influenced by the productio method. In the case of films made by PVD, the properties depend on vacuum an deposition conditions such as: residual pressure, its composition and spatia distribution, substrate temperature, deposition rate, angle of incidence, substrate surface topography, and so on.

Figure 26a shows the spectral reflectance measured at normal light incidence of some typical evaporated opaque metal films which are used in various optica applications.

Figure 26b shows the spectral reflectance obtained with linear polarized ligh at 45° incidence. It is remarkable that the ratio between R′ and R″ deviates more from unity the smaller the term ½ (R′ + R″).

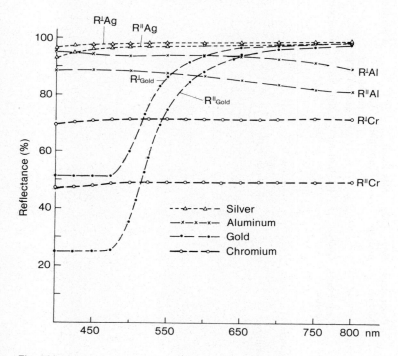

Fig. 26b

Spectral reflectance of opaque metal films at 45° incidence of linear polarized light. (Films produced and measured by Carl Zeiss, FRG).

For reliable production it is a priority to obtain films with reproducible optical constants. Typical of the problems in producing films with optimal properties are the different optical constants of evaporated aluminium films, shown in Table 9, which were obtained by Hass in the period 1945 to 1961 [207].

TABLE 9

OPTICAL CONSTANTS OF VACUUM-DEPOSITED ALUMINIUM FILMS PREPARED
AND MEASURED BY HASS IN 1946 AND 1961

λ (nm)	(1946) n	(1961) n	(1946) k	(1961) k
644	1.12	–	6.26	–
650	–	1.30	–	7.11
578	0.89	0.93	5.68	6.33
546	0.76	0.82	5.50	6.99
491	0.57	–	5.20	–
492	–	0.64	–	5.50
435	0.40	–	4.16	–
436	–	0.47	–	4.84
380	–	0.37	–	4.25

As a consequence of an improved technology (higher and cleaner vacua, higher deposition rate), the newer films, possibly containing less oxygen, clearly show better optical values, especially in the blue and near ultraviolet region. Also, many other metal films are very sensitive to oxidation, and even small amounts of oxides produce a degradation of their optical properties.

As can be seen in Fig. 26a, the reflectivity of aluminium is lower than that of silver in the visible and infra-red regions. However, silver films deteriorate rapidly on exposure to sulphur-containing atmospheres, whereas aluminium films retain their reflectivity, because of the glasslike protective oxide film Al_2O_3 which forms on their surface. Pure aluminium films have a high reflectance in the ultra-violet region. From the visible range down to about 200 nm, the Al reflectance remains practically constant. Below this wavelength, however, it decreases continuously and reaches zero at about 80 nm. Even small amounts of incorporated oxygen reduce the high ultraviolet reflectivity, so that Al films with good uv reflectance must be deposited very quickly in clean high vacuum.

The relatively high reflectivity of pure Al films deposited on glass can also be obtained using lacquer or plastic substrates, providing both are chosen to avoid contamination of the vacuum system. It is interesting to note that evaporated silver does not tarnish as rapidly as chemically deposited silver, possibly because the former has greater purity. The reflectivity of aluminium also decreases with age but the greatest reduction occurs during the inital period of oxide formation. After one month the aluminium oxidation is complete and the films retain a reflectivity of about 89 % for a long time in normal atmospheres.

The essential feature of dielectric optical film materials is their very low absorption ($\alpha < 10^3$ cm^{-1}) in the relevant region of the spectrum. The films are primarily characterized by transparency and refractive index.

Figure 27 shows schematically the curve of transmission versus wavelength of an optical thin-film material. The desired region of high transmittance (region II) is situated between the short-wavelength absorption edge, region I (which depends on the electronic structure of the material), and the long-wavelength limit, region

III (which is determined by lattice vibrations or, in the case of semiconductors, by free carrier absorption). The extent and quality of region II depends strongly on the material – on its stoichiometry and purity.

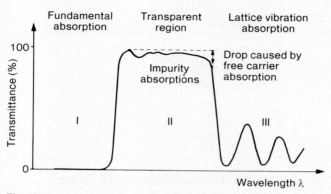

Fig. 27
Schematic curve of transmission versus wavelength of an optical material.

The refractive index of a material at optical frequencies is mainly determined by the polarizability of the valence electrons. In elements with increasing atomic weight, the electron population is enlarged, which results in the shielding of the positive charge of the nucleus, so that the polarizability of the valence electrons and consequently the refractive index are increased; for example, for Si, $n = 3.4$ and for Ge, $n = 4.0$. In compounds, the type of bonding also influences the indices. Compounds with predominant ionic bonding show lower refractive indices than compounds with a high degree of covalent bonding; for example, for ZnO, $n = 2.08$ and for ZnS, $n = 2.87$ and for ZnTe, $n = 3.56$. A detailed survey about these relationships was published by Black and Wales [208].

The transparency of thin films is often slightly worse than that of bulk materials. For example, bulk single crystals of cubic ZnS have an absorption coefficient k_{515} of between 4.3×10^{-6} and 1.2×10^{-5} [209], whereas for evaporated films a value of 2.7×10^{-4} is obtained. Deposition experiments in high vacuum with ZnS showed that the absorption coefficient k, determined by a calorimetric method [210], does not depend on deposition rates lying between 0.1 nm s^{-1} and 10 nms^{-1}, nor on different film thicknesses, but is influenced by the amount of oxygen in the residual gas. A change in oxygen content from PO$_2$/P tot $= 0.11$ to PO$_2$/P tot $= 0.55$ increases the absorption coefficient k_{1060} from 4×10^{-6} to 1.3×10^{-5}. Table 10 shows the absorption and extinction coefficients of well prepared sulfide and fluoride films. Similar experiments have also been performed for single oxide films prepared by reactive deposition, with and without excitation (partial ionization) of the oxygen gas. Typical results are presented in Table 11. Deposition in the presence of excited oxygen yields films with smaller k values. The effect is more pronounced near the short-wavelength absorption edge. One

TABLE 10

ABSORPTION COEFFICIENT k AND EXTINCTION COEFFICIENT k' OF WELL
PREPARED SULFIDE AND FLUORIDE FILMS
(Depositions were carried out by evaporation and condensation in vacuum)

Film	Ref.	k (1060 nm)	k (515 nm)	k' (633 nm)
ZnS	[226]	4.1×10^{-6}	2.7×10^{-4}	–
ZnS	[227]	–	–	3.5×10^{-5}
CdS	[271]	–	2.9×10^{-3}	1.7×10^{-3}
ZnSe	[271]	–	3.4×10^{-3}	2.9×10^{-3}
ZnTe	[271]	–	6.7×10^{-3}	2.6×10^{-3}
ThF$_4$	[226]	2.1×10^{-6}	5.0×10^{-6}	–
MgF$_2$	[226]	6.0×10^{-6}	9.0×10^{-5}	–
CeF$_3$	[227]	–	–	1.4×10^{-5}
NdF$_3$	[227]	–	–	1.2×10^{-5}

reason for the increased extinction of the films is often the true absorption due to
small deviations from stoichiometry, and contamination.

Another reason is the scattering of light by surface and volume imperfections
[211-215]. Such imperfections include surface roughness, rough internal boundaries, and density fluctuations which stem from crystallinity, porous microstructure,
pinholes, cracks, splashes, microdust, etc.

The contribution of light scattering to optical losses (L) is given by:

$$L = A + S, \tag{20}$$

$$1 = R + T + L, \tag{21}$$

where A is absorption, S is scattering, R is reflectance, and T is transmittance. It
can often be of the same order or larger than the true absorption, and it depends
strongly on cleanliness, deposition method, and deposition conditions.

Sputtered films often show lower scatter losses than evaporated films, because
of their dense homogeneous microstructure and smooth surface. For optimal
performance of a multilayer, the extinction coefficients of the individual single
films should not exceed 10^{-4}. For example, in order to obtain negligible losses in
multilayer antireflection coatings, in a film 100 nm thick, only loss values of less
than 0.01 % can be tolerated, corresponding to a k' value of 0.44×10^{-4} or an
α' value of 10 cm^{-1} ($\lambda = 550$ nm) [216].

In the case of planar waveguides in integrated optics, the quality requirements
are much more stringent because of the long path inside the film [217]. Low loss
values reported here are listed in Table 12. Determinations of the light attenuation
in waveguides enable loss measurements of 0.1 dB cm^{-1}, corresponding to an
extinction coefficient of 1×10^{-7} or an α' value of 0.02 cm^{-1} ($\alpha = 633$ nm) [218].

TABLE 11

ABSORPTIVITY A, ABSORPTION COEFFICIENT k, AND EXTINCTION COEFFICIENT k' OF SOME OXIDE FILMS DEPOSITED UNDER DIFFERENT REACTIVE EVAPORATION CONDITIONS

Film (starting material)	Ref.	O_2 pressure (mbar) (ionization of O_2)	1060 nm A%	1060 nm k	515 nm A%	515 nm k	633 nm k'
TiO_2 (TiO)	[226]	4×10^{-4} (No)	0.290	2.5×10^{-4}	1.510	7.5×10^{-4}	–
TiO_2 (TiO)	[226]	4×10^{-4} (Yes)	0.090	7.7×10^{-5}	1.250	5.5×10^{-4}	–
TiO_2 (Ti)	[226]	6×10^{-4} (Yes)	0.098	7.9×10^{-5}	2.520	1.1×10^{-3}	–
ZrO_2 (ZrO_2)	[226]	2×10^{-4} (No)	0.020	2.0×10^{-5}	0.480	1.6×10^{-4}	–
ZrO_2 (ZrO_2)	[226]	2×10^{-4} (Yes)	0.025	2.1×10^{-5}	0.490	1.8×10^{-4}	–
ZrO_2 (ZrO_2)	[226]	7×10^{-5} (No)	–	–	–	–	1.2×10^{-5}
Al_2O_3 (Al_2O_3)	[226]	2×10^{-4} (No)	0.011	8.0×10^{-6}	0.073	2.3×10^{-5}	–
Al_2O_3 (Al_2O_3)	[227]	7×10^{-5} (No)	–	–	–	–	2.3×10^{-5}
	[228]		–	–	–	–	3.5×10^{-5}
SiO_2 (SiO_2)	[226]	6×10^{-5} (No)	0.005	2.0×10^{-6}	0.060	1.5×10^{-5}	–
SiO_2 (SiO_2)	[226]	6×10^{-5} (Yes)	0.005	2.0×10^{-6}	0.050	1.3×10^{-5}	–
CeO_2 (CeO_2)	[227]	7×10^{-5} (No)	–	–	–	–	2.3×10^{-5}
MgO (MgO)	[227]	1×10^{-4} (No)	–	–	–	–	2.8×10^{-5}

TABLE 12

LOW-LOSS FILMS AS WAVEGUIDES IN INTEGRATED OPTICS

Film material	Ref.	Wave-length (nm)	dB (cm^{-1})	Losses k'	α' (cm^{-1})	Preparation
Ta$_2$O$_5$	[219]	633	0.9	1×10^{-6}	0.21	Ta sputtering oxidation in air
CAS 10 (Planer, GB)	[229]	633	1.1	1.2×10^{-6}	0.25	Electron-beam evaporation in high vacuum
Organosilicon	[218] [230]	633	0.04	5×10^{-8}	0.01	RF discharge in gaseous organosilicon compounds

By the application of proper conditions, volume losses can be distinguished from surface and interface losses [219, 220]. Advances in measurement techniques for small absorption coefficients are given by Hordvik [221].

Some years ago, Baumeister and co-workers [222, 223] made substantial progress in the deposition of low-loss optical films of HfO$_2$, La$_2$O$_3$, Al$_2$O$_3$, Y$_2$O$_3$, LaF$_3$, MgF$_2$, and SiO$_2$ for the near ultraviolet. Such films are required in dielectric mirrors for lasers, for example, at 351 nm and 248 nm [224].

The complex nature of the mechanism of laser damage by high-energy radiation in single and multilayer films has not yet been fully understood. However, there is no doubt that, among other possible mechanisms, absorption contributes to the deterioration. Higher damage thresholds of films seem to be associated with shorter ultraviolet absorption edges. More details on resistance to high-energy radiation are given in Ritter's review article [216], in the annual reports of the Boulder Damage Symposia on laser-induced damage in optical materials e.g. [225a] and in a recently published paper by Guenther and Humphreys [225b]. The refractive index of a dielectric film often differs from that of the bulk material. Generally, lower values are obtained. For example, bulk cubic ZnS has a mean refractive index of $n_{633} = 2.35$ (see [209]), but in polycrystalline cubic films of 100 nm thickness an index of $n_{633} = 2.27$ was determined [191]. Thinner films show lower values, whereas the indices of thicker films approach the bulk value. The observed refractive indices depend on the deposition method and on process-parameter-sensitive variations in structure, microstructure, chemical composition, and gas content of the films. Thus, for example, TiO$_2$ films show a strong dependence of their refractive index on the substrate temperature during deposition because between 20°C and 400°C they can be amorphous or can consist of anatase or rutile or a mixture of these phases [23, 28]. As a consequence, the refractive index of TiO$_2$ films varies between $n_{550} = 1.9$ (20°C) and $n_{550} = 2.6$ (400°C). Preferred crystallographic orientations (textures) may also influence the refractive index of thin films. The substrate temperature and other weaker parameters influence the microstructure of the film. The substrate temperature supports the tendency to crystallization.

Amorphous, glasslike, and crystalline deposits have a higher average film density (ρ_f) if deposited onto heated substrate or if they are annealed afterwards.

As already shown in a preceding section, many polycrystalline films consist of fairly well developed columnar or prismatic crystals of different diameters which can be arranged on the substrae in a close-packed structure. This means that the films are composed of relatively well developed crystalline aggregates, grain boundaries, and vacant places such as intermediate gaps and pores. A higher substrate temperature increases the mobility of the film atoms or molecules and favours the formation of larger and more closely packed crystals (an increase in crystal size and a decrease in grain boundary area and vacancies occur).

Similarly, a densification through a decrease in vacancies can be brought about even in non-crystalline amorphous or glasslike films by suitable heat treatment. Ion bombardment during or after film formation also produces densification.

Density and refractive index of isotropic materials are interrelated by the Lorentz-Lorenz equation. The mean refractive index can be calculated for the general case from the expression

$$\bar{n}_f = \frac{2\,\bar{\rho}_f\ R + M}{M - \bar{\rho}_f\ R} \qquad (22)$$

if the mean film density is known, R is the molar refraction, which can be calculated from single crystal data, and M is the molecular weight. With thin films, it is convenient to use the concept of the packing density (p) which is defined to be the ratio of the average film density (ρ_f) and the bulk density (ρ_m):

$$p = \rho_f / \rho_m \qquad (23)$$

Rather than being directly a ratio of densities, p is defined better with reference to the space within the film as the ratio: volume of the solid part of the film (i.e. columns) / total volume of the film (i.e. columns + voids).

Equation (24) shows the correlation between the packing density p and the refractive index [231]:

$$p = \frac{\rho_f}{\rho_m} = \frac{n_f^2 - 1}{n_f^2 + 2}\ \frac{n_m^2 + 2}{n_m^2 - 1} \qquad (24)$$

A small packing density causes a reduction of the film refractive index. the refractive index of a porous film measured in vacuum is therefore low. On exposure to the atmosphere, however, a markedly increased value due to the sorption of water vapour, with $n_{H_2O} = 1.33$ is observed especially in the case of films of low refracting materials, such as magnesium fluoride or cryolite. MgF_2

films deposited onto cold substrates have a relatively low packing density of $p = 0.74$. The influence of water vapour and substrate temperature on the refractive index of MgF_2 is demonstrated in Table 13 [232]. Similar, but often less pronounced, effects have also been observed with other dielectric films.

TABLE 13

REFRACTIVE INDICES OF MgF_2 FILMS DEPOSITED UNDER VARIOUS EVAPORATION CONDITIONS

Residual gas pressure (mbar)	Gas composition	Substrate temperature (°C)	Refractive index of the film, $\lambda = 633$ nm	
			in vacuum	in air
2×10^{-9}	Free of water	30	1.326	1.383
1×10^{-5}	90% N_2	30	1.328	1.385
1×10^{-5}	90% H_2O	30	1.341	1.385
1×10^{-7}	40% H_2O	280	1.384	1.385
3×10^{-5}	Dry air	30	1.32*	1.39*
5×10^{-6}	Dry air	30	1.33	–

*$\lambda = 550$ nm

Optical shifts [233 – 239] as well as mass changes [236, 45] produced by water sorption, are used to determine the packing density of thin films.

Many films deposited on heated substrates (250–350°C) possess packing densities larger than 0.9, so that the changes in refractive index on exposure to the atmosphere are very small and can often be neglected. The same effect can be observed with films produced by ion-assisted deposition techniques, such as sputtering and ion plating.

Considerations of Maxwell-Garnett [273, 274] and Bragg and Pippard [275] based on Lorentz-Lorenz polarization theory have been put into the form for dielectric films composed of cylindrical columnar crystals by Harris, Macleod et al. [276]:

$$n_f^2 = \frac{(1-p) n_v^4 + (1+p) n_v^2 n_s^2}{(1+p) n_v^2 + (1-p) n_s^2} \tag{25}$$

In the formular n_f is the index of the composite film, n_s is the index of the solid material, that is the columns (bulk value is often assumed) and n_v is the index of the voids in the film (in case of vacuum this index is unity).

An alternative expression for the index of a composite film based simply on an interpolation betweeen two limits is that of Kinosita and Nishibori [277]:

$$n_f = (1-p) n_v + p n_s \tag{26}$$

The accuracy of these expressions has been investigated by Harris, Macleod et al. using a finite difference approach to calculate the dielectric permittivity of a hexagonal array of columns. The linear interpolation turned out to be surprisingly

TABLE 14

OPTICAL PROPERTIES OF NON-ABSORBING AND ABSORBING METAL-OXIDE FILMS PRODUCED BY DIP COATING
(Ref. [28, 36] of Chapter 6)

Starting material	Film material	Refractive index n 550	Onset of absorption (nm)	Remarks
Al-sec-butylate, $Al(NO_3)_3\,9H_2O$	Al_2O_3	1.62	250	forms mixtures with other oxides
$Y(NO_3)_3$	Y_2O_3	1.82	$\leqslant 300$	–
$La(NO_3)_3$	La_2O_3	1.78	220	–
$Ce(NO_3)_3\,6H_2O$	CeO_2	2.11	400	forms mixtures with other oxides
$Nd(NO_3)_3$	Nd_2O_3	inhomogeneous	–	attenuation of transmittance by absorption bands of Nd^{3+} between 500 and 600 nm
$In(NO_3)_3$	In_2O_3	1.95	420	semiconductor
$Si(OR)_4$	SiO_2	1.455	205	forms mixtures with other oxides
$Ti(OR)_4$ $TiCl_4$	TiO_2	2.3	380	forms mixtures with other oxides
$ZrOCl_2$	ZrO_2	1.72	340	–
$HfOCl_2\,8H_2O$	HfO_2	2.04	220	traces of Cl in the layers
$ThCl_4$ $Th(NO_3)_4$	ThO_2	1.93	220	–
$SnCl_4$	SnO_2	inhomogeneous	350	semiconductor
$Pb(OOCCH_3)_2$	PbO	inhomogeneous	380	diffuses into glass at 500°C
$TaCl_5$	Ta_2O_5	2.1	310	–
$SbCl_5$	Sb_2O_5	1.9	340	–
$Cu(NO_3)_2\,3H_2O$	CuO	–	–	–
$VOCl_2$	VO_x	2	0.01	optical properties strongly dependent on preparation conditions
$Cr(NO_3)_3\,9H_2O$ $CrOCl$	CrO_x	–	–	–
$Fe(NO_3)_3\,9H_2O$	Fe_2O_3	2.38	0.14	–
$Co(NO_3)_2\,6H_2O$	CoO_x	2.0	0.16	–
$Ni(NO_3)_2\,6H_2O$	NiO_x	–	–	–
$RuCl_3\,H_2O$	RuO_x	–	–	semiconductor
$RhCl_3$	RhO_x	–	0.2	–
$UO_2(OOC\,CH_3)_2$	UO_x	1.95	0.015	–

TABLE 15

OPTICAL PROPERTIES OF SPUTTERED OXIDE FILMS

Starting material	Sputtering method	Film composition	Film structure	Useful transmittance range (nm)	Refractive index (visible range)	Ref.
Si (or SiO_x)	reactive HF sputtering Ar/O_2	SiO_2	glasslike	$<200-8000$	1.45–1.50	[197, 198, 202–204, 206]
Ge (or GeO_2)	–"–	GeO_2	–	280–8000	1.6 –2.12	[199, 204]
Ti (or TiO_2)	–"–	TiO_2	poly-crystalline	380–8000	2.1 –2.7	[197–202, 204, 206]
Zr (or ZrO_2)	–"–	ZrO_2	poly-crystalline	240–8000	2.15	[198, 203, 204]
Hf	Ar/O_2 reactive sputtering Ar/O_2	HfO_2	–	220–8000	2.0	[204]
Ta (or Ta_2O_5)	reactive HF sputtering Ar/O_2	Ta_2O_5	glasslike	350–8000	2.03–2.09	[199, 201, 204]
Nb (or Nb_2O_5)	–"–	Nb_2O_5	glasslike	320–8000	2.20	[202, 204]
Y (or Y_2O_3)	–"–	Y_2O_3	–	220–8000	1.90	[204, 206]
La (or La_2O_3)	–"–	La_2O_3	–	250–8000	1.76	[204]
Al (or Al_2O_3)	–"–	Al_2O_3	poly-crystalline (glasslike)	$<200-8000$	1.67–1.78	[197, 203, 204]
Ce (or CeO_2)	–"–	CeO_2	–	400–1200	2.5	[198, 206]
Th	reactive sputtering Ar/O_2	ThO_2	–	300–	2.04	[197]
MgO	reactive HF sputtering Ar/O_2	MgO	poly-crystalline	225–8000	1.69–1.72	[199, 204]
V_2O_5	reactive HF sputtering Ar/O_2	V_2O_5	–	–	2.43	[199]
ZnO	–"–	ZnO	–	400–	1.91–2.0	[199]
SnO_2	–"–	SnO_2	–	400–	1.95–2.07	[199]

good for low index films, but for high index films it was poor. The finite difference approach was useful for films with values of packing density, lower than 0.8. Many important thin films, however, have packing densities in the range 0.8 to 0.95. For such materials the finite difference approach does not work well and a rather better arrangement that has much in common with the finite difference approach, is that of finite elements. Using this method investigation of a film with $n_s = 2.35$, $n_v = 1$ and p between $0.7 - 0.95$ resulted in refractive index values showing a gradual change from Bragg and Pippard's eqn. (25) to Kinosita and Nishibori's eqn. (26) in the range between $p = 0.75$ and 0.85 [278]. A review paper

TABLE 16

OPTICAL PROPERTIES OF SPUTTERED MIXED OXIDE AND OXIDIC COMPOUND FILMS

Starting material	Sputtering method	Film composition	Film structure	Useful transmittance range (nm)	Refractive index (visible range)	Ref.
ZrO_2 CaO	reactive HF sputtering	ZrO_2 CaO	polycrystal-line	250–8000	2.10	[204]
HfO_2 Y_2O_3	reactive HF sputtering Ar/O_2	HfO_2 Y_2O_3	crystalline	220–8000	2.0	[204]
$(Y_2O_3)_2$ Al_2O_3	reactive HF sputtering Ar/O_2	$Y_4Al_2O_9$	–	220–8000	1.76	[204]
MgO Al_2O_3	–"–	$MgAl_2O_4$	–	225–8000	1.61	[203, 204]
InSn $(In_2O_3$ $SnO_2)$	reactive (HF) sputtering Ar/O_2	ITO $In_{1.9}Sn_{0.1}O_3$	–	400–>1000	1,8 –2.0	[201, 202, 204]
x(SiO) y(TiO_2)	reactive HF sputtering Ar/O_2	x(SiO) y(TiO_2)	amorphous	400–1200	1.5 –2.4	[206]
x(SiO_2) y(CeO_2)	–"–	x(SiO_2) y(CeO_2)	–"–	400–1200	1.5 –2.4	[206]
$LiNbO_3$	reactive HF sputtering Ar/O_2	$LiNbO_3$	–	–	1.94–2.0	[199]
Mg_2SiO_4	–"–	Mg_2SiO_4	–	–	1.64–1.65	[199]
Ca SiO_3	–"–	Ca SiO_3	–	–	1.60–1.62	[199]
Ba SiO_3	–"–	Ba SiO_3	–	–	1.60–1.66	[199]
Zr SiO_4	–"–	Zr SiO_4	–	–	1.89–1.98	[199]
Ca TiO_3	–"–	Ca TiO_3	–	–	1.90	[199]
Sr TiO_3	–"–	Sr TiO_3	–	–	1.92–2.01	[199]
Ba TiO_3	–"–	Ba TiO_3	–	–	2.04–2.17	[199]
Pb TiO_3	–"–	Pb TiO_3	–	–	2.46–2.53	[199]

TABLE 17

OPTICAL PROPERTIES OF SPUTTERED ZINC AND CADMIUM
CHALCOGENIDE FILMS

Starting material	Sputtering method	Film composition	Film structure	Useful transmittance range (nm)	Refractive index (visible range)	Ref.
ZnS	reactive HF sputtering Ar/H_2S	ZnS	polycrystalline (cubic and hexagonal)	400–14000	2.3	[205]
CdS	–"–	CdS	polycrystalline	600–11000	2.5	[204]
CdTe	reactive HF sputtering	CdTe	–"–	900–11000	2.70	[204]
ZnSe	–"–	ZnSe	–"–	500–20000	2.6	[204]

TABLE 18

OPTICAL PROPERTIES OF SPUTTERED NITRIDE AND CARBIDE FILMS

Starting material	Sputtering method	Film composition	Film structure	Useful transmittance range (nm)	Refractive index (visible range)	Ref.
Si, Si_3N_4	reactive HF sputtering Ar/N_2	Si_3N_4	amorphous	250–9000	1.95	[204]
Ge, Ge_3N_4	–"–	Ge_3N_4	–	800–9000	2.11	[204]
B, BN	–"–	BN	–	250–9000	1.67	[204]
Ti, TiN	–"–	TiN	polycrystalline	500–8000	2.13	[204]
Si, SiC	reactive HF sputtering Ar/C_2H_4	SiC	polycrystalline	1500–>11000	2.95	[204]

concerning this topic was recently published by Macleod [279]. These investigations offer interesting possibilities to calculate the film refractive index from microstructural data.

To obtain thin films with stable values of refractive index, there must be no variation in the structure or composition of the film. However, variations in structure may happen during crystal growth with increasing film thickness. Unwanted variations in the chemical composition can occur with less stable compounds through decomposition, dissociation, and fractionation during the evaporation process, and through insufficient recombination during the film formation.

As regards mixed films, it is known that in general mixtures of various compounds evaporate non-uniformly. Exceptions are some specially prepared two- and multicomponent mixtures that evaporate uniformly in a predefined ratio

TABLE 19

OPTICAL PROPERTIES OF HIGH VACUUM EVAPORATED FLUORIDE FILMS [192]

Starting material	Deposition method[a] evaporation temperature	Film composition	Film structure, packing density (p), and substrate temperature (T_s)	Transmittance range (μm) $\alpha < 10^3\,cm^{-1}$	Refractive index (n_1, n_2),[b] wavelength (μm), and substrate temperature (T_s)	Mechanical and chemical film properties
NaF	B (988°C)	NaF	Crystalline	0.2 –	1.29–1.30 (0.55 μm)	Soft, soluble in water
LiF	B (870°C)	LiF	Crystalline	0.11– 7	1.3 (0.55 μm)	Soft, hygroscopic
CaF$_2$	B (1280°C)	CaF$_2$	Crystalline p=0.57	0.15–12	$n_1 = 1.23$ (0.55 μm) $n_2 = 1.23$–1.46 (0.55 μm)	Fairly hard, low tensile stress
Na$_3$AlF$_6$	B (1000°C)	NaF Na$_3$AlF$_6$ NaAlF$_4$	Crystalline p=0.88 (T_s= 30°C) p=0.92 (T_s=190°C)	0.2 –14	1.32–1.35 depending on film composition (0.55 μm)	Soft, tends to recrystallization, low tensile stress
AlF$_3$	B ( )	AlF$_3$	Amorphous p=0.64 (T_s= 35°C)	0.2 –	$n_1 = 1.23$ (0.55 μm) $n_2 = 1.38$ (0.55 μm)	Soft, low tensile stress
MgF$_2$	B (1270°C)	MgF$_2$	Crystalline p=0.72 (T_s= 30°C) p=0.98 (T_s=300°C)	0.11– 4	1.32–1.39 in vacuum depending on T_s (0.55 μm) 1.38–1.40 in air, depending on T_s	Hard at high T_s, resistant against humidity, high tensile stress (cracking)
ThF$_4$	B (1100°C)	ThF$_4$	Amorphous (x-ray)	0.2 –15	$n_1 = 1.50$ (0.55 μm) (T_s= 35°C) $n_2 = 1.52$ (0.55 μm) (T_s= 35°C)	Soft, medium tensile stress Radioactive!
LaF$_3$	B (1490°C)	LaF$_3$	Crystalline p=0.80 (T_s= 30°C)	0.25– 2	1.55 (0.55 μm) (T_s= 30°C) 1.65 (0.55 μm) (T_s=300°C)	Fairly hard

a B=boat   b n_1 = in vacuum, n_2 = in air

Starting material	Deposition method[a] evaporation temperature	Film composition	Film structure, packing density (p), and substrate temperature (T_s)	Transmittance range (μm) $\alpha < 10^3 \, cm^{-1}$	Refractive index (n_1, n_2),[b] wavelength (μm), and substrate temperature (T_s)		Mechanical and chemical film properties
NdF_3	B (1410°C)	NdF_3	Crystalline $p=0.80$ ($T_s = 30$°C)	0.25–	1.61	(0.55μm) ($T_s = 300$°C)	Fairly hard
CeF_3	B (1360°C)	CeF_3	Crystalline $p=0.80$ ($T_s = 30$°C)	0.3 – 5	1.63	(0.55μm) ($T_s = 300$°C)	Hard, high tensile stress
PbF_2	B (850°C)	PbF_2	$\beta\,PbF_2$ $p=0.91$ ($T_s = 30$°C)	0.25–	1.98 1.75	(0.3 μm) ($T_s = 30$°C) (0.55μm) ($T_s = 30$°C)	Soft <100nm compressive >100nm tensile stress

a B = boat   b n_1 = in vacuum, n_2 = in air

TABLE 20

OPTICAL PROPERTIES OF HIGH VACUUM EVAPORATED OXIDE FILMS [192]

Starting material	Deposition method[a] evaporation temperature	Film composition	Film structure, packing density (p), and substrate temperature (T_s)	Transmittance range (μm) $\alpha < 10^3\,cm^{-1}$	Refractive index (n), wavelength (μm), and substrate temperature (T_s)	Mechanical and chemical film properties
SiO_2	E (~1600°C)	SiO_2	Amorphous p=0.9 (T_s= 30°C) p=0.98 (T_s=150°C)	0.2 – 9	1.45–1.46 (0.55μm)	Hard and resistant by deposition on heated substrate, compressive stress
SiO	B,R (~1300°C)	Si_2O_3	Amorphous	0.4 – 9	1.55 (0.55μm) (T_s= 30°C)	Hard by deposition on heated substrate
Al_2O_3	E (2050°C)	Al_2O_3	Amorphous p=1 (T_s= 30°C) p=1 (T_s=300°C)	0.2 – 7	1.54 (0.55μm) (T_s= 40°C); 1.63 (0.55μm) (T_s=300°C)	Hard and resistant
MgO	E (2800°C)	MgO	Crystalline	0.2 – 8	1.7 (0.55μm) (T_s= 50°C); (0.55μm) (T_s=300°C)	Hard and resistant
Nd_2O_3	B,R (1900°C)	Nd_2O_3	–	0.4 –10	2.05 (0.55μm) (T_s=260°C)	Hard
Gd_2O_3 ThO_2	B,R (2200°C) E (3050°C)	Gd_2O_3 ThO_2	– –	0.32–15 0.3 –	1.8 (0.55μm); 1.95 (0.30μm); 1.86 (0.55μm) (T_s=250°C)	Fairly hard; Radioactive! Hard
Y_2O_3	E,R (2400°C)	Y_2O_3	Amorphous, at higher T_s partially crystalline	0.3 –12	1.89 (0.33μm); 1.87 (0.55μm) (T_s=250°C); 1.83 (0.9μm)	Hard

a B = boat   E = electron beam   R = reactive deposition in O_2

Starting material	Deposition method[a] evaporation temperature	Film composition	Film structure, packing density (p), and substrate temperature (T_s)	Transmittance range (μm) $\alpha < 10^3\,cm^{-1}$	Refractive index (n) wavelength (μm), and substrate temperature (T_s)	Mechanical and chemical film properties
Sc_2O_3	E, R (2400°C)	Sc_2O_3	Amorphous, at higher T_s partially crystalline	0.35–13	1.90 (0.30μm) / 1.89 (0.55μm) / 1.86 (0.9 μm) (T_s=250–300°C)	Fairly hard
La_2O_3	B or E, R (1500°C)	La_2O_3	Amorphous	0.3 –	1.98 (0.33μm) / 1.9 (0.55μm) (T_s=300°C)	Hard
Pr_6O_{11}	B	Pr_6O_{11}	Amorphous	0.4 –	1.92–2.05 (0.55μm)	Fairly hard
ZrO_2	E, R (2700°C)	ZrO_2	– / p=0.67 (T_s= 30°C) / p=0.82 (T_s=250°C)	0.34–12	1.97 (0.55μm) (T_s= 30°C) / 2.05 (0.55μm) (T_s=200°C) / 2.11 (0.40μm) (T_s=250°C)	Hard and resistant
HfO_2	E, R	HfO_2	–	0.22–12	2.15 (0.25μm) (T_s=250°C) / 1.95 (0.55μm)	Fairly hard
SiO	B (~1300°C)	SiO	Amorphous	0.7 – 9	2.0 (0.7 μm) (T_s= 30°C)	Hard, tensile stress
Ta_2O_5	E, R (2100°C)	Ta_2O_5	Amorphous	0.35–10	2.25 (0.40μm) / 2.1 (0.55μm) (T_s=250°C)	Hard and resistant
ZnO	B (1100°C)	ZnO	Crystalline	0.4 –	2.1 (0.55μm) (T_s= 30°C)	Soft

a B = boat   E = electron beam   R = reactive deposition in O_2

Starting material	Deposition method[a] evaporation temperature	Film composition	Film structure, packing density (p), and substrate temperature (T_s)	Transmittance range (μm) $\alpha < 10^3\,\text{cm}^{-1}$	Refractive index (n) wavelength (μm), and substrate temperature (T_s)		Mechanical and chemical film properties
CeO_2	B or E (1600°C)	CeO_2	Crystalline	0.4 – 12	2.2	(0.55μm)	Fairly hard
TiO	B,R (1750°C)	TiO_2	Amorphous ($T_s = 30$°C) Crystalline ($T_s \geq 100$°C)	0.4 – 3	1.9 2.3 2.55	(0.55μm) ($T_s = 30$°C) (0.55μm) ($T_s = 220$°C) (0.55μm) ($T_s = 260$°C)	Hard and resistant, tensile stress
PbO	B (900°C)	PbO	Crystalline	0.53	2.6	(0.55μm)	Soft

[a] B = boat   E = electron beam   R = reactive deposition in O_2

TABLE 21

OPTICAL PROPERTIES OF HIGH VACUUM EVAPORATED NON-OXIDE
CHALCOGENIDE AND SEMICONDUCTOR FILMS [192, 271]

Starting material	Deposition method[a] (evaporation temperature)	Film composition	Film structure, packing density (p), and substrate temperature (T_s)	Transmittance range (μm) $\alpha < 10^3$cm^{-1}	Refractive index (n), wavelength (μm), and substrate temperature (T_s)	Chemical and mechanical film properties
ZnS	B (1100°C)	ZnS	Crystalline p ≥ 0.94 (T_s=35°C)	0.4 –14	2.3–2.4 (0.55μm) (T_s=35°C)	Soft, medium compressive stress
CdS	B (800°C)	CdS	Crystalline	0.55– 7	2.5 (0.6μm)	Soft
ZnSe	B (950°C)	ZnSe	Crystalline	0.55–15	2.57 (0.6μm) (T_s=30°C)	Soft
ZnTe	B (1000°C)	ZnTe	Crystalline	–	2.8 (0.55μm)	Soft
Sb$_2$S$_3$	B (370°C)	Sb$_2$S$_3$	–	0.5 –10	3.0 (0.55μm)	Soft
Ge$_{30}$As$_{17}$– Te$_{30}$Se$_{23}$	E	Ge$_{30}$As$_{17}$– Te$_{30}$Se$_{23}$	Amorphous	–	3.1 (10.6μm)	Toxic
InSb	B[b]	InSb	–	7 –16	4.3	Toxic
InAs	B[b]	InAs	–	3.8 – 7	4.5	Toxic
PbTe	B (850°C)	PbTe	–	3.5 –20	5.6 (5.0μm)	Soft, toxic
Si	B or E (1500°C)	Si	Amorphous up to T_s=300°C	1 – 9	3.4 (3.0μm)	Hard
Ge	B or E (1600°C)	Ge	Amorphous up to T_s=300°C	2 –23	4.4 (2.0μm) (T_s=30°C)	Fairly hard

[a] B = boat, E = electron beam   [b] Controlled evaporation from two sources

yielding fairly homogeneous films with refractive indices that lie between those of the pure components, e.g., CeO$_2$-SiO$_2$, CeF$_3$-ZnS, CeO$_2$-CeF$_3$ [240 – 242] and Ge$_{30}$As$_{17}$Te$_{30}$Se$_{23}$ [243]. Mixed films are often easier to prepare by co-sputtering, for instance from two rf diode sources. In this way films with predetermined index of refraction between the ones of the original materials have been prepared reproducibly from the couples CeO$_2$ – SiO$_2$ and TiO$_2$ – SiO$_2$ [206]. Evaporated films with perfect optical homogeneity and isotropy are rare. Many examples of inhomogeneous films are known, for example, MgF$_2$ films [45, 244, 245], cryolite films [19, 22, 246, 247], CaF$_2$ films [248] and ZnS films [191, 249, 250]. One must distinguish between inhomogeneities with gradual increase or decrease in refractive index with increasing film thickness and inhomogeneities with fairly

380

abrupt changes. The first type has been observed more frequently than the second. Abrupt changes are frequently associated with accidental changes in deposition parameters (pressure, rate, temperature). Generally speaking, inhomogeneities limit the utility of thin films. Exceptions to the above are films with special index profiles which offer interesting alternatives to some optical applications. Past work on the controlled production of special inhomogeneous films is described in [251 – 255].

The unisotropic columnar microstructure of many films strongly suggests that they should be birefringent. Films evaporated at normal incidence are expected to be uniaxial birefringent with optic axis normal to the plane of vapour incidence whereas films deposited at oblique incidence should be biaxial birefringent with the optic axes in the plane of the inclined columns. Bousquet [280, 281] detected uniaxial birefringence in CaF_2 films with $(n_o-n_e) = 2.5 \times 10^{-3}$. In the meantime anisotropy in thin films was investigated by various authors [282–284]. Recently birefringence could be found also in obliquely deposited TiO_2 and ZrO_2 films by Macleod and coworkers [285–287].

The Tables 14–21 present collected optical data of dielectric films deposited by wet chemical methods and by various PVD technologies. Finally some data concerning optical properties of plasma polymerized films are reported. The properties depend on the starting material and on deposition parameters.

Most films obtained from organic starting materials have refractive indices between 1.3 and 2.0. Many films deposited from hydrocarbons are coloured ranging from practically colourless through yellow to brownish depending on the deposition power density. The higher the deposition power the deeper the colour in the film. Nitrogen, when present in the film, can increase the index of refraction

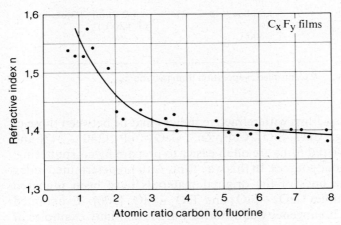

Fig. 28
Refractive index of plasma deposited films as function of the fluorine content [271].
The refractive index was determind by ellipsometry, the carbon fluorine ratio by Auger electron spectrometry.

from $n = 1.5$ to $n = 1.9$ depending on concentration. Films deposited from fluorocarbon containing starting materials can have a lower refractive index of $n = 1.4$ with no absorption in the visible range [270]. The refractive index of fluorine containing films vary with the amount of fluorine as can be seen in Fig. 28.

8.7 ELECTRO-OPTICAL MATERIALS AND THEIR PROPERTIES

The decrease in electrical resistance of selenium during exposure to light was first observed more than hundred years ago. Later a similar behaviour was also found with a number of other materials, but at that time the potential possibilities of the photo-electrical effect were not utilized practically. Today, however, the situation has changed and, for example, energy conversion with photo-voltaic solar cells has become an important technology.

In opto-electronics, a classification can be made similar to that in conventional semiconductor electronics, for example, in light emitting diodes and phototransistors, in hybride components such as alphanumeric displays and in integrated elements such as phototransistor mosaics with integrated localizier logic.

TABLE 22

EXAMPLES OF LIGHT-EMITTING MATERIALS AND CORRESPONDING DETECTOR MATERIALS

Light emitters	Approximate range of emission and absorption λ max (μm)	Light detectors
GaN GaP, sun GaAsP GaAsP, GaAlAs, GaP	0.4 –0.85 blue green yellow red	eye CdS Si
GaAs	0.85–1.0	Si
YAG:Nd laser, electric bulb	1 – 1.5	Ge
air glowing	1.5 –2.0	Ge,PbS
troposphere	2 –3	PbS
InAs	3 –4	InSb
human being, CO_2 laser, PbSnTe	9 –10	CdHgTe

All the components, if possible, are made of silicon. The next important materials are germanium and gallium arsenide as well as some other III-V compounds such as GaP, InSb, InAs and GaN and their ternary mixed compounds. Furthermore, binary and ternary IV-VI compounds like PbSnTe have been found to be very useful [256, 257].

In Table 22, some examples are given for light emitters at various wavelengths and the corresponding detector materials.

In combination with glass substrates coated with transparent electrodes, the electro-optical properties of liquid crystal films, which themselves emit no light, offer some interesting possibilities [258, 259]. Liquid crystals have the mobility of liquids and the optical properties of solids. The molecular structure lies between the liquid and solid state. Liquid crystals are organic compounds of relatively long molecules compared with their diameter, and often contain polar groups and multiple bonds.

There are two broad classes of so-called thermotropic liquid crystals which are termed nematic and smectic liquids. These two classes can be subdivided into some groups, which, however, will not be discussed here. The typical molecular arrangements of liquid crystals are shown in Fig. 29.

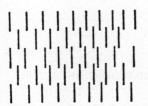

a) Classical nematic structure

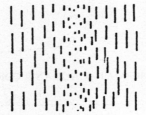

b) Twisted nematic structure (cholesteric)

c) Smectic structure A

Fig. 29
Schematic representation of the molecular arrangements of various types of liquid crystals:
a) Uniformly oriented nematic liquid.
b) Uniformly oriented twistet nematic liquid (cholesteric).This structure possess an infinite-fold screw axis.
c) Molecular packing in a smectic A liquid. This structure has an infinite-fold symmetry axis.

Optical investigations of smetic liquids indicated a behaviour of uniaxial or biaxial crystals depending on the special type of material. They are usually positive birefringent, which means that with transmitted light the ordinary beam has a lower refractive index. The nematic liquid is optically positive.

In both classes of liquid crystals, the velocity of transmitted light is higher in the direction of the long axis of the molecules than perpendicular to it.

The cholesteric liquid, which is a spontaneously twisted nematic, behaves like a negative uniaxial crystal, so that light vibrating perpendicular to the molecular layers shows maximum velocity. Linearly polarized light transmitted perpendicular to the molecular layers shows rotation of its electrical vector along a helical path.

Cholesteric liquids can rotate polarized light to a large degree, such as some thousand degrees per 1 mm layer thickness for visible light [258]. Because of different absorption of the two polarized components in the cholesteric liquid, the material shows dichroism. Most of the effects are applied practically. The colour change in reflection with temperature of a cholesteric liquid can be used for very sensitive temperature measurements of 0.001°C! It is even possible to construct infrared/visible image converters with cholesteric liquid crystals.

Of considerable significance to electro-optical application as non-emissive display is the dynamic scattering effect observed with nematic liquids [260, 261]. A thin film of a nematic liquid crystal, 5-25 μm, between two clear but conducting glass plates is transparent in the absence of an electric field, because the molecules are equally arranged under the influence of the plate walls. The action of a low-frequency ac or a low dc voltage of about 10 volts corresponding to a field of ~ 10 kV cm^{-1} causes the liquid to become opaque because the polar and long molecules become turbulent and scatter light. In the case of twisted nematic effect display cells, the long molecular axes of the nematic liquid crystals are oriented parallel to the conducting plates, and so the orientation of one plate is made perpendicular to the other. The unit-vector field describing preferred molecular orientation turn continuously through 90° across the film thickness. For the function of such a display cell polarizers are required.

Display devices can also be constructed using the field effect, the cholesteric memory effect and the cholesteric-nematic phase change effect [259, 262]. The recognition of the useful electro-optical properties of liquid crystals has stimulated efforts in synthesis of new mesomorphic materials. Today, more than 6000 compounds are available but an ideal liquid crystal is still elusive.

A further electro-optical effect which is studied for applications in display is electrochromism. Electrochromism is characterized as a reversible colour change induced in some organic and inorganic materials by an applied electric field or current. According to Chang [263], the physical mechanisms responsible for these phenomena are electronic and electrochemical in nature. A schematic electrochromic reaction can be formulated as follows:

$$M^n + e^- + A^+ \rightleftarrows A^+M^{n-1} \tag{27}$$

colourless   coloured

where M denotes an anion of the electrochromic material which can exist in different valency states and A a mobile cation such as H$^+$ or an alkali. An example is the formation of tungsten bronze.

$$WO_3 + (H^+ + e^-)_x \rightleftarrows H_x WO_3 \tag{28}$$

transparent   blue

384

Well known examples for the electrochemical reaction studied with respect to solid state display applications are WO_3 [264-267] and the organic viologen salts [268, 269].

A simple electrochromic display consists of a battery with a visible state of charges. When unwanted side reactions can be avoided, the open display retains its electrical charge and hence its colour. Such displays generally have a memory. The required voltage is about 1 V and the induced absorption is proportional to the accumulated charge. Development is still in progress.

REFERENCES

[1]   Ch. Weissmantel and C. Hamann (Eds.), Grundlagen der Festkörperphysik, Springer, Berlin, 1979, p. 125 ff.
[2]   G. B. Bokij, B. K. Vainshtein and A. A. Babarsko, Izv. Akad. Nauk SSSR, Otd. Khim. Nauk, 6 (1951) 667.
[3]   A. N. Lobachev, Acta Crystallogr., 7 (1954) 690.
[4]   Z. G. Pinsker, IZV Akad. Nauk SSSR. Ser. Fis., 13 (1940) 473; Doklad. Akad. Nauk SSSR, 73 (1950) 107; Trudy Inst. Kristallogr. SSSR, 10 (1954) 38.
[5]   G. N. Tishchenko and Z. G. Pinsker, Doklad. Akad. Nauk SSSR, 100 (1955) 913.
[6]   B. K. Vainshtein, Doklad. Akad. Nauk SSSR, 68 (1949) 301, 73 (1950) 103, 83 (1953) 227, 99 (1954) 81; Zhur. Fiz. Khim., 26 (1952) 1774; Zhur. Fiz. Khim., 29 (1955) 327; Trudy Inst. Kristallogr. SSSR, 10 (1954) 49.
[7]   B. K. Vainshtein, Doklad. Akad. Nauk SSSR, 85 (1952) 1239.
[8]   J. M. Cowley, Nature, 171 (1953) 440; Acta Crystallogr., 6 (1953) 53, 516, 522, 846; Acta Crystallogr., 9 (1956) 391.
[9]   J. M. Cowley and J. A. Ibers, Acta Crystallogr., 9 (1956) 421
[10] American Society for Testing Materials: X-ray Department Philadelphia, Pa., USA.
[11] C. A. Neugebauer, Structural disorder phenomena in thin films. In G. Hass (Ed.), Physics of Thin Films, Vol. 2, Academic Press, New York, 1954.
[12] H. Widmer, Appl. Phys. Lett., 5 (1964) 108.
[13] K. L. Chopra, M. R. Randlett and R. H. Duff, Phil. Mag., 16 (1967) 261.
[14] L. Wegmann, BALZERS Eldigraph KD4, Chem. Rundschau, 22 (1969) 487, 514.
[15] R. Niedermayer, Dissertation TH Clausthal (BRD), 1963.
[16] H. Hilbrand, Dissertation, University of Innsbruck, 1966.
[17] (a) H. K. Pulker and E. Ritter, Optical properties and structure of vapour deposited fluoride films. In M. Auwärter (Ed.), Ergebnisse der Hochvakuumtechnik und der Physik dünner Schichten, Vol. II, Wiss. Verlagsges. mbH., Stuttgart, 1971, p. 244.
      (b) H. K. Pulker, Habilitation paper, University of Innsbruck, Austria, 1973.
[18] O. S. Heavens and D. S. Smith, J. Opt. Soc. Am., 47 (1957) 469.
[19] E. Bauer, in M. Auwärter (Ed.), Ergebnisse der Hochvakuumtechnik u. d. Physik dünner Schichten, Vol. 1, Wiss. Verlagsges. mbH., Stuttgart, 1957, p. 39.
[20] E. Bauer, Z. Kristallogr., 107 (1956) 72, 290.
[21] H. G. Coleman, A. F. Turner and O. A. Ullrich, J. Opt. Soc. Am., 37 (1947) 521.
[22] H. K. Pulker and Ch. Zaminer, Thin Solid Films, 5 (1970) 421.
[23] B. Dudenhausen and G. Möllenstedt, Z. Angew. Physik, 27 (1969) 191.
[24] E. Suito and M. Shiojiri, Jap. Academy, 41 (1965) 455.
[25] M. Shiojiri, J. Phys. Soc. Japan, 21 (1966) 355.
[26] E. T. Fitzgibbons, K. J. Sladek and W. H. Hartwig, J. Electrochem. Soc.: Solid State Sci. and Technol., 119 (6) (1972) 735.
[27] G. Hass, Vacuum, 2 (1952) 331.
[28] H. K. Pulker, G. Paesold and E. Ritter, Appl. Opt., 15 (1976) 2986.
[29] E. Wäsch, Kristall und Technik, 8 (9) (1973) 1005.
[30] W. T. Pawlewicz, P. M. Martin, D. D. Hays and I. B. Mann, in R. I. Seddon (Ed.), Proc. SPIE, Vol. 325, Optical Thin Films, 1982 p. 105.
[31] Kirk-Othmer Encyclopedia of Chemical Technology, Vol. 20, 2nd edn. 1969, pp. 390 – 395.

[32] G. Skinner, H. L. Johnston, C. Beckett, Titanium and Its Compounds, H. L. Johnston Enterprises, Columbus, Ohio, USA, 1940, pp. 21 28.

[33] W. A. Pliskin, H. S. Lehmann, J. Electrochem. Soc., 112 (1965) 1013.

[34] W. A. Pliskin, P. P. Castrucci, Electrochem. Technol., 6 (1968) 85; J. Electrochem. Soc., 112, (148 c) (1965).

[35] S. Krongelb, Electrochem. Technol., 6 (1968) 251.

[36] S. Krongelb, T. S. Sedgwick, J. Electrochem. Soc., 113 (63 c) (1966).

[37] A. Cachard, J. A. Roger, J. Pivot and C. H. S. Dupuy, Phys. Stat. Solidi, 5a (1971) 637.

[38] W. A. Pliskin, J. Vac. Sci. Technol., 14 (1977) 1064.

[39] E. Ritter, J. Vac. Sci. Technol., 3 (1966) 225.

[40] T. H. Allen, in R. I. Seddon (Ed.), Proc. SPIE, Vol. 325, Optical Thin Films, 1982, p. 93.

[41] a) Ch. Weissmantel, R. Lenk, W. Forker and D. Linke (Eds.), Kleine Enzyklopädie Struktur der Materie, VEB Bibliogr. Institut Leipzig, 1982, p. 272 – 281.

b) W. Brügel, Einführung in die Ultrarotspektroskopie, 4th. edn., D. Steinkopff, Darmstadt, 1969.

c) L. J. Bellamy, The Infrared Spectra of Complex Molecules, Chapman & Hall, London, Vol. 1, 3rd. edn., 1975; Vol. 2, 1968.

d) R. Szeto and D. W. Hess, J. Appl. Phys., 52 (1981) 903.

[42] a) D. W. Pashley, Advances in Physics, 14 (55) (1965) 327.

b) G. Zinsmeister, Proc. Internat. Symp. Basic Problems in Thin Film Physics, Clausthal-Göttingen 1965, Vandenhoeck & Ruprecht, Göttingen, 1966, p. 33; Vacuum, 16 (1966) 259; Kristall und Technik, 5 (1970) 207; Thin Solid Films, 7 (1971) 51; Vak. Techn., 22 (1973) 85.

c) H. Poppa, J. Vac. Sci. Technol., 2 (1965) 42; J. Appl. Phys., 38 (1967) 3883; J. Vac. Sci. Technol., 9 (1971); Thin Solid Films, 32 (1976)223 and 229.

d) A. Barna and J. F. Pocza, Thin Solid Films, 4 (1969) 212.

e) K. Hayek, Diffusion und Keimbildung bei der Kondensation an festen Oberflächen, Veröffentl. der Universität Innsbruck No 78, Österr. Kommissionsbuchhandlung, Innsbruck, 1973.

f) B. Lewis and J. C. Anderson, Nucleation and Growth of Thin Films, Academic Press, New York, 1979.

g) G. Ehrlich, Proc. 9. Internat. Vac. Congr., Madrid, Spain, 1983, p. 3.

h) J. A. Venables, Proc. 9. Internat. Vac. Congr., Madrid, Spain, 1983, p. 26.

[43] C. Kooy and J. M. Nieuwenhuizen, in R. Niedermayer and H. Mayer (Eds.), Grundprobleme der Physik dünner Schichten, Vandenhoeck und Ruprecht, Göttingen, 1966, S. 181.

[44] H. K. Pulker and K. Hayek, in R. Niedermayer and H. Mayer (Eds.), Grundprobleme der Physik dünner Schichten, Vandenhoeck und Ruprecht, Göttingen, 1966, S. 204.

[45] H. K. Pulker and E. Jung, Thin Solid Films, 9 (1972) 57.

[46] S. Mutze and W. Gloede, Experim. Techn. Phys., XIX (1971) 413.

[47] Y. Namba, Thin Solid Films, 9 (1972) 459.

[48] N. G. Nakhodkin and A. I. Shaldervan, Thin Solid Films, 10 (1972) 109.

[49] J. P. Pearson, Thin Solid Films, 6 (1970) 349.

[50] H. K. Pulker and K. H. Günther, Vakuum-Technik, 21 (1972) 201.

[51] J. M. Nieuwenhuizen and H. B. Haanstra, Philips Techn. Rev., 27 (1966) 87.

[52] A. G. Dirks and H. J. Leamy, Thin Solid Films, 47 (1977) 219.

[53] H. K. Pulker, Proc. EUREM 80, Den Haag Netherlands, Electron Microscopy, 1980, Vol II, p. 788

[54] B. A. Movchan and A. V. Demchishin, Fiz. Met. Metalloved, 28 (1969) 653.

[55] H. A. MacLeod, in R. I. Seddon (Ed.), Proc. SPIE, Vol. 325, Optical Thin Films, 1982, p. 21.

[56] H. J. Leamy, G. H. Gilmer and A. G. Dirks, The microstructure of vapour deposited thin Films. In E. Kaldis (Ed.), Current Topics in Materials Sciences, Vol. 6, North Holland, 1980, pp. 390 – 444.

[57] D. Henderson, M. H. Brodsky and P. Chaudhari, Appl. Phys. Lett., 25 (1974) 641.

[58] K. H. Günther and H. Leonhard, Thin Solid Films, 90 (1982) 76.

[59] R. Messier, private communication to H. A. MacLeod (see [55]).

[60] P. Swab, S. V. Krishuaswamy and R. Messier, J. Vac. Sci. Technol., 17 (1980) 362.

[61] R. C. Ross and R. Messier, J. Appl. Phys., 52 (1981) 5329.

[62] C. H. Morrison (Ed.) Trace Analysis, Physical Methods, Intersciences Publ. Inc., New York, 1965.

[63] R. K. Williardson, Mass Spectrographic Analysis of Thin Films. In B. Schwartz and N. Schwartz (Eds.), Measurement Techniques for Thin Films, Johnson Reprint Corp., New York, 1968, p. 58.

386

[64]   D. L. Malm, RF Spark Source Mass Spectrometer for the Analysis of Surface Films. In E. M. Murt and W. G. Guldner (Eds.), Progress in Analytical Chemistry, Vol. 2, Plenum Press, New York, 1969, p. 148.
[65]   H. E. Beske, Recent Developments in Spark Source Mass Spectrometry, Abstracts, Keynote/3, 9th Int. Mass Spectrometry Conference, Vienna, Austria, 30.8. – 3.9.1982.
[66]   S. J. B. Reed, Electron Microprobe Analysis, Cambridge University Press, London, 1975.
[67]   A. J. Tousimis and L. Marton (Eds.), Electron Probe Microanalysis, Academic Press, New York, 1969.
[68]   K. F. J. Heinrich (Ed.), Quantitative Electron Probe Microanalysis, Natl. Bur. Stand, U. S. Spec. Publ. 298 (1968).
[69]   C. C. Chang, Surf. Sci., 25 (1971) 53; 48 (1975) 9.
[70]   P. H. Holloway, Adv. Electron. Electron Phys., 54 (1980) 241.
[71]   P. M. Hall and J. M. Morabito, Crit. Rev. Solid St. Mater Sci., 8 (1978) 53.
[72]   P. W. Palmberg, Anal. Chem., 45 (1973) 549A.
[73]   L. E. Davis, N. C. MacDonald, P. W. Palmberg, G. E. Riach and R. E. Weber, Handbook of Auger Electron Spectroscopy, Physical Electronics, Eden Prairie, 1976.
[74]   G. McGuire, Auger Electron Spectroscopy Reference Manual, Plenum Press, New York, 1980.
[75]   M. P. Seah, Surf. Interface Anal., 2 (1980) 222.
[76]   B. Feuerbacher and B. Fitton, Photoemission Spectroscopy. In M. B. Ibach (Ed.), Electron Spectroscopy for Surface Analysis, Springer, Berlin, 1977, p. 151.
[77]   A. W. Czanderna (Ed.), Methods of Surface Analysis, Elsevier, Amsterdam, 1975;
        P. F. Kane and G. B. Larabee (Eds.), Characterisation of Solid Surfaces, Plenum Press, New York, 1974;
        D. Briggs (Ed.), Handbook of X-ray and UV Photoelectron Spectroscopy, Heyden, London, 1977.
[78]   A. Benninghoven, Surf. Sci., 28 (1971) 541; 35 (1973) 427; 53 (1975) 596 and Crit. Rev. Solid St. Sci., 6 (1976) 291.
[79]   W. H. Werner, Surf. Interface Anal., 2 (1980) 56.
[80]   E. Zinner, Scanning, 3 (1980) 57.
[81]   K. Wittmaack, Surf. Sci., 89 (1979) 668.
[82]   H. Liebl, J. Vac. Sci. Technol.,12 (1975) 385.
[83]   W. K. Chu, J. W. Mayer, M. A. Nicolet, T. M. Buck, G. Amsel and F. Eisen, Thin Solid Films, 17 (1973) 1; 19 (1973) 423.
[84]   W. K. Chu, J. W. Mayer and M. A. Nicolet, Backscattering Spectrometry, Plenum Press, New York, 1978.
[85]   J. A. Borders, Thin Solid Films, 64 (1979) 403.
[86]   J. W. Mayer and J. M. Poate, Depth Profiling Techniques. In J. M. Poate, K. N. Tu and J. W. Mayer (Eds.), Thin Films, Wiley, New York, 1978, p. 119.
[87]   J. M. Walls, Thin Solid Films, 80 (1981) 231.
[88]   K. H. Günther, Appl. Opt., 20 (1981) 3487.
[89]   H. W. Werner, Mater. Sci. Eng., 42 (1980) 1.
[90]   M. L. Tarng and D. G. Fisher, J. Vac. Sci. Technol., 15 (1978) 50.
[91]   C. Lea and M. P. Seah, Thin Solid Films, 75 (1981) 67.
[92]   J. P. Chubb, J. Dillingham, D. D. Hall and J. M. Walls, Met. Technol., 7 (1980) 293.
[93]   J. M. Walls, D. D. Hall and D. E. Sykes, Surf. Interface Anal., 1 (1979) 204.
[94]   V. Thompson, H. E. Hintermann and L. Chollet, Surf. Technol., 8 (1979) 421.
[95]   E. N. Sicakfus, Ind. Res./Dev., 20 (1980) 127.
[96]   V. I. Derzhiev, G. I. Ramendik, V. Lieblich and H. Mai, Int. J. Mass Spectrom. Ion Phys., 32 (1980) 345.
[97]   L. Papagno, R. Scarmozzino and F. Simoni, Thin Solid Films, 67 (1980) 157.
[98]   P. Sigmund, Phys. Rev., 184 (1969) 384.
[99]   P. Sigmund, in N. H. Tolk, J. Tully, W. Heiland and C. W. White (Eds.), Inelastic Ion Surface Collisions, Academic Press, New York, 1977, p. 121.
[100]   W. D. Hoffman, Surf. Sci., 50 (1975) 29.
[101]   W. D. Hoffman, I. S. T. Tsong and G. L. Power, J. Vac. Sci. Technol., 17 (1980) 613.
[102]   I. S. T. Tsong, G. L. Power, D. W. Hoffman and C. W. Magee, Nucl. Instrum. Meth.,18 (1980) 399.
[103]   P. Sigmund and A. Gras-Marti, Nucl. Instrum. Meth., 168 (1980) 389.
[104]   A. Benninghoven, Z. Phys., 230 (1971) 403.
[105]   S. Hofman, Appl. Phys., 13 (1977) 205.
[106]   K. Röll and C. Hammer, Thin Solid Films, 57 (1979) 209.

[107] J. Erlewein and S. Hofman, Thin Solid Films, 69 (1980) L39.
[108] T. Ishitani and R. Shimizu, Appl. Phys., 6 (1975) 241.
[109] P. S. Ho, J. E. Lewis, H. S. Wildman and J. K. Howard, Surf. Sci., 57 (1976) 393.
[110] N. T. Chou and M. W. Shafer, Surf. Sci., 92 (1980) 601.
[111] R. D. Webber and J. M. Walls, Thin Solid Films, 57 (1979) 201.
[112] D. E. Sykes, D. D. Hall, R. E. Thurstans and J. M. Walls, Appl. Surf. Sci., 5 (80) 103.
[113] S. Hofman, Surf. Interface Anal., 2 (1980) 148.
[114] K. H. Günther, Appl. Opt., 20 (1981) 48.
[115] K. H. Günther, Thin Solid Films, 77 (1981) 239.
[116] H. Bach, Radiat. Eff., 28 (1976) 215.
[117] H. Bach, Radiat. Eff., 22 (1974)73.
[118] W. Weisweiler and R. Neff, Glastech. Ber., 53 (1980) 282.
[119] K. Röll, W. Losch and C. Achete, J. Appl. Phys., 50 (1979) 4422.
[120] F. Ohuchi, M. Ogino, P. H. Holloway and C. G. Pantano, Jr., Surf. Interface Anal., 2 (1980) 85.
[121] R. Buhl, Balzers AG, Project Report 39 171, unpublished, 1978.
[122] K. W. Raine, Thin Solid Films, 38 (1976) 323.
[123] D. L. Malm, M. J. Vasile, F. J. Padden, D. B. Dove and C. G. Pantano, Jr., J. Vac. Sci. Technol., 15 (1978) 35.
[124] A. Zalar, Mikrochim. Acta, 29 (1980) 435.
[125] H. W. Werner and A. E. Morgan, J. Appl. Phys., 47 (1976) 1232.
[126] G. Slodzian, Ann. Phys. Paris, 9 (1964) 591.
[127] A. Nelson and A. K. Green, J. Vac. Sci. Technol., 17 (1980) 855.
[128] H. Bach, Balzers Report BB 800 012 DE (8110).
[129] H. Bach, Z. Angew. Physik, 28 (1970) 239.
[130] R. W. Hoffman, Phys. Thin Films, 3 (1966) 211.
[131] K. Kinosita, Thin Solid Films, 12 (1972) 17.
[132] E. J. Mills, Proc. Roy. Soc. London, 26 (1877) 504.
[133] G. C. Stoney, Proc. Roy. Soc. London, A 32 (1909) 172.
[134] W. Buckel, in C. A. Neugebauer. J. B. Newkirk and D. A. Vermilyea (Eds.), Structure and Properties of Thin Films, Wiley Inc., New York 1959, p. 53.
[135] W. Oswald, Z. Phys. Chem., 22 (1879) 289.
[136] G. Günther and W. Buckel, in R. Niedermayer and H. Mayer (Eds.), Basic Problems in Thin Film Physics, Vandenhoeck Ruprecht, Göttingen, 1966, p. 231.
[137] H. Horikoshi and N. Tamura, Jpn. J. Appl. Phys., 2 (1963) 328.
[138] R. Carpenter and D. S. Campbell, J. Mater. Sci., 2 (1967) 173.
[139] D. B. Dove, J. Appl. Phys., 35 (1964) 2785.
[140] R. M. Hill, Nature 204, (1964) 35.
[141] J. D. Wilcock, D. S. Campbell and J. C. Anderson, Thin Solid Films, 3 (1969) 123.
[142] K. L. Chopra and L. C. Bobb, in M. H. Francombe, H. Sato (Eds.), Single Crystal Films, Pergamon, New York, 1964, p. 373.
[143] K. L. Chopra, J. Appl. Phys., 37 (1966) 2249.
[144] R. W. Hoffman et al., A. E. C. Techn. Rep. 18 (1961); 64 (1970); 76 (1971); 79 (1972); 82 (1975); 83 (1975); Case Western Reserve University, Cleveland, Ohio.
[145] J. D. Wilcock, Stress in thin films, Thesis, Imperial College, University of London, London, (1967).
[146] J. Priest, H. L. Caswell and Y. Budo, Vacuum, 12, (1962) 301.
[147] A. Brenner and S. Senderorff, J. Nat. Bur. Stand., 42 (1949) 105.
[148] a) Campbell, in L. I. Maissel and R. Glang (Eds.), Handbook of Thin Film Technology, Ch. 12, McGraw-Hill Inc., New York, 1970.
      b) J. P. Seidel, Thesis, 1981, Institute of Experimental Physics, University of Innsbruck, Austria.
[149] R. W. Hoffmann, in Physics of Thin Films, Vol. 3, Academic Press, New York, 1966, p. 211.
[150] H. K. Pulker, R. Buhl and J. Mäser, Proc. 7th Int. Vacuum Congress, Vienna, 1977, p. 1761.
[151] S. M. Wong, Thin Solid Films, (1978) 65.
[152] J. N. Zemel, J. D. Jenson and R. B. Schodar, Phys. Rev., 140 (1965) 330.
[153] R. L. Coren, J. Appl. Phys., 33 (1962) 1168S.
[154] C. Lu and A. A. Milgram, J. Appl. Phys., 38 (1967) 2038.
[155] C. A. Pela, G. E. Barberis, J. F. Suassuna and C. Rettori, Phys. Rev., B 21 (1980) 34.
[156] W. B. Ittner, III, in G. Hass (Ed.), Physik of Thin Films, Vol. 1, Academic Press, New York, 1963, p. 233.
[157] A. F. Turner, Thick Thin Films, Bausch and Lomb, Techn. Report, 1951.

388

[158]   A. E. Ennos, Appl. Opt., 5 (1966) 51.
[159]   Pulker et al., Thin Solid Films, (a) 58 (1979) 371; (b) 59 (1979) 65; (c) Proc. 8th Int. Vac. Congr., Cannes, Sept. 1980.
[160]   H. K. Pulker, in, R. I. Seddon (Ed.), Proc. SPIE, Vol. 325, Optical Thin Films, 1982, p. 84.
[161]   Y. Nakajima and K. Kinosita, Thin Solid Films, 5 (1970) R 5.
[162]   R. Abermann, R. Koch and R. Kramer, Thin Solid Films, 58 (1979) 365.
[163]   E. Klokholm, J. Vac. Sci. Technol., 6 (1969) 138.
[164]   M. Laugier, Thin Solid Films, 79 (1977) 169.
[165]   A. Kubovy and M. Janda, Thin Solid Films, 42 (1977) 169.
[166]   R. Berger and H. K. Pulker, Abstract PB4, in E. Ledinegg (Ed.), Öster. Physikal. Gesellschaft Jahrestagung 1980, TU.-Graz, 22-26 Sept. 1980, Internal Publicaton; and in J. R. Jacobsson (Ed.), Proc. SPIE, Vol. 401, Thin Film Technologies, 1983 p. 69.
[167]   H. P. Martinz and R. Abermann, Proc. 8th Int. Vac. Congress, Vol. 1, Thin Films, Sept. 22-26, 1980, Cannes, France, Suppl. Le Vide, les Couches Minces, No. 201 (1980) 311.
[168]   R. Carpenter and D. S. Campbell, unpublished calculation 1964, (see [148]).
[169]   A. Kinbara and S. Baba, Proc. 8th Int. Vac. Congress, Vol. I, Thin Films, Sept. 22-26, 1980, Cannes, France, Suppl. Le Vide. les Couches Minces, No. 201 (1980) 323.
[170]   H. Schröder and G. M. Schmidt, Z. angew. Phys., 18 (1964) 124.
[171]   K. Kinosita et al., Jpn. J. Appl. Phys., 4 (1965) 340.
[172]   J. Seidel, Thesis, Inst. Experimentalphysik, University of Innsbruck, Austria, 1981.
[173]   K. C. Ludema, W. A. Glaeser and S. K. Rhee (Eds.), Wear of Materials, 1979, Amer. Soc. Mech. Eng., New York.
[174]   R. F. Smart and J. C. Moore, Wear, 56 (1979) 55.
[175]   E. F. Finkin, Mat. Eng. Appl., 1 (1979) 154.
[176]   T. Spalvins, Thin Solid Films, 53 (1978) 285.
[177]   J. L. Jellison, IEEE Transact., PHP-13 (1977) 132.
[178]   Handbook of Chemistry and Physics, 54th edn., CRC Press, 1973/74.
[179]   J. E. Hiller, Grundrisse der Kristallchemie, W. de Gruyter, Berlin, 1952.
[180]   Schott Informationen, Heft, 4 (1967) 15.
[181]   Werkstofftabellen der Metalle, Kröner Verlag, Stuttgart, 1972.
[182]   R. D. Wales and W. Hortinuiko, Plating and Surf. Finishing, 64 (1977) 30.
[183]   B. W. Mott, Micro-Indentation Hardness Testing, Butterworth Sci. Publ., 1956.
[184]   M. Nishibori and K. Kinosita, Thin Solid Films, 48 (1977) 325.
[184a] H. Bangert, A. Wagendristel and H. Aschinger, Thin Solid Films, 89 (1982) 131.
[184b] H. Markhof, FRG, Offenlegungsschrift 2735340 (8 Februar 1979).
[185]   L. S. Palatnik, G. V. Federov, A. I. Prokhavulov and A. I. Ferenki, Fisika Metall., 20 (1965) 574.
[186]   B. A. Movchan and A. V. Demchishin, Fizika Metall., 28 (1969) 83.
[187]   J. Colen and R. F. Bunshah, J. Vac. Sci. Technol., 13 (1976) 536.
[188]   L. Holland and E. W. van Dam, J. Opt. Soc. Am., 46 (1956) 773.
[189]   Th. Kraus, BALZERS Fachbericht 5 (1965).
[190a] H. Ebel, A. Wagendristel and H. Judtmann, Z. Naturf., 23a (1968) 1863.
[190b] H. Ebel and F. Hengstberger, Z. Naturf., 25a (12) (1970) 1984.
[191]   A. Preisinger and H. K. Pulker, Jpn. J. Appl. Phys. Suppl., 2 (1) (1974) 769.
[192]   H. K. Pulker, Appl. Opt., 18 (1979) 1969.
[193]   F. M. D'Heurle, Metallurg. Trans., 1 (1970) 725.
[194]   R. W. Hoffman, Thin Solid Films, 89 (1982) 155.
[195]   H. T. G. Hentzell, B. Andersson and S. E. Karlsson, Thin Solid Films, 89 (1982) 116.
[196]   H. K. Pulker, Thin Solid Films, 89 (1982) 191.
[197]   L. Hiesinger and H. König, Optik und Struktur kathodenzerstäubter Schichten unter besonderer Berücksichtigung der Reaktion mit dem Restgas, in K. Ruthardt (Ed.), Wissenschaftl.-Techn. Festschrift 100 Jahre, W. C. Heraeus GmbH., 1951.
[198]   W. J. Coleman, Appl. Opt. 13 (1974) 946.
[199]   R. H. Deitch, E. J. West, T. G. Giallorenzi and J. F. Weller, Appl. Opt., 13 (1974) 712.
[200]   W. T. Pawlewicz and N. Laegreid, Proc. SPIE, Vol. 140, Optical Coatings, Application and Utilization II, March 1978, Washington, DC.
[201]   W. T. Pawlewicz and R. Busch, Thin Solid Films, 63 (1979) 251.
[202]   W. T. Pawlewicz, R. Busch, D. D. Hays, P. M. Martin and N. Laegreid, Battelle Report PNL-SA-8012 (1979), Pacific North West Lab., Richland, Washington.
[203]   W. T. Pawlewicz, D. D. Hays and P. M. Martin, Thin Solid Films, (1980) 169.

[204] W. T. Pawlewicz, P. M. Martin, D. D. Hays and I. B. Mann, in R. I. Seddon (Ed.), Proc. SPIE, Vol. 325, Optical Thin Films, 1982, p. 105.
[205] a) P. Bugnet, J. Deforges and S. Duromd, J. Phys. D.: Appl. Phys., 6 (1973) 1986.
      b) b)E. Krikorian, Abstract in J. Vac. Sci. Technol., 15 (1978) 260.
[206] a) C. Misiano, M. Varasi, C. Mancini and P. Sartori, Proc. 4eme Coll. Int. Plasmas et Pulverisation Cathodique, CIP 82, IUT Nice, 13-15 Sept. 1982. Suppl. no 212 de la revue Le vide les couches minces, Juillet 1982, p. 149.
      b) C. Misiano, Vuoto, 24 (1983) 35.
      c) M. Varasi, C. Misiano and L. Lasaponara, Proc. Internat. Ion Engineering Congr. – ISIAT 83 and IPAT 83, Kyoto, 1983, p. 1041., publ. by Institute of Electrical Engineers of Japan, Tokyo, T. Takagi (Ed.).
[207] G. Hass, Optik, 1 (1946) 8;
      G. Hass and M. Waylonis, J. Opt. Soc. Am., 51 (1961) 719.
[208] O. W. Black and J. Wales, Infrared Phys., 8 (1968) 209.
[209] S. J. Czyzak, D. C. Reynolds, R. C. Alten and C. C. Reynolds, J. Opt. Soc. Am., 44 (1954) 864.
[210] H. Ahrens, Thesis, Technical University, Hannover, West Germany, 1974; H. Ahrens, H. Welling and H. E. Scheel, Appl. Phys. 1 (1973) 69.
[211] G. Hass, J. B. Ramsey and R. Thun, J. Opt. Soc. Am., 48 (1958) 324.
[212] W. P. Barr, J. Sci. Instrum., 2 (1969) 1112.
[213] K. H. Günther, Thesis, University of Innsbruck, Austria, 1974.
[214] K. H. Günther, H. L. Gruber and H. K. Pulker, Thin Solid Films, 35 (1976) 363.
[215] F. Flory, Thesis, University D'Aix-Marseille France, 1978.
[216] E. Ritter, in Physics of Thin Films, Vol. 8, Academic Press, New York, 1975.
[217] J. E. Goell and R. D. Standley, Proc. IEEE, 58 (1970) 1504.
[218] P. K. Tien, G. Smolinsky and R. J. Martin, Appl. Opt., 11 (1972) 637.
[219] D. H. Hensler, J. D. Cuthbert, R. J. Martin and P. K. Tien, Appl. Opt., 10 (1971) 1037.
[220] H. P. Weber, F. A. Dunn and W. N. Leibolt, Appl. Opt., 12 (1973) 755.
[221] A. Hordvik, Appl. Opt., 16 (1977) 2827.
[222] P. Baumeister and O. Arnon, J. Vac. Sci. Technol., 14 (1977) 195; Appl. Opt. 16 (1977) 439.
[223] D. C. Smith, M. S. Thesis, University of Rochester, 1978.
[224] C. E. Chase, Electro-Opt. Syst. Des., 9 (1977) 59.
[225] a) H. E. Bennett, A. J. Glass, A. H. Guenther and B. E. Newnam (Eds.), Laser Induced Damage in Optical Materials: 1980 (Boulder Damage Symposium) NBS Special Publ. 620; ASTM, STP 759; US Government Printing Office, Washington, DC 20402: 1981.
      b) A. H. Guenther and T. W. Humphreys, in J. R. Jacobsson (Ed.), Proc. SPIE, Vol. 401, Thin Film Technologies, 1983, paper 28.
[226] H. K. Pulker, Thin Solid Films, 34 (1976) 343.
[227] R. T. Kersten, H. F. Mahlein and W. Rauscher, Thin Solid Films, 28 (1975) 369.
[228] P. K. Tien, R. Ulrich and R. J. Martin, Appl. Phys. Lett. 14 (1969) 291.
[229] R. T. Kersten and W. Rauscher, Opt. Commun., 13 (1975) 189.
[230] G. Smolinsky, J. Vac. Sci. Technol., 11 (1974) 33.
[231] G. Bauer, Ann. Phys., 19 (1934) 434.
[232] H. K. Pulker, Optik (Stuttgart), 32 (1971) 496.
[233] A. P. Bradford, G. Hass and M. McFarland, Appl. Opt., 11 (1972) 2342.
[234] H. Koch, Phys. Stat. Solidi, 12 (1965) 533.
[235] W. Heitmann and G. Koppelmann, Z. Angew. Phys., 23 (1967) 221.
[236] E. Ritter and R. Hoffmann, J. Vac. Sci. Technol., 6 (1969) 733.
[237] A. C. S. van Heel and W. Van Vonno, Appl. Opt., 6 (1967) 793.
[238] J. T. Gruyters and N. A. Krijzer, in Proc. Int. Conf. Thin Films, Vol. II., 1972, p. 299.
[239] S. Ogura, N. Sugawara and R. Hiraya, Thin Solid Films, 30 (1975) 3.
[240] H. Vonarburg, Optik, 20 (1963) 43.
[241] S. Fujiwara, J. Opt. Soc. Am., 53 (1963) 1315.
[242] S. Fujiwara, J. Opt. Soc. Am., 53 (1963) 880.
[243] A. D. McLachlan and W. E. K. Gibbs, in A. J. Glass and A. H. Guenther, Appl. Opt., 17 (1978) 2386, 2398.
[244] J. F. Hall, Jr., J. Opt. Soc. Am., 46 (1956) 1013.
[245] K. Nagata, Jpn. J. Appl. Phys., 7 (1968) 1181.
[246] G. Koppelmann, K. Krebs and H. Leyendecker, Z. Phys., 163 (1961) 557.
[247] R. P. Netterfield, Appl. Opt., 15 (1976) 1969.
[248] A. Bourg, N. Barbaroux and M. Bourg, Opt. Acta, 12 (1965) 151.

390

[249] W. R. Oliver, Phil. Mag., 21 (1970) 1229.
[250] W. R. Oliver, Opt. Acta, 17 (1970) 593.
[251] R. Jacobsson, J. Phys. Paris, 25 (1964) 46.
[252] R. Jacobsson and J. O. Martensson, Appl. Opt., 5 (1966) 29.
[253] H. Anders and R. Eichinger, Appl. Opt., 4 (1965) 819.
[254] S. K. Sharma, V. N. Yadova and K. L. Chopra, Jpn. J. Appl, Phys. Suppl. 2, (1) (1974) 685.
[255] R. Jacobsson, in G. Hass, M. H. Francombe and R. W. Hoffman (Eds.) Physics of Thin Films, Academic Press, New York, 1975.
[256] D. E. Charlton, J. Cryst. Growth, 59 (1982) 98.
[257] R. F. C. Farrow, J. Vac. Sci. Technol., 129 (1981) 150.
[258] B. Hampel, Laser, 3 (1971) 53.
[259] G. H. Brown, J. Opt. Soc. Am., 63 (1973) 1505.
[260] G. H. Heilmeier, Appl. Phys. Lett., 13 (1968) 46.
[261] M. Schadt and W. Helfrich, Appl. Phys. Lett., 18 (1971) 127.
[262] E. P. Raynes, Cholesteric texture and phase change effects, in A. R. Kmetz and F. K. von Willisen (Eds.), Non-emissive Electrooptic Displays, Plenum Press, New York, (1976), p. 25.
[263] I. F. Chang, Electrochromic and electrochemichromic materials and phenomena, in A. R. Kmetz and F. K. von Willison (Eds.), Non-emissive Electrooptic Displays, Plenum Press, New York, 1976, p. 155.
[264] S. K. Deb, Phil. Mag., 27 (1973) 801.
[265] S. K. Deb and J. A. Chopvarian, J. Appl. Phys., 37 (1966) 4818.
[266] S. K. Deb, Appl. Opt Suppl., 3 (1969) 192.
[267] B. W. Fanghnan, R. S. Crandall and P. M. Heyman, RCA Rev., 36 (March) (1975) 177.
[268] C. J. Schoot, J. J. Ponjee, H. T. van Dam, R. A. van Doom and P. R. Bolwjis, Appl. Phys. lett., 23 (1973) 64.
[269] H. T. van Dam and J. J. Ponjee, J. Electrochem, Soc., 121 (1974) 1555.
[270] G. Kaganowicz, P. Datta and J. W. Robinson, Proc. of the 4th Internat. Symp. on Plasma Chemistry, Zürich, 1979.
[271] K. Sasaki, Y. Kudo, H. Watanabe and O. Hamano, Thin Solid Films, 89 (1982) 297.
[272] R.P. Netterfield and P.J. Martin, Proc.Internat.Ion Engineering Congr.–ISIAT 83 and IPAT 83, Kyoto, 1983, p.909, publ. by Institute of Electrical Engineers of Japan,Tokyo, T.Takagi (Ed.).
[273] J. C. Maxwell-Garnett, Roy. Soc. Phil. Trans., 203 (1904) 385.
[274] J. C. Maxwell-Garnett, Roy. Soc. Phil. Trans.. 205 (1906) 237.
[275] W. L. Bragg and A. B. Pippard, Acta Cryst., 6 (1953) 865.
[276] M. Harris, H. A. Macleod, S. Ogura, E. Pelletier and B. Vidal, Thin Solid Films, 57 (1979) 173.
[277] K. Kinosita and M. Nishibori, J. Vac. Sci. Technol., 6 (1969) 730.
[278] M. Harris, M. Bowden, G. Hayden and H. A. Macleod, «Refractive index calculations for thin film with columnar structure». Paper presented at the Institute of Physics Meeting on Developments in Optical Thin Film Coatings, Imperial College, London, March 27, 1980 (Summary only).
[279] H. A. Macleod, «Some Effects of Microstructure on the Properties of Optical Thin Films». Paper presented at Journees d'Etude sur les Multicouches pour Application Optique, 8 – 9 November 1983, Ecole Nationale Superieure de Physique, Univ. Marseille. To be published in Le Vide, les Couches Minces, 1984.
[280] P. Bousquet, «Theoretical study of the optical properties of thin transparent films», Ann. Phys. (Paris), Vol. 2 pp. 5 – 15, 1957.
[281] P. Bousquet and Y. Delcourt, «Evidence of the birefringence of calcium fluoride thin layers obtained by vacuum evaporation», J. Phys. Rad., Vol. 18, pp. 447 – 452, 1957.
[282] P. Bousquet and P. Rouard, «Constantes optiques et structure des couches minces», J. Phys. Rad., Vol. 21, pp. 873 – 892, 1960.
[283] P. Rouard, «Determination of the optical constants of thin films», J. Opt. Soc. Am., Vol. 46, p. 370, 1956 (abstract).
[284] R. J. King and S. P. Talim, «A comparison of thin film measurement by guided waves, ellipsometry and reflectometry», Optica Acta, Vol. 28, pp. 1107 – 1123.
[285] F. Horowitz and H. A. Macleod, «Forum birefringence in thin films», Paper presented at the Los Alamos Conference on Optics, Santa Fe, April, 1983.
[286] F. Horowitz, «Structure-induced optical anisotropy in thin films», PhD Dissertation, University of Arizona, Tucson, Optical Sciences Center, 1983.
[287] I. J. Hodgkin, F. Horowitz, H. A. Macleod, M. Sikkens and J. J. Wharton, «Birefrigence in optical coatings», Paper to be presented at the OSA Fall Meeting, New Orleans, 1983.

CHAPTER 9

9. APPLICATION OF COATINGS ON GLASS

9.1 GENERAL CONSIDERATIONS

Thin films are applied to effect a desired change in chemical and/or physical properties of glass and plastic surfaces [1-6]. Our first example of technical surface coating is important in the container industry: glass containers such as non-returnable and returnable bottles are coated by CVD and spray processes with various metallic oxides such as tin oxide, and given topcoats such as non-toxic silicones, various waxes and glycols, etc. The thin invisible oxide film is deposited generally by the application of a hot-end coating after the containers leave the forming machine (550 – 680°C) and before they enter the annealing lehr. The organic coating, known as a cold-end coating, is applied over the oxide film at a lower temperature (100 – 200°C) when the glass containers emerge from the lehr. The coating increases their strength and abrasion resistance, and provides sufficient dry lubrication to ensure a smooth flow in the high-speed procedures of inspection, automatic filling and packing [7-9]. Protection from friction damage is essential in modern filling lines where speeds of up to 800 bottles min^{-1} are used. In the case of returnable containers, an additional requirement is that the coating is retained during about 20 alkaline hot washing operations and that it prevents the formation of opacity of the glass bottles [8]. Other examples are the deposition of thin films to increase the strength and the chemical and environmental stability of various sensitive glass types. Then there is the coating of hollow and flat glass substrates with perfect dense and glossy coloured enamel films by a special thermal spraying process [10]. Thin vitreous coatings of SiO_2, TiO_2 and ZrO_2 deposited by precipitation from solutions, and suitable for a variety of technical glass applications can also be used to protect ancient glasses against weathering. This was demonstrated by weathering tests after coating glass objects dating from Roman times and from the late Renaissance [11].

Decorative coatings produced by sputtering, which are used for consumer product containers, are of increasing importance. Containers produced from glass or other inexpensive materials often lack the pleasing aesthetic appearance desired by the marketing profession. Gold, silver and bronze coatings as well as other coloured coatings are known to provide this marketing appeal, as well as to add functional benefits to the containers, such as wear resistance and humidity resistance [12]. Electrical thin film applications often require ceramics or single crystalline materials rather than glass or plastic substrates, and will therefore only be considered in some special cases in this monograph. One exception is the use of insulating Langmuir-Blodgett films of Cd-stearate with tunnelling dimensions (1-

3 nm) in Schottky barrier MIS structures with hydrogenated amorphous silicon, a-Si:H. Such configurations can be used for solar cells. With thicker insulators (typically 100 nm), they can be used for insulated gate field-effect transistors [13]. By contrast, glass and plastics are typical substrate materials for the deposition of optical coatings. Compared with other technical coating applications, the production of optical thin-film components is relatively old, pioneered by Smakula, Strong, Geffcken, Turner, Hass and Auwärter, among others [4], and has grown continuously. Optical coatings can split light into transmitted, reflected and absorbed fractions in a prescribed manner. Many technical problems, both functional and decorative, can be solved using these possibilities. Newer components often combine optical and other desired film characteristics, e.g. electrical, to achieve an optimized operational possibility for special purposes, such as coatings in display techniques and energy-related optical coatings [15].

Moisture-sensitive optical components such as alkali metal halide materials require protection against exposure to high humidity but in addition need antireflective surface properties. With plasma polymerization, pinhole-free, highly adherent, mechanically intact fluorocarbon-type polymer films in the thickness range between 500 and 1500 nm can be obtained which combine moisure resistance and antireflection properties, see [14]. Optical thin-film components are used in the visible as well as in the infrared and ultraviolet region. It seems that even neutron and soft X-ray optical applications may become relevant [16]. In an increasing number of applications, optical systems consisting of films with the lowest possible optical losses are required, promoting the use of highest purity starting materials and the development of clean and clearly definded vacuum coating technologies.

Plastic optical components like lenses and spectacle lenses are generally coated with scratch-resistant layers and/or antireflection films, e.g. [17]. With the versatility of design and expanding usage of many plastics, vacuum metallizing of such parts has received considerable attention in various industries. The automobile industry offers many opportunities with the wide range of plastic parts used on cars [20, 21]. Many different functional and decorative components such as head lamp and side marker bezels, tail-light trims, instrument knobs, lock buttons, door pulls, mirror holders, arm rests, grilles, etc. are made of coated plastics. In addition to the classical plastics such as polystyrene, PVC, plexiglass, and polycarbonate, over the last 15 years the chemical industry has developed plastic combinations, such as ABS polymers or Celcon Acetal copolymers [18, 19], which are cheap, possess the desired mechanical properties and can be vacuum metallized.

Generally, plastic parts are varnished before vacuum coating. The influence of the base material is decreased by the varnish, for example degassing, and by using a correctly selected varnish, faults and slight impurities are smoothed and trapped. Also, the tendency of stress cracking of some metal films such as Cr films can be decreased by deposition onto base-laquered plastics. In selecting a varnisch for the base-laquering procedure, its adhesive properties and suitability for use in high vacuum must be carefully considered [36]. Varnishing is generally carried out by spraying, dipping or flow coating. After drying in an oven, the varnished parts are placed in the vacuum coating system and metallized. Normally, a second coat of an ultraviolet-resistant varnish is applied after metallizing to protect the deposited metal layer from abrasion and wear.

Metallizing with gold and copper is a well known technique of enriching plastics. However, gold can be replaced by the cheaper CuAl and CuZn alloys. Aluminium and chromium are the most often used metal films. Multicomponent alloy films, e.g. stainless steel and various hard metal films, are generally deposited by cathode sputtering. In metallizing plastics with mirror films or decorative layers, it is usually sufficient for them to have metallic reflection and the required brilliance. These can be simply obtained by quantitative evaporation. Particularly good decorative effects can often be obtained by coating the rear side of transparent plastic parts. In metal coating of plastics for electrical applications, however, special measuring devices must be provided to reproduce a definite surface resistance.

Aluminizing lamp reflectors dates back to 1937 when Daniel Wright at General Electric, USA, made the first sealed beam lamp [22]. Now more than 500 types of light sources with aluminized reflectors are produced. The 200 nm thick aluminium reflector film is generally protected by evaporated silicon oxide or by plasma-polymerized organic coatings.

Plasma polimerization of organic and inorganic films has become an important practical technique for the deposition of thin coatings of excellent physiycal and chemical properties. As already mentioned, pinhole-free conformal and water-resistant films can be obtained. The use of highly crosslinked organic films such as protective or dielectric layers has been described by many authors, e.g. [23-27]. The starting materials are numerous and diverse. Styrene is very popular but also butadiene, acetylene, various olefines and propylene oxide are frequently used. The properties of plasma-polymerized films can be varied substantially and can often be tailored towards a desired application. Inorganic plasma polymerized films such as Si_3N_4, SiO_2 and oxynitrides are used in the semiconductor industry as insulating films or diffusion masks [28, 29]. Plasma polymerization is also used for the preparation of high quality laser light guides in integrated optics [30]. Recently, coloured coatings, 300-500 nm thick, have been developed [31], and can be applied to a variety of substrates by a codeposition process of Au, Cu or Al and plasma polymerization of hexamethyldisiloxane. To date, for the production of these metal colloid type films, gold has been used to produce red and pink coatings, copper for yellow and green, and aluminium for blue coloured coatings [31]. Theory, i.e. Maxwell-Garnett or Mie theories, and experiment have been found to be in good agreement. The coatings are solvent-resistant and are believed to be as thermally stable as other plasma-polymerized deposits. The technology of plasma polymerization has been improved recently to allow efficient coating of large substrates of 180×180 cm in area [31].

Thin-film products have always been deposited on substrates of relatively small dimensions. Exceptions have been the wet chemical silvering for mirror production and perhaps the making of special oxidic heat protection layer systems by dip coating. A few other rare examples for large area coating can be given in film deposition on large astronomical telescope mirrors and on 1.2 m diameter optical components for the lasers used in nuclear fusion experiments.

The more traditional way of coating small pieces has been complemented today by coating large panes of flat glass used in building and automotive industries and by coating of various plastic foils, see for instance [32]. Within the building industry, solar protection glasses using a variety of metallic or oxymetalliy coatings, made by dip and spray coating or by CVD and PVD technologies, are used. For the same purpose, plastic foils, in particular polyethylene, polypropylene, polyester, polyamide and polycarbonate are coated with solar protection films in various types of roll-coaters using reactive and non-reactive physical vapour deposition processes [33, 34]. However, plastic foils are also metallized, e.g. with aluminium for decorative and electrotechnical applications. Metallized plastic foils are particularly suitable for electric condensors and replace the paper strip formerly used, as a result of which many of the difficulties associated with paper have been eliminated. The coating of long lengths under vacuum can be performed by uncoiling from specially constructed roll coaters. This is possible because the gas release from suitable plastic foils is considerably smaller than was the case with paper. Metallization of plastic webbing is also important. The conditions are generally similar to those with the foils. For video-recorder tape production, long plastic foils are coated with magnetic films and are subsequently cut into strips.

In special plants with special locks, aluminium coatings are deposited on optical phono and/or video disks [35a and b] made of plastic or glass. The next generation of optical disks is programmable and has been developed using as the optical storage medium, either vacuum deposited tellurium, its alloy and compound films or other alloy films . Laser diodes are used as writer and reader to record video signals and play them back in real time mode [35b – 35d]. Recently even an erasable and reuseable type of optical disk using a doped tellurium suboxide thin film has been announced [35e – 35g]. In this case recording is achieved by transforming the highly reflecting crystalline phase into the low reflecting amorphous phase by irradiation with an 830 nm laser. The recorded signal is erased by inducing recrystallization at a laser wavelength of 780 nm. This phase transition can be repeated up to a million times.

In the transport industry, the most important application is the de-icing and de-misting of large extended surface areas by transparent and electrically conductive coatings and there are, in addition, some solar control applications.

A great number of optical thin-film components increase or decrease light intensity, split light beams or act as light guides. These are made by the deposition of metal and/or dielectric films, as well as by combinations of metals and dielectrics or of dielectric multilayers taking advantage of their properties of reflection, multiple reflections and interference in the layers. Such coatings are produced mainly by evaporation under vacuum and other PVD techniques. In some cases, however, wet chemical processes such as dip coating and various CVD technologies are also used. As such films are very important in coating glass, the principles responsible for the optical effects obtained will be treated briefly in the next section.

9.2 CALCULATION OF OPTICAL FILM SYSTEMS

Films with no absorption are very important for most optical thin film applications. Following Anders [4] and Macleod [5], some simple statements have to be accepted in the qualitative and quantitative understanding of the performance of optical thin film devices. These statements for the interaction of light with non-absorbing films are the following.

1) The films under consideration are homogeneous, i.e. the optical constants do not depend on position within the film.

2) With non-absorbing films, $A = 0$, and for reflectance and transmittance the relation $R + T = 1$ is valid. It is therefore sufficient to know either R or T. Reflectance R is considered in the following treatment.

3) At any boundary between two different media, the amplitude of reflectance of light is $r = (1-\rho)/(1+\rho)$, where ρ denotes the ratio of refractive indices at the boundaries, e.g. $n_0 = $ air, $n_1 = $ film, $n_2 = $ substrate.

4) When reflection occurs in a medium of lower refractive index than the adjoining one, there is a phase-shift of 180°. In the reverse case, the phase-shift is zero.

5) The two split beams formed by reflection at the top and at the bottom of a film can interfere destructively or constructively. The first is the case if the resultant amplitude is the difference of the amplitudes of the two components when the relative phase-shift is 180°. Constructive interference occurs when the relative phase-shift is zero or a multiple of 360°.

Various mathematical ways exist of evaluating how light , incident on one or more thin films, is reflected. According to Anders [4], three different attempts to solve this problem have been tried.

1) For each special problem, the Maxwell equations are used with their limiting and transition conditions and numerical solutions of these equations are tried.

2) Analogous considerations to the well known quadrupole theory of electricity are made with optical systems.

3) The well known solutions for every single boundary between two optical media, the Fresnel equations, are applied. The reflected portions of the waves are added with respect to amplitudes and phases. Only one way of calculating the reflectance of multilayers applying the Fresnel formula will be demonstrated here.

If a light wave from a medium with refractive index n_0 falls vertically on a thin film n_1 which is on a glass substrate n_3, the light wave undergoes a reflection according to Fresnel equations upon passing through both boundary surfaces of the film:

$$r_1 = \frac{n_0 - n_1}{n_0 + n_1}, \quad r_2 = \frac{n_1 - n_2}{n_1 + n_2} \tag{1}$$

These two waves, also enhancend by repeated reflected fractions of r_2, indicate a path difference of a multiple of $2\,n_1\,t_1$. The addition of the waves according to phase and amplitude gives for the reflected amplitude:

$$r\,e^{i\varepsilon} = \frac{r_1 + r_2\,e^{-2i\delta_1}}{1 + r_1\,r_2\,e^{-2i\delta_1}} \tag{2}$$

$$\delta_1 = \frac{2\pi}{\lambda}\,n_1\,t_1\,\cos\varphi_1 \;\ldots\; \text{phase thickness}$$

$$\varepsilon \;\ldots\; \text{phase angle}$$

and for the reflected intensity, the reflectance:

$$R = \frac{r_1{}^2 + r_2{}^2 + 2r_1\,r_2\,\cos2\delta_1}{1 + r_1{}^2\,r_2{}^2 + 2r_1\,r_2\,\cos2\delta_1} \tag{3}$$

If the values r_1 and r_2 obtained from n_0, n_1 and n_2 are applied for various angles 2δ in a Gaussian plane, the values from the reflection amplitude r and phase ε of the numerator (N) and denominator (D) of eqn. (3) are obtained (r_N, ε_N, r_D, ε_D) and from this $\varepsilon = \varepsilon_N - \varepsilon_D$ and $R = (r_N/r_D)^2$.
The reflection shows an extremum value at

$$2\delta_1 = \frac{4\pi}{\lambda}\,n_1 t_1$$

$$= \pi, 3\pi, \ldots \quad (\text{for } \cos\varphi_1 = 1)$$

i.e. $n_1 t_1 = \lambda/4, \;\; 3\lambda/4, \ldots$, this term is called the phase condition.

Regarding the amplitude condition, it is valid that if amplitudes r_1 and r_2 have the same sign, the film shows a reflection minimum; however, if they have different signs, a reflection maximum occurs. In the case of oblique indicidence of light at an angle φ_1 to the perpendicular, the path difference changes by $\cos\varphi_1$ to $(4\pi/\lambda)n_1 t_1 \cos\varphi_1$. This means that the reflected amplitude is split into a vertical ($r_1^{\perp}$, $r_2^{\perp}$) and a parallel component ($r_1^{\parallel}$, $r_2^{\parallel}$) and $n_1 t_1 \cos\varphi_1 = \lambda/4, 3\lambda/4$, i.e. the maximum of the reflectance is shifted to short wavelengths in the case of oblique incidence of light. If several films are considered for the purpose of an increase in reflection, then for 2 films, eqn. (2) extends to:

$$r\,e^{i\varepsilon} = \frac{r_1 + r_2\,e^{-2i\delta_1} + r_3\,e^{-2i(\delta_1+\delta_2)} + r_1\,r_2\,r_3\,e^{-2i\delta_2}}{1 + r_1\,r_2\,e^{-2i\delta_1} + r_1\,r_3\,e^{-2i(\delta_1+\delta_2)} + r_2\,r_3\,e^{-2i\delta_2}} \tag{4}$$

$$n_0 \quad n_1 \quad n_2 \quad n_3 \; ; \quad r_1 = \frac{n_0 - n_1}{n_0 + n_1}, \quad r_2 = \frac{n_1 - n_2}{n_1 + n_2}, \quad r_3 = \frac{n_2 - n_3}{n_2 + n_3}$$

air   films   glass

$$\delta_1 = (2\pi/\lambda)\,n_1 t_1, \quad \delta_2 = (2\pi/\lambda)\,n_2 t_2$$

For three, four and more films, eqn. (1) is analogously further developed. The strongest reflection increase arises then if r_1, r_2, r_3 ... are as high as possible as a result of high differences in refractive indices and the amplitudes fit together almost linearly by appropriate selection of the phase angle. A further increase in reflection of a glass substrate for the selected λ_0 is achieved if in a triple layer a low refracting intermediate film is deposited between the two high refracting films. In addition to the high reflectance band, the width and quality of which depends on the refractive index difference and the number of films, secondary bands already occur.

By consequently following this synthesis principle of alternating non-absorbing films with large refractive index difference, it is possible to build up dielectric mirrors with more than 99% reflectance. These simple dielectric $\lambda/4$ multilayer systems are the basis of a large number of optical thin film products fabricated with special film thickness variations, a few examples of which will be mentioned.

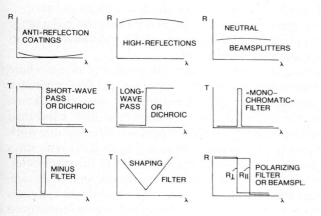

Fig. 1
Typical applications of thin films and film systems in optics.

Suitable introduction of one or more $n\,t = \lambda/2$ films produces zero positions in the reflection and, depending on the construction, monochromatic narrow pass band or broad pass band interference filters with varying transmittance characteristics are obtained. The side bands of a quarterwave stack can be modified with suitable smoothing or correcting films so that various types of high-pass and low-pass filters are obtained. By series arrangement of two or more single all dielectric mirror systems which are coupled by interference, broad band, full reflecting or partial transmitting dielectric mirror systems are obtained. The various types of optical thin film products are shown schematically in Fig. 1. Theoretical treatment of optical properties of multilayer systems is far more advanced than the practical realisation. It should be stated that the current situation of film system design could not have been achieved without the existence of high performance digital computers. Some completely automatic design techniques have been developed, see [37–40], which do not require any initial design as a starting point. Nevertheless, generally pure synthesis is seldom used because of the vast volume and complexity of the calculations which would sometimes overload the capacity of even modern computers. It also depends on wether or not a fast computer with sufficient storage capacity is available for the individual film system designer. To overcome these difficulties, in practice a computer technique called refinement is often applied. It involves the gradual improvement of a simple starting design, e.g. a quarterwave multilayer system by varying refractive indices, thicknesses and number of the films and by comparing the spectral characteristic of any one design with that of the proposed modification until a satisfactory solution is obtained [41 – 46].

Computer-aided optical film system design is to a large extent computer work and will therefore not be treated here in more detail. For practical application, a medium-sized or even small table model computer in which the system design refinement program – which may be commercially available – is recorded, is fed with the optical data of the substrate, the film and the adjoining medium as well as with the desired values of the spectral characteristics of, say, a reflection curve.

After calculation, the best approximation to the desired spectral reflectance curve and the required optical or physical thicknesses of the individual films in correct sequence forming the system are printed out.

In the design process, multilayer stacks are often obtained with layers where the film thickness bears no simple relationship with quarter waves of one reference wavelength. For monitoring the deposition of such multilayers, optical thickness monitoring systems analysing the reflected and/or transmitted light combined with a rate meter, or the oscillating quartz crystal thickness and rate monitor work best. If a quartz crystal monitor is used, it is advantageous that its sensitivity stays practically constant. Furthermore, film systems for application in the ultraviolet, visible and infrared regions can be monitored with the same quartz crystal device.

A prerequisite for optimum production of the calculated multilayer system is that the required refractive indices for every used film material can be reproduced with adequate accuracy. This can be performed by careful experimental determination of the parameter values and their safety limits for proper and reproducible film deposition. Arrangements for highly uniform thickness distribution must be adapted and the complex conditions for low losses in the dielectric films must be maintained.

9.3 ANTIREFLECTIVE COATINGS

9.3.1. SINGLE-LAYER ANTIREFLECTION COATING

When a beam of light travelling in air, refractive index $n_0 = 1$, passes through a glass plate, refractive index n_2, a part of the incident light will be reflected on the front and rear surfaces, as shown schematically in Fig. 2a. The reflectance R of a glass/air interface, definded as the ratio of the reflected intensity to the incident intensity, is a function of the difference in the refractive indices of the media and may be expressed for normal incidence as:

$$R = (\frac{n_0 - n_2}{n_0 + n_2})^2 \tag{5}$$

With an ordinary glass plate, $n_2 = 1.52$, the original beam of light is decreased in intensity by this reflected amount, approximately 4.1 % per surface, making a total of 8.2 %. If appropriate thin films, refractive index n_1, are deposited on the surfaces, these reflections can be reduced, resulting in an effective improvement in the transmission of the glass. This process, shown schematically in Fig. 2b, is known as antireflection coating of the glass plate. It depends for its operation on almost complete elimination of the light reflected at the upper and lower of the two surfaces of the thin film n_1. For complete elimination of the two beams, the

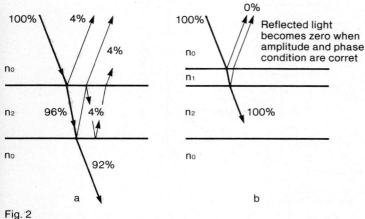

Fig. 2

Reflection of light incident on a glass surface (a) and on a glass surface coated with a single film (b).

reflected intensities at the upper and lower boundaries of the film should be equal, which implies that the ratio of the refractive indices at each boundary should be equal:

$$\frac{n_0}{n_1} = \frac{n_1}{n_2} \qquad \text{or} \qquad (6)$$

$$n_1 = \sqrt{(n_0\, n_2)} \qquad (7)$$

This is the amplitude condition which shows that the film index n_1 should be intermediate between the indices of air and glass.

Part of the incident light will be reflected at the top and bottom surfaces of the antireflection film, and in both cases the reflection will take place in a medium of lower refractive index than the adjoining medium. Thus, to ensure that the relative phase shift is 180°, the optical thickness of the film should be made one quarter wavelength so that the total difference in phase between the two beams will correspond to twice one quarter wavelength, that is 180°. This is the phase condition:

$$n_1\, t_1 = \lambda/4, \quad 3\lambda/4, \qquad (8)$$

Both conditions must be fulfilled to get an ideal single-layer antireflection coating [47, 48, 51]. The most widely used antireflection film is a $\lambda/4$ layer of magnesium fluoride $n_1 = 1.38$. It can be coated on either glass or acrylic substrates.

Curve S in Fig. 3 shows the effect of a single-layer antireflection (a.r.) coating on normal mirror glass ($n_D = 1.52$).

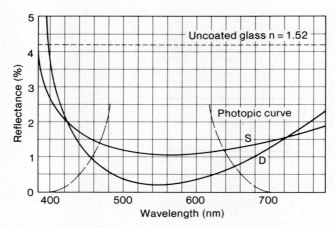

Fig. 3
S = Single layer antireflection coating on glass.
D = Double layer a.r.-coating TRANSMAX®.

Without antireflection coatings, surfaces of higher refracting glasses give higher reflections than those of lower refracting glasses. Higher refractive indices of the glass substrates, however, increase the effectiveness of the single layer MgF_2.

TABLE 1

VARIOUS GLASSES UNCOATED AND COATED WITH A MgF_2 SINGLE LAYER

Refractive index n_D of the glass	% R uncoated	% R at the minimum with a quarter wave thick film
1.52	4.2	1.2
1.67	6.3	0.4
1.75	7.4	0.25
1.90	9.6	0.0

In general, the centering point of the antireflection coating and therefore the minimum on the reflection curve is set at 550 nm to allow for the spectral sensitivity of the eye. For photographic instruments the a.r. centering point is normally 500 nm.

Since magnesium fluoride transmits from approximately 200 nm up to 7μm, in special applications the centering point can be selected at an optional wavelength between 250 nm and 5 μm. One major advantage of the single layer for antireflection is that even for wavelengths far outside the selected centering point, the coating cannot have the effect of increasing reflection.

Generally the single magnesium fluoride layer has very good adhesion to its substrate and provides a wear- and scratch-resistant coating. The single layer is also able to withstand the effects of moisture, changeable climate, salt water, organic solvents, etc. Most available coatings satisfy the requirements of national standards, e.g. the German Amt für Wehrtechnik (Ministry of Defence) (VTL 6600 – 002) and the US Military Acceptance Regulation MIL-Std 810 A.

9.3.2 DOUBLE-LAYER ANTIREFLECTION COATINGS

The refractive indices of most glasses are too low for optimum efficiency of a single layer antireflection coating. If a more effective reduction in reflectance for various glass types is required, this can be achieved with a minimum of two films. Deposition of a high-index material on the substrate as a first film enables the use of magnesium fluoride as the second layer. Such systems have been studied by Catalan [49] and other authors e.g. [50, 51, 4]. With ordinary crown glass, $n_3 = 1.52$, the reflectance will be zero at one wavelength using materials of the following sequence of indices and thicknesses: $n_0 = 1$, $n_1 = 1.38$, $n_2 = 1.70$, $n_3 = 1.52$.

$$n_2 = (n_1/n_0) \sqrt{n_3} \tag{9}$$

$$n_1 t_1 = \lambda/4$$
$$n_2 t_2 = \lambda/4$$

(10

Resignation to use films with quarter-wave optical thickness always yields a zero position in reflectance R with $|r_2| + |r_3| > |r_1|$. The following examples are given in [4]:

$n_0 = 1$	$n_0 = 1$
$n_1 = 1.38$, $t_1 = 177$ nm	$n_1 = 1.38$, $t_1 = 97.8$ nm
$n_2 = 2.15$, $t_2 = 35.9$ nm	$n_1 = 2.15$, $t_2 = 238$ nm
$n_3 = 1.52$	$n_3 = 1.52$

Generally the antireflection double layer can be matched to the refractive index of the glass substrate which is to be used. Hence, a particularly low residual reflection is achieved in the antireflection centering point. Glasses with refractive indices of $n = 1.45$ up to 1.82 can be antireflection coated in this way. The double layer is superior to the single layer over a narrow spectral region adjacent to the centering point, but outside that region it has a somewhat higher reflectance, as can be seen in Fig. 3. It is obvious that antireflection with a double layer having that V-shaped curve of residual spectral reflectance with a characteristic minimum is beneficial for all visual optical instruments.

Commercially available a.r. double layers have mechanical and chemical characteristics which are equivalent to those of single layers. There are also double layers which have been developed for reducing the reflection of organic substrates such as acrylic glass and Perspex. These special double layers can also be used for reducing the reflection of temperature-sensitive optical components made of inorganic glass. Such a coating of inorganic films on organic glasses improves the mechanical durability, and on Perspex its resistance to climatic effects is superior to that of the single layer. Double layer a.r. coatings on glass can even be made by plasma polymerization starting with, for example, hexamethyldisiloxane and tetraethyl-tin. A $\lambda/4$, $\lambda/4$-design with

$$n_0 = 1, \quad n_1 = 1.5, \quad n_2 = 1.6, \quad n_3 = 1.516$$

gives reasonable results [52].

Such coatings are excellent in protecting the substrate from water vapour attack, but their hardness is rather low.

9.3.3 MULTILAYER ANTIREFLECTION COATINGS

When a low reflection is required over a wider range of the spectrum, triple or multiple layer films may be applied consisting of combinations of low- and high-index materials in various thicknesses on the glass. Calculations for the design of such broadband a.r. coatings have been performed by Geffcken in 1944 [53] and later by Brauer [54] Cox, Hass and Thelen [55], Thetford [56] and by many other authors listed in Ref. [5]. However, many existing designs used in industrial film production remain unknown because they have never been published. The most popular design showing two zero positions in reflection are the various $\lambda/4$, $\lambda/2$, $\lambda/4$ a.r. coatings which can be made on different glass types. A possible solution for $n_{glass} = 1.52$ is e.g.:

$$
\begin{aligned}
n_0 &= 1 \\
n_1 &= 1.39 & n_1 t_1 &= \lambda/4 \\
n_2 &= 2.43 & n_2 t_2 &= \lambda/2 \\
n_3 &= 1.71 & n_3 t_3 &= \lambda/4 \\
n_4 &= 1.52
\end{aligned}
$$

Since the effectivity of all a.r. coatings decreases with increasing angle of incidence, special designs have been developed for such applications. An example of a triple layer for use with 45° incidence is e.g. [57]:

$$
\begin{aligned}
n_0 &= 1 \\
n_1 &= 1.4 & n_1 t_1 &= \lambda/4 \\
n_2 &= 1.75 & n_2 t_2 &= \lambda/2 & & \text{matched for 45° incidence.} \\
n_3 &= 1.58 & n_3 t_3 &= \lambda/4 \\
n_4 &= 1.5
\end{aligned}
$$

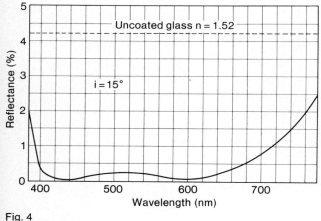

Fig. 4

Broadband antireflection coating IRALIN®.

Modern broadband antireflection coatings generally consist of three to seven layers. The following four layer a.r. coating design on crown glass was created by Shadbolt [58]:

$$n_0 = 1$$

$n_1 = 1.38$	$n_1 t_1 = \lambda/4$	(MgF_2)
$n_2 = 2.3$	$n_2 t_2 = \lambda/2$	(ZnS)
$n_3 = 1.63$	$n_3 t_3 = \lambda/4$	(CeF_3)
$n_4 = 1.38$	$n_4 t_4 = \lambda/4$	(MgF_2)
$n_5 = 1.52$		

The multiple coating Iralin® for instance shown in Fig. 4, from Balzers, gives antireflection over a very wide spectral range. Iralin is used on glasses with refractive indices between $n = 1.5$ and $n = 1.85$. The multiple-coating Iralin satisfies the optical conditions of the US-Military Acceptance Regulation MIL-C-14806 dated November 1968. Iralin has good adhesion and is wear-resistant. The coating system is very resistant to climatic effects, humidity, salt water, organic solvents and cleaning agents. In addition to fulfilling the requirements of the German Ministry of Defence, the coating also complies with the more precisely specified US-Military Acceptance Regulations MIL-STD-810 and 810A dated June 1962 and June 1964. Similar coatings are also available from other thin film companies. Some multilayer broadband a.r. coatings with special optical characteristics are shown in Fig. 5. Antireflection coatings for infrared

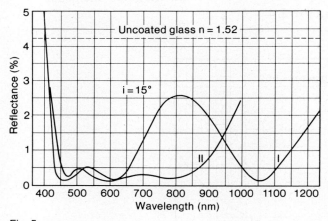

Fig. 5
I = Antireflection coating DUOLIN® VIS + 1060 nm (special).
II = Antireflection coating TRIOLIN®.

materials such as Si, Ge, InAs and InSb have been designed extensively by Cox and various coworkers [59 – 61]. A commercially available multilayer broadband a.r. system for Ge is shown in Fig. 6. For ultraviolet applications, only MgF_2 single layers or double layers are used because of the lack of suitable high-index film materials which would satisfy the requirements for half wave films.

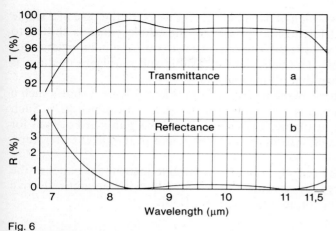

Fig. 6

Measured curves for germanium samples coated with the broadband antireflection films INFRALIN 211® (Balzers).
a) Transmittance of a 1 mm thick germanium plate, both sides coated.
b) Reflectance per surface, measured on a germanium wedge of 7°.

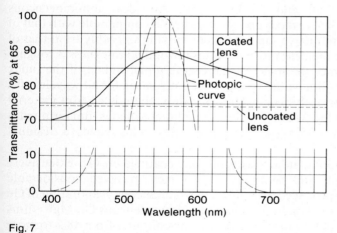

Fig. 7

Transmittance for unpolarized light incident at 65° through an uncoated glass plate and the same sheet coated with an antireflection coating optimized for high transmittance at high angles of incidence [62].

9.3.4   ANTIREFLECTION COATINGS AT OBLIQUE INCIDENCE

Consideration of angles of incidence other than normal leads to changes in the optical chateristics of antireflection coatings. Generally, the observed deviations are small up to an angle of 20°. However, light is split into TE (s-polarization) and

TM (p-polarization) waves which behave in different ways. In the case of a single layer originally optimized for normal incidence, the whole curve is shifted to shorter wavelengths and it is impossible to obtain a zero position in reflectance for both polarisation directions. The behaviour of double and mulilayer coatings is similar to or even worse than that of single layers.

Special designs have to be made to solve this problem. Figure 7 shows such a special design of an a.r. coating for ordinary glass at 65° incidence [62]. Such coatings are commercially available. Turbadar [57, 63] has published designs for oblique incidence. Additional information can be found in the patent literature, e.g. [64 – 67].

Veiling glare caused by reflection of an automobile instrument panel in the windshield can be reduced by coating the windshield with an antireflective multilayer dielectric coating for oblique incidence [156].

9.3.5 INHOMOGENEOUS ANTIREFLECTION COATINGS

The theoretical treatment of inhomogeneous films started even in 1913 with calculations made by Försterling [68], followed later by the work of Kofink and Menzer [69], Schröder [70] and Geffcken [71]. An ideal inhomogeneous antireflection coating is a layer whose optical thickness is comparable with that of a single layer but with a refractive index gradient varying smoothly from that of the substrate to that of the incident medium. Thus the optical thickness is roughly the physical thickness times the mean of the two terminal indices if the index variation is linear. Such a film would be a perfect a.r. coating for all wavelengths shorter than twice its optical thickness. Unfortunately, there is no film material available with an index as low as unity so that any inhomogeneous film terminates with an index of about 1.35 (MgF_2 n $_{infrared} \simeq$ 1.35). This yields a residual reflectance of 2.2%.

For infrared applications, inhomogeneous a.r. coatings on Ge, extensively investigated by Jacobsson and Martensson [72] have received considerable attention. The films were produced by controlled simultaneous evaporation of Ge and MgF_2. A Ge plate coated on both sides with inhomogeneous Ge/MgF_2 films with a physical thickness of 1.2 μm increased the transmittance from 48% to more than 90% in the spectral range of 2 to 6 μm [72].

9.3.6 APPLICATIONS OF ANTIREFLECTION COATINGS

Antireflection coatings make up the majority of all optical coatings produced. They are used for example on lenses of photographic objectives, on ophthalmic glasses, and on the lenses and prisms of binoculars, microscopes, rangefinders, periscopes as well as on instrument cover glasses and on sights. Figure 8 shows the

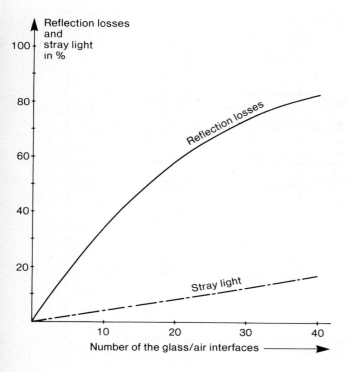

Fig. 8
Increase of the reflection losses and of stray light in an
optical instrument with increasing number of the glass / air
interfaces.
Normal incidence of light, R = 4 % per glass / air interface (Leitz,
Wetzlar, FRG).

increase of the reflection losses and of the stray light with the number of uncoated
glass/air interfaces. The deposition of antireflection coatings increases the transmission of light in optical lens systems and also increases the contrast by the elimination of ghost images and by the reduction of stray light. Only through use of
antireflection coatings did the design of modern high speed objectives for
photography become possible. The influence of a.r. coatings on the transmittance
of such objectives is shown in Fig. 9.

9.4 REAR SURFACE MIRRORS, SURFACE MIRRORS AND BEAM SPLITTER MIRRORS

These reflecting optical film components are almost as important as the
antireflection coatings for transmitting components. Figure 10 shows schematically the two principal constructions.

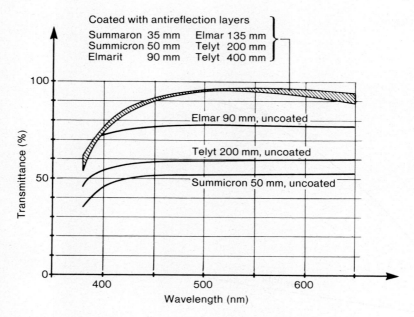

Fig. 9
Spectral transmittance of photo objectives uncoated and coated with anti-reflection layers (Leitz, Wetzlar, FRG).

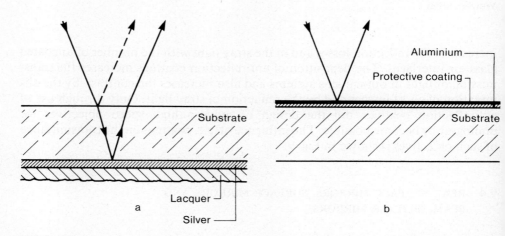

Fig. 10
a) Rear surface or second surface mirror.
b) Front surface or first surface mirror.

9.4.1 REAR SURFACE MIRRORS

The conventional mirror, a glass plate which is silver-plated on the rear surface by a chemical process, is a rear surface mirror. Radiation first passes through the glass substrate and is then reflected on its rear surface.

This construction is necessary because an exposed silver surface is not chemically or climatically resistant and therefore the silver film only remains permanently bright on the side adhering to the glass. Disadvantages of this mirror type are: There is an additional reflection from the front of the substrate which generally produces a second image. The radiation passes twice through the substrate material and is weakened by absorption or undergoes spectral changes. Also, with this double passage, deviations from parallelism of the two surfaces, bubbles and striae in the glass are particularly detrimental.

9.4.2 METAL FILM SURFACE MIRRORS

In the surface mirror, modern technology has enabled an improved method of construction: the reflecting metal film is coated under high vacuum on the side on which radiation is incident. Various metals are applied for the different types of surface mirrors. Their reflectance is shown in Fig. 26 of Chapter 8. The most frequently used metal for surface mirrors is aluminium. Its reflectance as a function of wavelength and its dependence on preparation conditions is shown in Fig. 11.

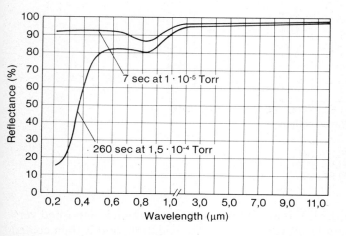

Fig. 11
Reflectance of two Al coatings prepared under extremely different evaporation conditions in the wavelength region from 0,22 to 11 microns.

410

In the same work cycle used to deposit the aluminium films, the mirror can be coated with a thin, optically transparent protection film by which means the mirror becomes more resistant to chemical and climatic effects and to mechanical wear.

Surface mirrors are used on the side on which the mirror is coated, the substrate providing only the surface for the mirror, and radiation does not pass through it. By these means, the above mentioned disadvantages of the rear surface mirror are eliminated in front surface mirrors by the method of construction.

Commercially available are various types of aluminium front surface mirrors to suit different requirements. For the visible spectral range, there are standard mirrors such as Alflex A®. If improved reflection is required, a multiple film mirror Alflex B® can be used. Both types of mirrors are provided with a hard and resistant dielectric protection coating. Such mirrors were first made by Hass et al. [73, 74]. The aluminium film on the surface mirror Alflex B is even protected by an interference film system, which also enhances the reflectance for the visible range. In the visible and infrared, the spectral curve of the reflectance is approximately the same for Alflex A as that of an unprotected aluminium surface. With a mirror type Alflex B, the increase in reflection in the visible, with a maximum at 550 nm, can be clearly seen in Fig. 12. If required, this maximum can also be shifted to other wavelengths in the visible spectrum.

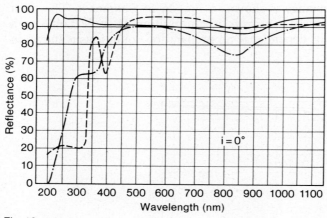

Fig. 12

——————— Metallic mirror ALFLEX UV®.
– – – – – Metallic mirror ALFLEX B®.
–·—·—·– Metallic mirror ALFLEX A®.

A surface mirror for the UV is an aluminium mirror which is deposited very fast in a very good vacuum to prevent detrimental oxidation, and is then coated immediately with a protective magnesium fluoride film [75 – 83] sometimes LiF is also used. This film combination gives excellent reflectance in the ultraviolet region of 200 nm up to 400 nm, as can be seen in Fig. 12.

The spectral reflectance clearly shows the improvement resulting from the interference effect of the MgF_2 protection film in the ultraviolet. Such mirrors also show very good reflection in the visible and the neighbouring infrared, and are therefore ideal for optical instruments, particularly spectral instruments of the entire wavelength range. The aluminium and magnesium fluoride coating has been examined in some detail by Canfield, Hass and Waylonis [81].

All coated surface mirrors have good adherence and can be carefully cleaned with cloth and organic solvents.

Aluminium surface mirrors are produced as plane mirrors for:
mirror reflex cameras, view-finders, telescopes, episcopes, microscopes, holography, projection enlargers, viewing tubes, bubble chambers, etc;
as spherical concave mirrors, parabolic and ellipsoidal mirrors for:
still pictures, moving pictures, cinema projectors, illumination equipment, sun simulators, head lamps, and special instruments;
as signalling mirrors for:
image frames and similar purposes.

The geometrical quality of reproduction of a plane mirror is determined by the precision of the surface of the glass substrate. If a test piece is compared with an ideal standard plane sample in an interferometer, the deviations of the test piece from the ideal sample will be visible as a pattern of interference fringes.

As with the relief lines on a map, these interference fringes give an accurate image of the geometrical arrangement and the extent of the deviations. In this way, the test piece can be classified.

The following classification of plane mirrors according to their surface quality shown in Table 2 may help to find a suitable compromise between requested quality and the costs of its manufacture. All quality standards are produced from films on common machine-drawn glass to optically finest polished glass.

TABLE 2

MAIN QUALITY GROUPS OF PLANE MIRRORS IN RELATION TO PLANITY

Class IV:	Not tested with an interferometer (indiscriminate, industrial quality glass). Application: non-directed lighting, no requirements for reproduction
Class III:	Interferometric check of random samples (e.g. before finishing). Irregular fringe pattern. Limitation of the maximum permissible fringes to a certain diameter. Application: directed lighting, simple reproduction requirements.
Class II:	100% interferometric check (individually). Irregular fringe pattern, limited number of fringes to a certain diameter. Application: higher reproduction requirements
Class I:	100% interferometric check (individually). Regular fringe pattern. In general, mirrors of this group are precision optically ground and polished. Plane substrates are used. Such mirrors are available with a surface accuracy to conform with DIN 3140.

Highest quality unprotected aluminium coatings are particularly required for the mirrors of large astronomical reflecting telescopes. Dust and dirt accumulating on the mirror surface cause a decrease in reflectance always reducing the amount of light which should be collected. Because such mirrors cannot be cleaned without damage, periodic recoating, e.g. once a year, must be performed in special coating plants which are installed in the observatory see, for example Fig. 61a, b of Chapter 6. Many investigations to coat astronomical mirrors have been performed by Strong [84]. His technique of evaporating Al rapidly from tungsten coils is still in use for this purpose. In most cases a reflective coating of uniform thickness is required. However, a technique also exists for converting a spherical mirror into a parabolic one. For parabolizing of telescope mirrors [84], a film of exact preselected non-uniformity is desired. With a spherically curved mirror substrate, a reflective film that is thicker around the edges and becomes thinner toward the center must be deposited. However, limitations on maximum usable film thickness limit the focal length producible by this method. It is further possible with this method to aluminize spherical surfaces to produce off-axis paraboloids [85]. Proper deposited Al mirror films consist of very small crystals and have rather smooth surfaces. The images obtained from Al coated telescope mirrors are therefore excellent. This is also true for rhodium coated mirrors [86] which,

a b

Fig. 13

Photographs of the moon obtained with a 30 cm telescope mirror coated with rhodium (a) and with silver (b). The photos were made in Berlin in 1938 [91]. As can be seen by observation of the craters a sharper and halo-free picture was obtained with the rhodium mirror as a consequence of the finer grained and smoother coating.

however, have a lower reflectance in the visible. In contrast to the above-mentioned two materials, silver mirror films often have a diffuse component in reflection and the images obtained may show some strong halos. This was clearly demonstrable by observations of the moon at the observatory in Berlin in 1938. The telescope used had a 30 cm mirror. The first image was taken using a rhodium mirror coating and the second image was obtained with a silver mirror. Both coatings were performed by Auwärter [87] at W.C. Heraeus. The quality of the picture obtained with the rhodium mirror was quite superior, as can be seen in the historical photographs in Fig. 13 [91]. Materials such as rhodium and also chromium offer in themeselves a high degree of hardness and abrasion resistance and of adherence to a glass substrate, but have only a reflectance of 78% and 60% respectively. The reflectance of about 60% of chromium is, however, sufficient to use this films as a coating for rear mirrors in automobiles. Rhodium is a mirror metal which retains a constant reflecting power for any length of time without the use of a protective layer. Furthermore, it can be kept even for long periods of time, at temperatures up to 430°C without any appreciable loss in reflectance. Because of these properties, rhodium films are best suited for use as standards for absolute intensity measurements, and in application as mirrors for medical purposes.

An interesting type of mirror on an epoxy resin base was described first by Hass and Erbe [88] and is now commercially abailable e.g. [89]. Such mirrors are made by a replication technique. Replication is a simple moulding process made possible by the ability of a liquid polymer to accurately reproduce the optical surface of a carefully polished master tool. After a parting compound has been applied to the precision master, the reflective film, usually aluminium, is deposited by vacuum evaporation. This is cemented together in a «sandwich» and the polymer is cured thermally, thus forming strong bonds between the reflective metal film and the organic substrate. When the master is separated, the substrate's surface duplicates the original. For series production, the process can be repeated using a fresh substrate blank. Replication frees designers from traditional constraints on the use of high quality optical surfaces in small or in large quantities. With this technique it became possible to produce aspherical, toroidal or any other normally difficult surface very economically in any quantity. It also means that the optical distortions typically induced by conventional polishing of thin parts can be eliminated. Accuracies of $\lambda/4$ are routine and $\lambda/10$ is possible. Typical applications of such products are nutating and rotating scan mirrors, choppers, integral mirrors and mount components, as well as low inertia mirrors.

In television tube fabrication, vacuum evaporated high reflecting aluminium coatings are used to increase the brightness of the pictures by more than 80%. For that purpose, the Al film is deposited on the rear inner surface of the phospor layer in a similar coating plant as shown in Fig. 67a, b of Chapter 6. Because of the relatively high electron permeability of aluminium in the form of very thin layers, practically no absorption of the electron beam is observed.

For infrared applications, special aluminium mirrors with a protection film which increases the reflection in the near infrared are used. Such mirrors are primarily used in the region above 900 nm. Surface mirrors of gold films which

414

are made to adhere to glass by using special contact layers can be used in the entire infrared spectral region above 600 nm. At approximately 600 nm, the reflectance reaches 90% and from 800 nm it remains constant above 97%.

A different type of reflector is formed by a finely ground glass surface coated with aluminium or gold. Such coated surfaces of special fineness of grinding are applied as scatter plates in infrared spectrometers [90]. By selective scattering of the short wavelengths and specular reflection of the longer wavelengths, such plates serve to remove short-wavelength components from the measuring beam.

9.4.3 BEAM SPLITTER MIRRORS

Another class of reflector coatings are beam splitters, also known as semi-transparent mirrors. These are mirrors which reflect part of the incident light and transmit the other part. This splitting is uniform over the entire surface of the mirror, enabling a beam splitter to be built into a beam path at any selected position without interfering with the optical image.

Spectrally selective and unselective or neutral beam splitters can be used according to the requirement. Examples of selective beam splitters are the colour splitter for colour printing or colour television. Non-selective beam splitters are used in binotubes for microscopes, in camera view finders with luminous frames, etc. Neutral beam splitters may be made of semi-transparent metallic films, but absorption inherent in such materials precludes their use when maximum light intensity is required. Figure 14 shows typical spectral reflectance and transmittance curves of cemented silver/dielectric neutral beam splitters. The shaded areas indicate losses by absorption. The coating which is protected by cementing withstands normal climatic conditions. For loss-free beam splitters, between 1 and 8 dielectric films without absorption are used. In the simplest case one high

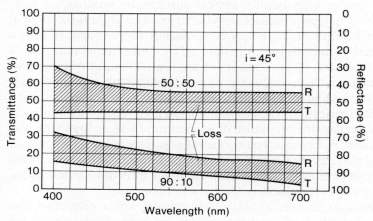

Fig. 14
Neutral beamsplitters SEMIMET® Ag (cemented).

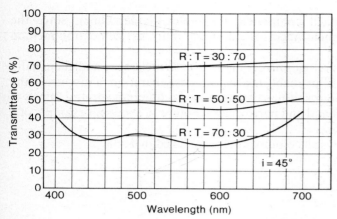

Fig. 15
Neutral beamsplitters TRANSFLEX® (uncemented).

refracting oxide film of quarterwave optical thickness at the desired angle of incidence is sufficient for $R:T = 30:70$. Figure 15 shows some commercially available all-dielectric neutral beam splitters of the Balzers Transflex series. As these coating systems are very scratch resistant and also resistant to climatic conditions, they can be used without cementing.

Beam splitters are single films or film systems, and can therefore be coated on any surface in the optical beam path. Beam splitters are therefore produced not only on plane surfaces but also on spherical surfaces such as lenses. The angle of light incidence can also be pre-selected to suit requirements, common values being 0° and 45°. At oblique incidence also polarization effects have to be considered e.g. [4]. It must then be decided whether the beam splitter on its substrate is to be used with air as the boundary surface or whether it is to be

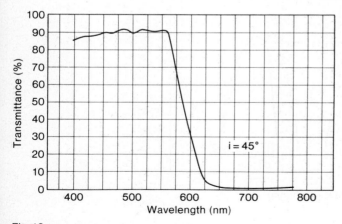

Fig. 16
Broadband colour separation mirror (red mirror).

416

cemented to another optical component (example: divider cube). Since the refractive index of the cement affects the dividing ratio, the cementing method must be agreed upon.

Dielectric colour-separating mirrors made of extremely hard and chemically resistant oxide multilayer films enable nearly loss-free separation or mixture of the primary colours. Such colour splitters can be made for different spectral ranges of the reflection band, broadband and narrowband. Figure 16 shows an example of a red reflector at 45° incidence.

Typical data of broadband colour splitters are listed in Table 3.

TABLE 3

COLOUR-SEPARATING BROADBAND MIRRORS

Blue	$T \leqslant 5\%$	400 – 450 nm
	$T \geqslant 80\%$	530 – 760 nm
Green	$T \geqslant 80\%$	400 – 480 nm
	$T \leqslant 10\%$	520 – 560 nm
	$T \geqslant 80\%$	630 – 760 nm
Red	$T \geqslant 80\%$	400 – 560 nm
	$T \leqslant 5\%$	640 – 760 nm

Typical spectral response curves for narrowband colour separators are shown in Fig. 17.

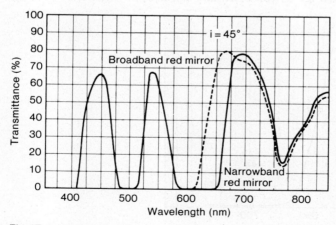

Fig. 17

Blue, green and red narrowband colour separators (mirrors/filters).

To enhance transmission outside the reflection band and prevent ghost images, rear surfaces of colour splitters can be antireflection coated. The tolerable thermal load is generally 250°C.

1) Colour separating coatings on prisms are widely used in colour TV cameras. the advantages of this approach are shorter optical paths, fewer air-glass interfaces, and good separation of the individual colour components. An example of a typical system is shown schematically in Fig. 18.

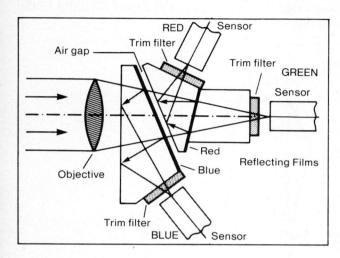

Fig. 18
Optical path through colour separating prisms.

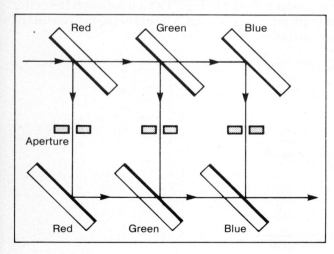

Fig. 19
Colour regulator for printers.

2) Another application for colour separators is film printing. In this case, the primary colours are separated by means of mirrors, which produce three colour channels. Apertures mounted in each channel then provide continuous regulation of the light intensity for proper colour balance and production, as shown in Fig. 19. Poor quality colour film copies sometimes result from the overlapping sensitivity ranges of film materials, shown in Fig. 20. This problem can be overcome by colour separators, which filter out the overlapping absorptions. With no loss of copier speed, colours that are true, bright, and entirely faithful to the original print are readily achieved.

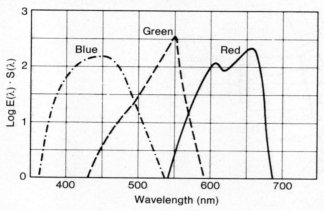

Fig. 20
Representative spectral response of colour film.

3) Finally a special beam splitter is shown in Fig. 21 which separates the entire visible spectrum from the near infrared by reflection. The beam splitter consists of a dielectric multilayer system on a silicon substrate.

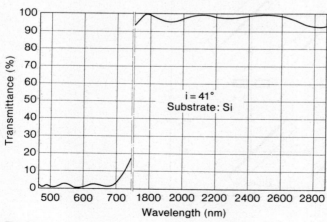

Fig. 21
Beam splitter R VIS/T IR (uncemented).

9.4.4 NEUTRAL DENSITY FILTERS

Thin film neutral density filters consist of single metallic films of various thicknesses on glass and cause a desired uniform attenuation of the passing light beams by reflection and absorption. Usually, rather than using the transmitted intensity I, the filter is characterized by the optical density: $D = \log I_0/I$, with $I_0 =$ incident intensity. For the production of the filters, metals such as Rh, Pd, W, Cr are used to some extent but the best results are obtained by evaporation of Ni/Cr alloys, e. g. [92]. In Fig. 22, some transmission curves of the neutral density filters of Ni_{80}/Cr_{20} on glass are shown. Such filters are also deposited on glass disks with a continuously changing optical density. Typical applications are found in various measuring and analyzing instruments.

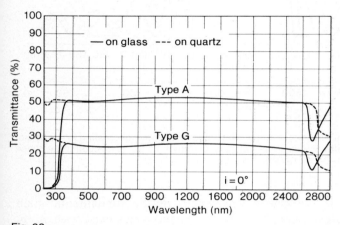

Fig. 22
Neutral density filters.

9.4.5 DIELECTRIC MIRRORS

As already mentioned in Section 9.2, a multilayer stack formed of alternating high and low refracting non-absorbing quarterwave films shows a high reflectance, because the reflected beams from all the interfaces have equal phase when reaching the top surface and therefore interfere constructively. The spectral characteristics of such multilayer stacks are shown in Fig. 23. As can be seen in the figure, the height of the high reflection zone increases with the number of the layers. Also the number of oscillations outside the high reflection zone increases with the number of the layers but their extension is limited. The width of the high reflection zone depends on the ratio in the refractive indices of the film materials used. The higher the ratio, the larger the spectral extension.

This is valid not only for the fundamental reflectance zone considered but also for all wavelengths for which the layers are an odd number of quarter

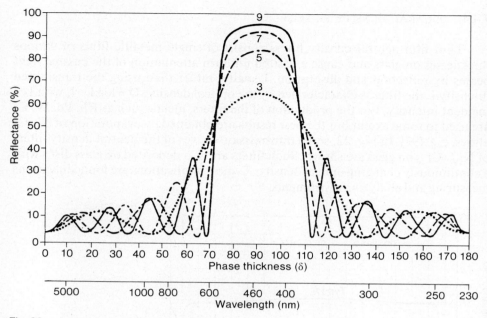

Fig. 23
Spectral characteristics for normal incidence of multilayer stacks formed of alternating $\lambda/4$ layers of high $n_H=2.3$ and low $n_L=1.38$ refractive index on a glass substrate $n_g=1.52$ as function of the phase thickness $\delta=2\pi nt/\lambda$ of wavelength λ for $\lambda_0=460$ nm [96].

wavelengths thick. If λ_0 is the centre wavelength of the fundamental high reflectance zone, then higher orders of high reflectance zones exist with $\lambda_0/3$, $\lambda_0/5$, $\lambda_0/7$, etc. However, at wavelengths for which the optical thicknesses of the layers are equivalent to an even number of quarter waves, which is the same as an integral number of half waves, the layers will all be absentee layers and consequently at the positions the reflectance is equal to that of the uncoated substrate. With λ_0 the centre wavelength of the fundamental high reflectance, this is the case for wavelengths λ equal to $\lambda_0/2$, $\lambda_0/4$, $\lambda_0/6$, etc. Depending on film material, such simple quarterwave high reflecting multilayer coatings which can be used as interferometer mirrors or, in higher quality, as laser mirrors can be designed for application in the ultraviolet, visible or infrared region. Deposition of highly transparent films yields multilayer stacks with very small residual absorption. If additionally a homogeneous microstructure of the individual films and a smooth surface and interface topography of the whole film system can be achieved, such mirrors may show very low optical losses, e.g. [93, 94, 95]. It is evident that the losses in the coatings are not only determined by the materials alone but are also influenced by the applied film deposition technique and by exactly how clean the deposition plant and the substrates are.

The relatively narrow spectral range over which a high reflectance can be achieved with a quarterwave multilayer is a stringent limitation for some practical applications. This is a consequence of the lack of materials with sufficient high refractive indices. To overcome this disadvantage, a number of attempts have been undertaken to extend the range of high reflectance by various means. As was found early on [96, 97], successive harmonic variations in the thicknesses of the individual films of a multilayer yielded broadening of the reflectance zone. Calculations of layer assemblies with thicknesses either in the arithmetic or in geometric progression were undertaken later [98]. It can be shown that with the same number of layers the geometric progression yields a slightly broader high reflection zone. Using films with $n_L = 1.39$ and $n_H = 2.36$ on glass $n_g = 1.53$ with a total number of 25 layers, the arithmetic approach yielded a high reflectance zone of more than 90% between 418 and 725 nm, whereas the geometric approach had an extension between 342 and 730 nm [98].

The simplest method, however, is to deposit a quarterwave stack for one wavelength on the top of another stack for a different wavelength. This procedure can be repeated by deposition of further shifted stacks until a sufficient spectral extension of high reflectance is achieved [99]. To avoid a transmission peak by Fabry-Perot filter formation in the overlapping area of two shifted mirrors (since the uppermost high refractive quarterwave film of system $1 = H_1$ and the first high refractive film of system $2 = H_2$ together form a halfwave film HH $(1+2)/2$ for an intermediate wavelength), a low refractive quarterwave film L is deposited between them [99] according to:
glass [0.8 (HLHLHLHLH)] [1.2 (HLHLHLHLH)] air . . . transmittance peak, but
glass [0.8 (HLHLHLHLH)] − L [1.2 (HLHLHLHLH)] air . . . without transmittance peak.

The latter design for extension of spectral reflection is often used in technical film production.

9.4.6 COLD LIGHT MIRRORS AND HEAT MIRRORS

Cold light mirrors have very high reflection characteristics for the visible light and a high transmission for infrared radiation. They are particularly suitable for illumination systems in which the illuminated object must be protected from thermal radiation. Their function is partially or fully based on the interference of multilayer systems. In early attempts to design such mirrors [100, 101], silicon or germanium films, which show good reflectance in the visible and high transmittance in the infrared, were deposited as a first layer on glass. Using additional multilayer overcoats, the spectral broadness and the height of the reflectance in the visible can be improved.

Another possibility of fabricating a cold light mirror is the deposition of a quarterwave stack with a high reflectance zone between 400 and 550 nm on the front side of a glass and to coat the rear side with a quarterwave stack which has its high reflectance between 550 and 750 nm. The relatively thick glass substrate prevents interference between the two multilayer systems and the total reflectance is obtained according to:

$$R = \frac{R_1 + R_2 - 2 R_1 R_2}{1 - R_1 R_2} \qquad (11)$$

R_1 denotes the front reflectance and R_2 that of the rear side. The most commonest method is, as already discussed, the superposed deposition of two or more quarterwave stacks, with shifted centre wavelength and suitable transmittance peak attenuation, on the front surface of the glass.

Commercially available mirrors are often made of ZnS/MgF_2, TiO_2/SiO_2 multilayers. Figure 24 shows a typical spectral transmission curve of a cold light mirror. The effective reduction of the temperature in reflected light depends to a great extent on the geometric arrangement and the proportion of the direct radiation. Good results are thus achieved with the use of cold light mirrors as deflection mirrors. It is efficient to design the interference system for the special angle of incidence (often 45°). For special applications the reflected range can be stretched into the ultraviolet or near infrared to a certain extent. Depending on the arrangement of the mirroring layers, the colour temperature or colour nuances of the reflected light can be influenced. So-called hard coating such as TiO_2/SiO_2 used on large cinema projector mirrors, can also be applied to small, highly curved reflectors. The advantage of these layers is their high thermal, mechanical and chemical stability and the long lifetime connected with these features. The generally high transmittance in the infrared, on average 80 % to 2.6 μm, means that heat is efficiently removed while the over 95 % reflectance in the visible ensures an optimum even light yield across the entire spectrum. Such coatings should withstand temperatures up to 400°C in continuous operation. The small mirrors are applied, for example, in film and slide projectors, microfilm readers, fibre optic or signal lighting systems.

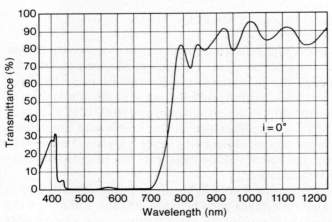

Fig. 24
Cold light mirror for strongly curved substrates.

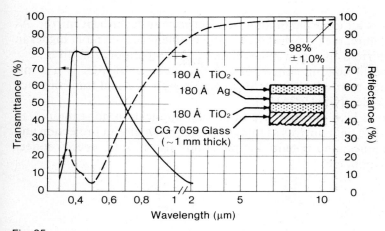

Fig. 25
Transmittance and reflectance of a transparent heat mirror
using a thin metal-dielectric film combination [62].

Transparent heat mirrors have optical properties which enable them to have high transmission of the visible radiation and high reflectivity of infrared radiation. There are basically two ways to achieve high visual transmittance simultaneously with high infrared reflectance. One is the use of the interference effect in all-dielectric multilayers, the other is the use of intrinsic optical properties of electrically conducting films such as Au, Ag, and others which have high infrared reflection with relatively low visual absorption. Their suitability as transparent heat mirror can be improved by antireflection coating for the visible. Figure 25 shows an example for such types of heat mirrors, according to Fan et al. [102].

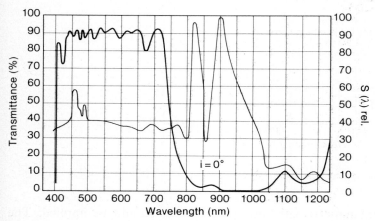

Fig. 26
Heat protection filter CALFLEX® / Light source spectrum: Xenon lamp.

Figure 26 shows an example of an all-dielectric multilayer heat mirror. As can be seen in the figure, the transmittance in the visible is considerably higher in that design. Such mirrors also show an excellent high-temperature stability.

A further type of coating suited for transparent heat mirrors is a semi-conducting film such as In_X-Sn_Y-O_Z, which combines good visual transmittance with high infrared reflectivity [103–105].

Such transparent heat mirrors have important application, for example, in combination with cold light mirrors as thermal radiation shields in projection and illumination techniques; and potential applications also in solar energy collection, window insulation, etc.

9.4.7 LASER COATINGS

Multilayer interference systems are ideally suited as reflection coatings, for the fully reflecting and the partially transmitting mirrors of lasers. Because of negligible absorption, reflection values of approximately 100 % are possible. Such reflectors give minimum attenuation of laser emission. They can be made of ZnS/ThF_4, TiO_2/SiO_2 and of some other oxide combinations.

There are two basic types of mirror, those for a single wavelength and those for a fairly broad spectral band. The single wavelength mirrors are simple quarterwave multilayers. The broadband mirrors are designed using methods as already described in Section 9.4.5.

Figure 27 shows the spectral reflectance curve of a broadband laser mirror.

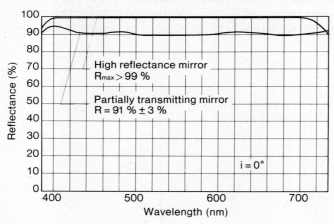

Fig. 27
Laser broadband mirror (e.g. for dye lasers).

Only film combinations with low optical losses should be applied. The surface quality of the laser substrates is one of the critical parameters in obtaining low light scattering values and optimum adhesion of the coating. Various substrate materials are used, e. g. glass, fused silica, glass ceramics, etc. Laser mirror coatings can be applied to substrate shapes such as lenses, disks, flat plates, prisms and cubes. For high power laser mirrors, special designs are used. The laser mirrors are produced for various angles of incidence such as e. g. 0°, 30° and 45°. Except for normal incidence, the reflection values of the two polarization components $R\,T_\perp$ and $R\,T_\parallel$ will be different.

Polarizing beam splitters are important thin film products required for many laser experiments. In their design, observation of the above-mentioned fact that a quarterwave stack has different width of the high reflectance zones for the two planes of polarization s and p is made. There are polarizers for single wavelengths, uncemented or cemented design, and for a broad wavelength range, only available in the cemented design, e. g. [106]. The following sequence of layers shows a theoretical design created by Thelen [107]:

air / [0.735 (L/2 H L/2)]² [0.84 (L/2 H L/2)]⁸ [0.735 (L/2 H L/2)]²/ glass with $n_L = 1.45$, $n_H = 2.35$, $n_g = 1.52$ and $\alpha = 56.7°$. Depending on type, the common angles are 45° and 57°.

In Figure 28, a commerically available uncemented laser polarizer for 57° incidence and for $\lambda = 1060$ nm is shown, e. g. [108].

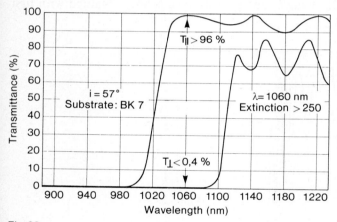

Fig. 28
Laser polarizer 1060 nm/57° (uncemented).

Most laser polarizers are made of hard, chemically and mechanically durable dielectric coatings, which generally meet or exceed applicable military specifications such as MIL-C-675a, MIL-C-14806, MIL-M-13508b, regarding adhesion, abrasion, humidity and temperature (MIL-STD-810-A).

Entrance/exit faces, depending on type, can be coated with a suitable antireflection coating.

Dichroic laser beam splitters are thin film products separating the incident light into two spectrally different beams. The commonest angle of incidence is about 45°, where the reflection values will be different for the two polarization components ($RT_{\perp}$ and $RT_{\parallel}$). The dichroic laser beamsplitters are made of hard, chemically and mechanically durable, dielectric coatings. The film systems also meet the above-mentioned US military specifications. The dichroic laser beamsplitters are distinguished by uncemented and cemented designs, e.g. [109]. Special designs such as high power dichroic laser beamsplitters are also made. Anti-laser protection coatings are used for safety purposes. One or more laser wavelengths will be suppressed. For laser protection filters, it is very important that, while the laser light is attenuated, the remaining part of the visible spectrum should be affected as little as possible. The attenuation can be achieved with three types of filters [110]:

– absorbing glasses or plastics
– all dielectric thin film reflectors and
– combinations of absorbing glasses with dielectric thin film reflectors.

Such components are usually used at or near an angle of incidence of 0°.

The protection coating with the spectral transmittance curve shown in Figure 29 is of the composite type [111]. Specifications for such protective coatings are found for example in [112]. A high damage resistance is only obtained with first surface reflection coatings.

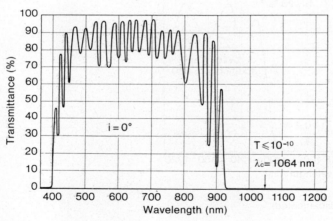

Fig. 29
Anti-laser protection filter 1064 nm (uncemented).

Special requirements must be met in the case of high power laser components e.g. [45b]. A complete assessment of laser fusion coatings was published by Apfel [113].

Because of the high power densities, damage may become a limiting factor and only more resistant highest quality materials can be used. Of the optical materials such as bulk substrates and thin films, as well as surfaces used in high power laser systems, thin film optical coatings are the least damage resistant and also the most complicated. A quantitative model for the mechanisms under which optical materials fail under intense irradiation is still incomplete. In particular, pulsed laser damage is inadequately understood. Also, with intrinsic damage, the relative importance of multiphoton absorption compared with avalanche ionization and the relative importance of self-focusing and self-defocusing in the damage process under various experimental conditions is also not well understood.

There is evidence that damage in thin films is not intrinsic, but arises from defects, impurities or absorbed materials which also depend in a complex manner on the applied preparation technique and the chosen parameters. It is therefore not surprising that attempts to find simple solutions to film damage have not yielded consistent results.

The deposition of a barrier layer sometimes work, but not always. With pulses of several nanoseconds or more the observed damage seems to be thermal, originating at highly localized sites. However, the identity and the origin of these sites is still not clear.

Information on progress in this important topic can be obtained from the publications (NBS Special Publications) of the annual Boulder Damage Symposia and the published Index of Papers of these symposia e.g. [114].

9.4.8 ARTIFICIAL JEWELS

Artificial jewels can be made of cut of pressed bodies of transparent materials like glass or plastic which are coated with metal and/or dielectric films. It is not surprising that the most desirably cut diamonds were imitated by cheaper materials. In its optical behaviour, it can be very effectively copied by new synthetic minerals such as fabulite ($SrTiO_3$) and zirconia cubic (a Y_2O_3 or CaO stabilized cubic ZrO_2). However, glass remains the much cheaper base material and coating by evaporation or sputtering can be performed very economically. Diamond is outstanding in its high index of refraction $n = 2.417$ and its high reflection of about 20% per facet. Together with total reflection, it is to this property that a diamond owes «fire» and «brilliance». Numerous attempts have been made to provide a jewel imitating the fire and brilliance of a diamond. Metallization of the rear-side facets of glass bodies and colouring by deposition of high reflecting all-dielectric multilayer systems, e.g. TiO_2/SiO_2, have not produced the sparkle and shimmering colour effects of a true diamond, which changes greatly with direction of observation. Many investigations to solve this problem have been made by Auwärter, Ross and Rheinberger. The best imitation was obtained by deposition of a multilayer antireflection coating for long wavelengths. In this way, a subjective impression of a bluish reflection colour is produced even under very flat or small acute angles of observation and the reflectance per facet is increased considerably [116]. Other artificial jewels, such as imitations of marcasite or topaz, can also be made of glass with deposited thin film of Cr and SiO [117] or by SiO alone [118].

Artificial jewels of glass and plastic with vapour-deposited interference layer systems present a great variety of very appealing colour nuances. Even opalescence effects can be produced by deposition of all-dielectric or dielectric-metallic multilayers onto rough gem surfaces [119] which are achieved by wet chemical etching or sand blasting of the cut glass bodies. Generally, it can be stated that high quality coated artificial jewels are always made from cut glass bodies. The finish of the facets achieved by cutting is quite superior to that obtained by pressing. The Swarovski Company in Wattens, Tyrol (Austria) have developed special machines to cut large quantities of glass jewels and to coat them economically. Swarovski thus became the most famous and important producer of coated artificial jewels in the world.

9.5. SEPARATION OF LIGHT BY FILTERS

Light separation by filters can be performed with coloured glass or dyed gelatine filters. The quality of separation and the thermal stability of the filters, however, have been improved considerably by the use of thin film interference systems, see [4, 5, 115].

9.5.1 LOW- AND HIGH-PASS EDGE FILTERS

These filters are characterized by producing an abrupt change between a region of rejection and a region of transmission. They are generally produced from all-dielectric multilayer systems with corrected side bands to increase transmittance on the short or long wavelength side, as is shown in Fig. 30, or on both sides of the rejection region. The filters are sometimes also combined with absorbing

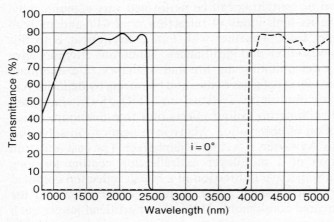

Fig. 30
IR longwave pass filter and IR shortwave pass filter.

coloured glasses to improve and/or extend the rejection zone. Another way to extend the rejection zone is to place a second stack in series with the first and to ensure that their rejection regions overlap.

In this category belong blocking or transmitting filters for separation of broad spectral ranges in the ultraviolet, the visible and the near infrared range. Such blockers can be used for suppression of the unwanted higher or lower orders of various narrow and broad band interference filters, or to block disturbing uv radiation and to limit the sensitivity range of a receiver. Some examples are shown in Figs. 31 and 32.

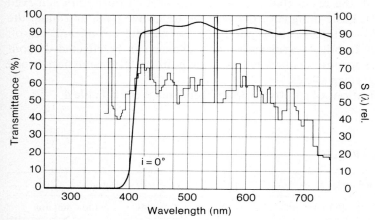

Fig. 31
UV blocking filter / Light source spectrum: Mercury lamp.

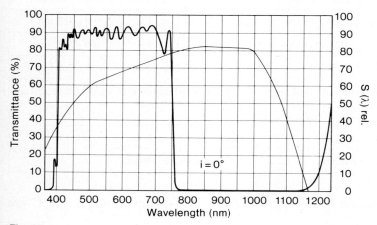

Fig. 32
NIR blocking filter / Sensitivity of receivers: Silicon cell.

Furthermore, the dichroic colour filters should be mentioned here. Generally, additive filters of this type are produced in the three primary colours: blue, green and red. Through a combination of dielectric, mostly oxide, coatings with coloured glasses, colour filters achieve a complete suppression from the ultraviolet range to the near infrared without causing a noticeable loss of tansmittance in the passband region. Because the coloured glass and the cement are thermosensitive, the maximum thermal load is, for about 100°C, relatively small. If, however, all-dielectric oxide systems are produced with the coatings deposited on both sides of the substrate, then the filter also has high transmittance in the pass band region and a broad suppression range in the visible. Cut-on and cut-off slopes are then relatively steep, with the result that colour outputs of high purity are achieved with these products. Basic designs of the red, the longwave pass, and blue, the shortwave pass, colour filters can be modified so that the cut-on and cut-off positions are shifted to other wavelengths. This change can be effected without altering colour filter spectral characteristics, namely its high transmission, broad blocking region, and steep slopes.

Typical specifications of additive colour filters are given in Table 4.

TABLE 4

ADDITIVE COLOUR FILTERS

With coloured glass:			all dielectric:		
Blue	T ⩾ 65%	400 – 450 nm	Blue	T ⩾ 80%	400 – 460 nm
	T ⩽ 1%	505 – 760 nm		T ⩽ 1%	505 – 760 nm
Green	T ⩽ 1%	400 – 475 nm	Green	T ⩽ 1%	400 – 485 nm
	T ⩾ 75%	525 – 550 nm		T ⩾ 75%	530 – 555 nm
	T ⩽ 1%	610 – 760 nm		T ⩽ 1%	590 – 760 nm
Red	T ⩽ 1%	400 – 575 nm	Red	T ⩽ 1%	400 – 575 nm
	T ⩾ 80%	625 – 760 nm		T ⩾ 80%	630 – 760 nm
Substrate:Coated glass laminated to coloured glass			Substrate:Heat-resistant TEMPAX coated on both sides		
Glass thickness: 2 – 4 mm			Glass thickness: 1 mm		
Thermal load: 100°C maximum			Thermal load: 400°C		

With such all-dielectric oxide systems, thermal loads up to about 400°C are readily feasible.

Balzers subtractive colour filters are dielectric interference filters whose passbands extend over the spectral region of two primary colours: yellow over the green and red regions, magenta over the blue and red, and cyan over blue and green. When subtractive colour filters are placed in the path of a light source, a primary colour results. Thus yellow and magenta filters yield red, yellow plus cyan produce green, and magenta and cyan provide blue. Through suitable combinations, the subtractive colour filters can produce every colour hue in any density to full saturation.

Typical specifications of subtractive colour filters are given in Table 5.

TABLE 5

SUBTRACTIVE COLOUR FILTERS

Yellow	T ⩾ 80%	530 – 760 nm
	T ⩽ 1%	410 – 475 nm
Magenta	T ⩾ 75%	400 – 460 nm
	T ⩽ 1%	530 – 560 nm
	T ⩾ 75%	650 – 730 nm
Cyan	T ⩾ 80%	420 – 565 nm
	T ⩽ 1%	630 – 720 nm
Substrate: Heat-resistant TEMPAX		
Glass thickness: 1 mm		
Thermal load: 400°C maximum		

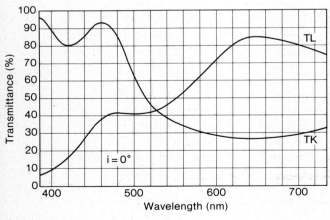

Fig. 33
Colour temperature conversion filters TL/TK.
(TL = day light, TK = artificial light)

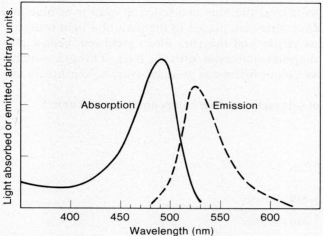

Fig. 34
Absorption and emission of FITC-protein conjugate as function
of wavelength.

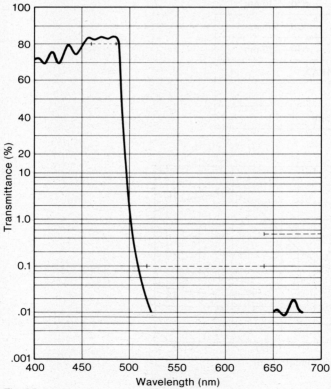

Fig. 35
Excitation filter for FITC immuno-fluorescence.

Applications for dichroic colour filters encompass TV cameras, film printers, telecine equipment, colour printers, colour enlargers, signal lighting, studio illumination, colour separating, and colour sorting. The stability and durability of these oxide coating products make them suitable for a wide range of environments in manufacturing, the laboratory, and the photographic darkroom.

As well as additive and subtractive colour filters, there are also colour-temperature conversion filters for natural and artificial light applications. Figure 33 shows an example of such filters.

Generally, in many types of long- and shortwave-pass filters, the steepness of the edge is not of critical importance. It is important, however, with filters applied in fluorescence microscopy where the excitation and emission bands of special fluorescent tracers may have such a small spectral distance, that they do overlap to a certain degree. This happens, for example, with fluorescein-iso-thio-cyanate FITC, a fluorochrome used in immunofluorescence. For excitation, the maximum absorption is at about 490 nm and the emission maximum is at 520-525 nm as can be seen in Fig. 34. When such an exceptional high degree of edge steepness is required, then the easiest way of improving it is to use more layers. Figure 35 shows an example of an excitation filter consisting of more than 31 TiO_2/SiO_2 layers inclusive the correcting layers in the stack. These filters show very high transmission of 80 % minimum in the area of excitation. In the blocking area starting at 510 nm, the transmission in lower than 5×10^{-3}. The transmission decreases towards longer wavelengths and is lower than 1×10^{-3} at the fluorescence maximum of $\lambda = 520$ nm. The red part of the spectrum is suppressed by combination with a special colour glass.

9.5.2 BAND PASS INTERFERENCE FILTERS

Band pass filters are generally characterized by a region of possibly high transmission, limited on either side of the spectrum by regions of rejection. Depending on the width of the transmission region, one may distinguish between narrow-band and broad-band filters.

9.5.2.1 NARROW-BAND FILTERS

A typical narrow-band filter is the metal dielectric Fabry-Perot filter basing on the Fabry-Perot interferometer. An interference filter of this type consists of two highly reflecting but partially transmitting mirror films spaced optically one half wavelength apart [120-122, 4, 5]. For the production of such filters, the vacuum deposition method has proved most successful. In the simplest case, the filter is made by depositing first a silver film onto a plane glass substrate then evaporating a $\lambda/2$ or a multiple of an absorption-free dielectric material following this with another film of silver. For symmetry, protection and stabilization of optical properties of the arrangement, another plane glass is added using, for

434

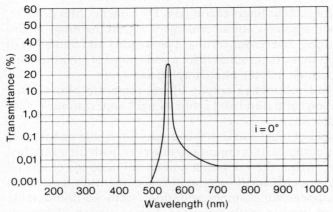

Fig. 36
Metal-dielectric interference filter FILTRAFLEX B-20 (Balzers).

example, a conventional optical cement. Figure 36 shows the spectral transmittance of a metal-dielectric F. P. Filter. The thickness of the dielectric intermediate or spacer layer determines the position of the transmission band within the spectrum and the degree of background transmission. The wavelength for which the internal multiple reflections in the forward direction are in phase will be strongly transmitted, other wavelengths are suppressed. First-order filters, that is, forward reflections differing in phase by only one wavelength, generally have a peak transmittance of 30 to 40% with a half width of the transmission curve of about 20 nm. Second-order filters, in which the forward reflections differ in phase by two wavelengths, may have the same peak transmittance as the first-order filters, but a half width of only about 10 nm. The undesired orders of the filter can be blocked by edge filters or coloured glasses.

An extensive mathematical treatment was performed by Hadley and Dennison [121, 122]. Only the most important formulae characterizing a Fabry-Perot filter are given here

The transmitted intensity I_T is:

$$I_T = \left[(1+\frac{A}{T})^2 + \frac{4R}{T^2}\sin^2\left\{\frac{2\pi}{\lambda}n_s t_s \cos(\varphi_s+\delta)\right\}\right]^{-1} \tag{12}$$

The maximum transmittance T_{max} is:

$$T_{max} = T^2/(1-R)^2 \tag{13}$$

The minimum transmission T_{min} outside the transmittance curve is:

$$T_{min} = T/(1+R)^2 \tag{14}$$

Finally the half width HW is given by:

$$HW \simeq (1-R)/X\,\pi\,\sqrt{R} \tag{15}$$

In the equations, T = transmittance of a single mirror layer, R = reflectance of a single mirror layer, A = absorption of a single layer, X = order of interference, $n_s t_s$ = optical thickness of the spacer layer and α = angle of radiation within the layers.

Examples of fields of application of such filters, producing monochromatic light, are in colourimeters, sensitometers, polarimeters, refractometers, interferometers, fluorimetrical measuring instruments, flash-photometers, microscopes, etc. Filters are manufactured for the near infrared and the visible range using Ag mirrors and for the ultraviolet using Al mirror films. The corresponding spacer layers are often made of cryolite, thorium fluoride, magnesium fluoride or lead fluoride. For infrared applications, materials such as germanium, silicon and tellurium and gold are also used. Most filters are designed for applications at room temperature and at normal incidence of light. Peak transmittance is shifted towards shorter wavelengths if either the temperature is decreased or the angle of incidence is increased e. g. [123]. The angle of incidence of the light determines the wavelength at which transmission is maximum. In fact, λ_{max} is shifted in the shorter wavelength direction with increasing angle of incidence and simultaneously the transmission is slightly reduced. Inclination of the filter therefore offers the opportunity to move λ_{max} within a small range. For light beams of considerable convergence or divergence incident at an angle, the effect is to broaden the transmission band. Where light is polarized, the displacement varies, depending upon whether the polarization is parallel or normal to the plane of incidence.

Fabry-Perot filters are also available commercially, with local continuously changing spectral transmittance. Such continuous filters are deposited on glass strips or on circular disks.

Vital to the operation of an interference filter is a very high reflectance of the mirror coatings adjacent to the spacer layer. The absorption of the metal mirrors can be reduced and thus the maximum transmittance of the filter increased if both metal layers are increased in reflectance by additionally deposited high reflecting dielectric multilayers [124]. In this way, with a first-order filter, a half width of 2 nm and a transmittance maximum of 41 % can be obtained [124]. If, however, the metal mirrors are completely replaced by absorption-free all-dielectric high reflecting multilayers, one can achieve a peak transmittance of greater than 75 % with a half width of only 2 nm, which is considerably better than the usual metal-spacer-metal system. Figure 37 shows an example of an all dielectric NIR-interference filter. The relation between half width and tenth width in a Fabry-Perot filter is 1 : 3. Such a transmission curve is not of ideal shape. To obtain a more rectangular shape of the transmittance band and to eliminate the disturbing influence of absorptions on the bandwidth, considerations analogous to coupling of tuned electric circuits led to acceptable results [125-131]. By coupling together tuned electric circuits, the resultant response curve is more rectangular than that of a single tuned circuit. This also happens when coupling together single Fabry-Perot filters. From the multiple half-wave filters the double half-wave (DHW) type of the form: // mirror / half wave spacer / mirror / half wave spacer / mirror // is often fabricated. The filters may be either

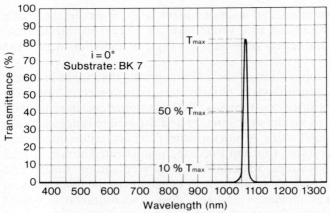

Fig. 37
All-dielectric narrowband interference filter λ = 1064 nm.

metal-dielectric or all-dielectric. The relation between half width and tenth width in a DHW filter is better than 1:1.8.

When using an interference filter, the uncemented mirror surface must always face the light source. Where filters are to be used above 40°C, the time of exposure should be kept as short as possible, and in no case should temperatures of 70°C be exceeded. Mechanical stresses and strains can destroy a filter. It is recommended that filters are stored in dry air at temperatures not exceeding 40°C, e.g. [132, 133]. For information on the influence of neutron bombardment of interference filters, see e.g. [134]. Each filter is guaranteed by most manufacturers for a period of one year. Other more special types of monochromatic filters are rather seldom produced and will therefore not be discussed further here. Details are given in:
reflection interference filters [121, 122, 124, 135, 136];
frustrated total reflection filters [124, 135, 137];
induced transmission filters [138]; and
phase dispersion filters [139 – 141].

9.5.2.2 BROAD BAND FILTERS

Broad band filters can be made by combining a coloured glass with an interference edge filter [4], or better by combining two interference edge filters such as a long-wave-pass and a short-wave-pass e.g. [5]. The two components can be deposited on opposite sides of a glass substrate. With some precaution, it is also possible to deposit both film systems on the same side of a substrate [142]. Care must be taken in the combining of the two stacks so that the transmittance in the pass band is a maximum and that no side transmission peaks occur in the rejection region. Such filters are suitable for filtering wide ranges of the visible or infrared spectrum, as can be seen in Fig. 38. It is thus possible, for example, to produce a trichromatic division of visible light, as is necessary for colourimetric purposes for

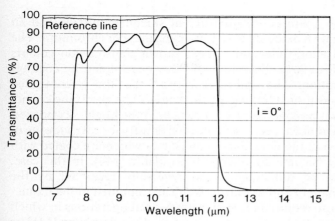

Fig. 38
IR broadband interference filter.

colour-film techniques and for colour television. In printing techniques, coloured photographs also come within the field of application. Compared with other coloured filters, these interference filters have the advantage of high transmission, steep edges on the long wavelength side of the transmission band and low heat absorption with high power light sources.

9.6   ABSORPTIVE COATINGS

Absorptive coatings are most commonly manufactured by films of metals, alloys, oxides and cermets, the transmittance being inversely proportional to the film thickness. Absorptive coatings may be rather thin even when approaching opacity. An aluminium film for instance with a transmittance of 10^{-5} is only 50 nm thick. Semi transparent films of nickel, chromium, platinum, palladium and rhodium are practically neutral, whereas that of silver and aluminium are blue, gold and copper are green or blue-green and that of antimony and titanium are brown and grey-brown respectively. Oxide films and especially suboxide films are often brown. the most neutral transmission is, however, obtained with alloy films such as Inconel and Nichrome. Cermet films, such as mixtures of silicon oxide and chromium, are brown in transmission.

Absorptive coatings have many technical applications, one of which, that for neutral density filters, has already been discussed in Section 9.4.4. Some other important applications will be reported in the following sections.

9.6.1   EYE PROTECTION COATINGS

Neutral filters and some green, blue and brown coloured films have achieved popular application as sunglasses in ophthalmic lens coating [143]. Most ophthalmic sun protection coatings are mixtures of silicon monoxide and an optically absorbing lower metal oxide such as iron, manganese or chromium. The absorbing film is generally anti-reflection coated. Brown transmission colours can be deposited with 15% to 90% absorption and grey neutral density coatings from 20% to 60% absorption. The optical density of the film is controlled photometrically during deposition. Grading the thickness of the absorbing cermet film on the sunglasses makes possible extra-high density in the bright-sky portion of the field of vision with a uniform decrease to normal density for easy viewing of objects in the lower portion of the field. Soon, sun glasses which are coated with electrochromic layers may become commercially available [144]. In this case, sunglasses absorption can be individually and reversibly regulated by the accumulated charge. The battery and a control switch are mounted in the frame.

Protection filters are also required for viewing industrial operations in which high-intensity radiation is emitted, such as during gas or electric welding [145] or when using lasers [112]. Even a short exposure to intensive ultraviolet radiation can induce severe irritation to the eyes so that protection is necessary in all welding operations and in particular when two specified wavelengths of 334 nm and 365 nm are in use. Also, intense infrared radiation is hazardous to the eye. To protect welders against harmful ultraviolet and infrared radiations and to provide increased visibility of the welding operation, special protection filters are used. Welding-protection filters constructed as eye pieces or as shaped eye protectors generally consist of two parts. One is made of coloured glass or dyed plastic. This part of the filter is combined with a chemically hardened glass cover plate which is coated on the front side with a semitransparent, absorptive and high reflecting thin layer such as gold.

Laser protection filters have already been discussed in Section 9.4.7.

9.6.2 PHOTO MASKS

Photo masks are high quality tools required for the reproducible generation of fine structured simple and complex patterns in various thin film applications. Photo masking is a process used today predominantly in microelectronics to etch different patterns on the oxide or on the metal deposit film of a silicon wafer. For this process, a photosensitive resin, the positive or negative photo-resist, is painted on the wafer, a photomask containing patterns is placed on this, and light is applied for sensitization of the resin. The two types of photoresists differ in their response to light and the solubility behaviour induced towards the developing solvent system. Resins which become less soluble by illumination yield a negative pattern of the mask and are called negative photo-resists. Resins which become more soluble when exposed to light and therefore yield a positive image of the mask are called positive photo-resists. Both types are utilized in practice. Selective solvents and special techniques are required to develop and finally remove the patterns. The result may be the creation of an oxide film corresponding to the patterns. Of great importance is the accuracy and stability of the photo mask used for this process. The soft photographic emulsion mask originally used was soon replaced by the very stable chromium mask. To produce such a mask, chromium is deposited onto a high quality polished flat glass or fused silica substrate under conditions which generate practically no pinholes. Generally the average density of all pinholes larger than 1.5 μm in a chromium blank is less than 0.1 inch^{-2}.

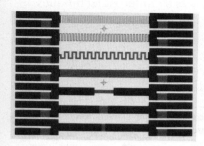

Fig. 39
Photo mask with etched test pattern.

Then a photo-resist is put over the chromium and a master bearing the desired pattern is applied. Subsequent exposure and development of the photo-resist exposes areas of the chromium which can be etched away by a suitable etchant. Plasma etching or any of the acid-based etchants can be used. After etching and removing the photo-resist, a chromium image is left on the glass plate. The chromium is considerably harder than a photographic emulsion and, therefore, is subject to wear and deterioration at a much slower rate than the emulsion masks. However, the process for making such masks is much more involved than the simple photographic process for making the emulsion masks, thereby resulting in much higher costs for chromium masks than for emulsion masks. The opaque areas of both types of masks are opaque to visible light as well as ultraviolet light. For the production of modern microelectronics, line widths between 3 and 2 μ are generally required and can easily be achieved with the chromium masks. Also, values smaller than 1 μm can be achieved. Disturbing thermal expansion effects are very small when using fused silica [146] or borosilicate low expansion glass as the substrate. Figure 39 shows a micrograph of a chromium mask. Photo masks of chromium are now established as a standard tool throughout the semiconductor industry. Good edge definition, narrow tolerances, ease of processing and long lifetime have lead to this general acceptance of such masks [147−150]. High quality coated glass plates (chromium blanks) for the manufacture of chromium masks which are designed to meet the different requirements of the semiconductor industry are commercially available, e.g. [151 − 153]. Also, «see through» low density iron oxide photo masks and masks of various other oxides are produced which shield ultraviolet rays necessary for the exposure of a photo-resist but transmit visible rays of long wavelength. Such masks aid processing due to the transparency in the visible range.

9.6.3 SCALES, RETICLES, APERTURES

In the production of dials on glass, enormous progress was achieved by replacing old techniques such as chemical etching and mechanical scratching by thin film deposition. Depending on the size of scale, mechanical masks or photo masks are used. The deposited metal is for the most part chromium. Figure 40a shows a comparison of the quality of scales produces by different techniques.

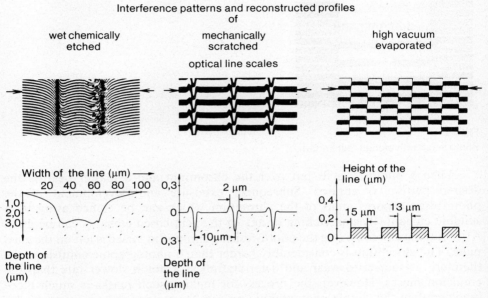

Fig. 40 a
Comparison of the quality of line scales produced with different methods (Leitz, Wetzlar).

9.6.4 PHASE PLATES

Annular phase plates are the principal items required for Zernike's phase contrast microscopy [154]. The phase plate, which consists in addition to the phase altering coating, of an absorbing film, is inserted into the rear part of the microscope objective. With the technique of phase contrast microscopy, invented in 1934, it became possible to make invisible phase objects visible like amplitude objects. This occurs because the phase of the undiffracted light contributing to the primary image is advanced or retarded by means of an annular phase plate so that it will interfere optically with the light beams diffracted by the specimen. In this way, a non-absorbing phase object can be made to appear lighter or darker than the surrounding medium. According to the positive or negative sign of the phase angle by which the fundamental wave is shifted in this special device, it is possible to distinguish between positive and negative phase contrast. In either case, however, the intensity of the undiffracted light is considerably greater than that of diffracted light, so that a neutral absorbing film must be applied to the phase ring to produce maximum optical contrast for the phase image.

The first commercially available phase contrast microscopes were fabricated by Carl Zeiss in 1944. The required phase plates were produced with a primitive technique using resin films and lacquer coatings. In 1950, however, Auwärter and Reichert [155] fabricated the first vapour-deposited positive and negative phase rings. As phase altering coatings, ZnS, MgF_2 or ThF_4 films can be used. These are deposited to a thickness producing a quarter-wavelength retardation. In Fig. 40b, the design of an annular phase plate for Zernike's phase contrast is shown schematically.

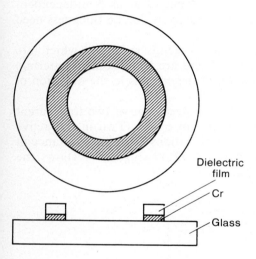

Dielectric
film

Cr

Glass

Fig. 40 b
Schematic representation of an annular phase
plate for Zernike's phase contrast microscopy.

9.7 TRANSPARENT CONDUCTIVE COATINGS

Transparent conductive coatings combine high optical transmission with good electrical conductivity. The existence of both properties in the same material is, from the physics point of view, not trivial and is only possible with certain semi-conductors like indium oxide, tin oxide, cadmium oxide, and with thin gold films, e. g. [157].

There are a number of interesting applications for this kind of coating, such as

– liquid crystal displays
– gas discharge displays

- electroluminescent devices
- electrochromic devices
- flat TV tubes
- fluorescent light sources and displays
- pockel's cells for laser Q-switch
- antistatic coatings for various meters (and combined with antireflection coatings)
- heated windows (aircraft, automobile, heating stages for microscopes, etc.).
- antennae for cars and radar protection for fighter planes.

The specifications for the various coatings depend on the specific application. The ranges are transmission from 60 to 90% and area resistance from $10 \, \Omega \, \square^{-1}$ to $100000 \, \Omega \, \square^{-1}$. (It is common practice to specify resistance in this case in $\Omega \, \square^{-1}$, which uses the fact that the resistance of areas is independent of absolute size. The advantage of using area resistance is that no thickness need to be specified.)

As already mentioned, liquid crystals are organic materials which are transparent when no electrical field is present and are translucent (scattering) when the field is on because of an orientation of the crystals in the direction of the electrical field, e.g. [157 – 160].

In order to utilize this effect for displays, one needs one or two transparent conductive electrodes consisting of, say, indium tin oxide. (Two in the active case and one in the passive case.) Required specifications are a transmittance of 80 to 90% and an area resistance less than $300 \, \Omega \, \square^{-1}$. Figure 41 shows the spectral transmittance of such electrode coatings.

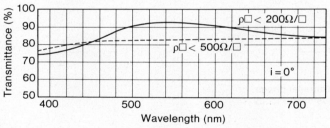

Fig. 41

Electrically conductive ITO coatings BALTRACON® (Balzers) which can be used as electrodes in various displays.

One of the crucial properties is the ability to resist the chemical and electrochemical impact of the liquid crystal material. Because of electrolytic processes, the use of dc fields is generally more critical than that of ac fields. In the passive case, normally opaque metal coatings are used as a second electrode. Here similar problems occur especially in the dc case.

Gas discharge displays, electrochromic and electroluminescent devices [161] are used for similar purposes as liquid crystals, but in place of the liquid crystal, a plasma (gas discharge like in a Ne tube) or solid state materials are used. The optical and electrical requirements are very similar. With gas discharge display electrodes, a high temperature stability is also required (450°C cycling).

The requirements for flat TV tubes and fluorescent displays are not well defined since these devices are still in development.

Pockels cells consist of potassium- or ammonium-dihydrogenphosphate which are very sensitive to humidity and temperature changes. The problem here is to get good adhesion to these difficult materials. The area resistance required is $1-10 \, k\Omega \, \square^{-1}$.

The transparent antistatic coating for meter windows is an old requirement and this coating has been used for many years on both glass and plastic substrates. Figure 42 shows the spectral characteristic of such a film. The surface resistance of approximately 10^{15} ohms/square of Perspex, is reduced to less than 10^5 ohms/square by the antistatic film. This resistance is not exceeded through ageing processes. The antistatic coating can be combined with an antireflection coating (e.g. conductive Iralin® (Balzers)). The area resistance lies in the range $10 - 100 \, k\Omega \, \square^{-1}$.

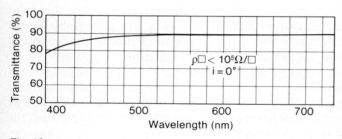

Fig. 42
Electrically conductive coating ANELL® (Balzers) which can be used as antistatic coating.

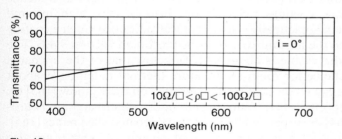

Fig. 43
Electrically conductive coating AURELL®(Balzers).

The transparent heating and antenna coatings require a very low area resistance of about $5 - 20 \, \Omega \, \square^{-1}$. Traditionally, special gold coatings have been used on both glass and plastics. The stability of the coating towards temperature changes has to be very good. The transmission generally lies between 60 and

80 % in the middle of the visible spectral range, as can be seen in Fig. 43. Fundamentally there is no reason why this requirement could not be met with semiconductor materials, such as semiconducting oxides, but no practical realisation by vacuum deposition is yet possible, the lowest values so far are around 60 $\Omega \square^{-1}$.

Most of the transparent conductive coatings are produced by vacuum evaporation and sputtering. Other techniques are hot spraying and chemical vapour deposition, as manufactured by Corning, Pittsbourgh Plate Glass (PPG) and others.

9.8 ENERGY–RELATED COATINGS

For this type of optical coating application, indium tin oxide films, the so-called ITO films, in particular have achieved remarkable importance.

Presumably, industrial use for these transparent and conductive layers first arose more than 40 years ago when a heating system was needed for de-icing aircraft windscreens. Nowadays, because of the large band gap in the range 3.5–4 eV, these layers are mainly used for optical windows for the solar spectrum, as well as for transparent conductors. They are very stable under typical environmental conditions and are resistant to chemical attack, adhere well to many substrates and have good mechanical properties.

Such coatings can significantly increase solar collector panel transparency and energy conversion efficiency. That same coating reflects inefficient infrared radiation to prevent photovoltaic cell overheating and boost the energy conversion ratio. ITO coated panels are excellent conductors, and diminish the sheet resistance typical of thick silicon cells.

They are also used as architectural coatings. The use of optical coatings can significantly improve the energy efficiency and performance of building systems, including windows, skylights, insulating materials and lighting systems. Indium tin oxide window coatings reduce room heat loss by high infrared reflectance without reducing practically visual light transmittance. Such a coating has a similar behaviour to the heat mirrors described before in Section 9.4.6. The mean transmittance in the visible is about 8 - 10 % below that of the uncoated glass, whereas the reflectance for room temperature thermal radiation is about 92 % [162]. ITO coatings can either be directly deposited on the window [163] or, more economically, deposited first onto plastic foils which are afterwards laminated to the glass [164] or suspended between two uncoated glass panes [165].

Other compound films such as cadmium tin oxide and $TiC+SnO_2$ are also applied and colouring effects with ITO/Ag/ITO and ITO/Cu/ITO have been investigated. Ag and Cu are used, replacing Au for infrared reflecting films [166]. The deposition of such functional long-life coatings must be applied by using high-rate deposition systems for large panes which are cost-effective for the intended application. For the deposition of flexible substrates, roll coaters, e.g. [33, 34], seem to be the optimum solution.

A further application of the indium tin oxide is a coating on low-pressure sodium lamps which enables higher operating temperatures resulting in more light output for the same power input [167]. There are also experiments with coated glass envelopes of incandescent lamps to reflect the heat back onto the filament, thus reducing the electrical power required to maintain the filament operating temperature [168].

For solar radiation collectors of the photothermal conversion type, which may consist of a coated glass or plastic tube system, e. g. [169a], a number of different film materials are used as selective solar absorbers. The applied film materials must exhibit a higher degree of solar radiation absorption α than their emission ε of long wavelengths at the practical temperatures at which the collector material will operate. Desired values are $\alpha \geqslant 0.9$ and $\varepsilon \leqslant 0.05$. For low temperature application (below 300°C), the most commonly used absorber coatings are black chromium and black nickel [169b and c]. Cermet films such as Cr/Cr_2O_3, Au/MgO [169d], Au/Al_2O_3 and Cu/Al_2O_3 [169e] which are made by sputtering, or Ni/Al_2O_3 [169g] which are made by coevaporation, are also able to produce similar results.

9.9 SOLDERABLE COATINGS

Solderable coatings are complex film systems of $Cr/Cu/Au$ which are vapour deposited, beginning with Cr, onto optical components made of glass, fused silica, ceramics or crystalline materials. In some cases the top gold film is electrolytically tin plated so that the whole coating becomes ideal for industrial high frequency brazing. The glass substrates coated with such a film system can be soldered with normal soft solder to wires, contact lugs or holders, etc. Because of the excellent adherence of the films, the soldered joint is mechanically durable, liquid and gas tight, e. g. [170]. Such coatings have various applications: for instance, sight glasses which are coated with heatable, transparent and antireflective films can be provided with the solderable film as bus bars. Instrument windows coated with an antistatic film can be provided with solderable films for electrically conducting electrodes. Finally it becomes possible to braze gas-tight instrument glasses edged with the solderable film into aircraft panel instruments.

9.10 INTEGRATED OPTICS

The transistor invented in 1947 started a revolutionary trend in the fields of communication, information processing and automatic control. Soon after, a large and still increasing number of miniaturized transistors and solid state devices were combined to integrated electronic circuits. Such IC's were not only compact, of small size and fast but also relatively cheap.

With the invention and development of continuously operating lasers [171] in 1961, a formal union between electronics and optics could be achieved using optical circuits. A new technology of integrated optics proposed by S. E. Miller from Bell Laboratories in the USA [172] in 1969 was based on the concept that optical circuits can be constructed in loose analogy to the above mentioned electronic integrated circuits. Whereas in electronic circuits current flows in thin strips of metal conductors, in the proposed optical circuits, the information was to be carried by modulated light beams which are guided along thin film or strips of an optically transparent extremely low loss material. The thin films, however, form miniature lasers, lenses, prisms, light switches and light modulators. Since the frequency of light is some 10^4 times higher than the highest frequency of an electronic device, the amount of information that can be carried by a light signal is correspondingly greater. Optical circuits are also in principle considerably faster than electronic circuits. Integrated optical circuits are thought to be used in connection with optical transmission lines in the form of glass fibres with circular cross section. The coded optical signal emitted by a first optical circuit is transmitted by means of a linked glass fibre. On the other end of the transmission line, a second optical circuit will decode the information carried by the light pulses. Such a system is shown schematically in Fig. 44 [173]. The wave guiding glass fibres consist of a core with refractive index n_1 and a cylindrical cladding with index n_2, so that $n_1 > n_2$. A difference of 1 % in the indices is sufficient. There are step-index and graded-index fibres which are either of single-mode or of multimode type. In the case of multimode fibres, the core diameter is 50 μm for the step-index type and some hundred μm for graded-index fibres. With single-mode fibres, the core diameter is only some microns to 10 μm.

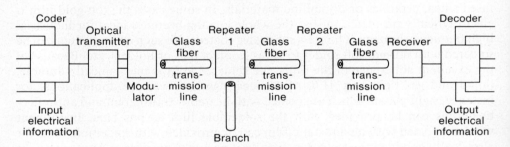

Fig. 44
Block diagram of an optical communication system using glass fiber transmission lines [173].

Light guidance in the fibres as well as in the planar wave guides occurs in a zigzag path by total reflectance. Extremely low losses of only 0.5 dB km^{-1} have been achieved in special fibres for near infrared radiation [174]. This would mean that the distances between repeater stations could be extended to about 30 km. This value is, however, not yet an industrial fabrication standard! The present standard is about 3-7 dB km^{-1} [175]. All advanced high bit rate systems, which have been realized recently use single-mode fibres [176]. Integrated monomode optics is considered to be a proper supplement for advanced single-mode fibre systems.

High efficiency coupling devices are required to link optical thin film circuits and glass fibres. For interconnection between fibres, snap-in fibre optic links are available, e.g. [177].

Integrated optics offer realizable goals of low cost, high band width, extreme ruggedness and great versatility. But it is equally evident that the problems to be solved in this case are numerous and the technology is quite difficult and expensive. The field of integrated optics is years away from full realization. In the meantime, however, there are many possible uses of partially integrated systems to transfer technical progress in the form of hybrid circuits into practical application.

9.10.1 INTEGRATED OPTICS COMPONENTS

In integrated optics experiments, planar wave guides are used which consist of various thin dielectric films deposited on fused silica or glass substrates. These transparent films must have very low optical losses, their thickness is generally between 0.1 and 10 μm and their refractive index is slightly higher than that of the

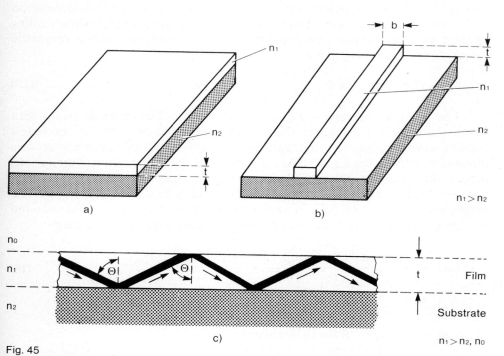

Fig. 45

Two dimensional planar wave guide (a), one dimensional strip shaped wave guide (b) and schematic representation of a wave guided by total reflection inside the thin film wave guide (c). [222].

transparent substrate. Inorganic and organic film materials are used [30, 178-187]. Photolithographic techniques are applied for defining the structures; and roughness of the sidewalls of the wave guides has to be avoided to keep losses small. Figures 45a and b show the two types of planar waveguides. The light waves are guided by total reflection as in the optical fibres. If the thickness t of the planar waveguide is large compared with the wavelength λ of the light, a geometric optical treatment can be applied [188]. With n_0 = refractive index of air, n_1 = that of the waveguide and n_2 = that of the substrate, so that $n_1 > n_2$, n_0 and $n_2 > n_0$ the zigzag propagation of light is given by:

$$\theta_{max} = \sin^{-1}(n_2/n_1) \tag{16}$$

In the dielectric waveguide, various waves can propagate with different angles θ_i between $90° \leqslant \theta_i < \theta_{max}$.

The propagation of light in waveguides with film thicknesses which are comparable with the wavelength of light must be treated by wave optical methods e. g. [189].

Mathematical treatment of the corresponding wave equation under consideration of all boundary conditions yields a number m of discrete solutions of possible waves. These possible waves, the so-called modes, which depend on refractive indices n_0, n_1, n_2 and on the film thickness t are distinguished according to their different propagation constant $ß_m$. The connection with a geometric optical consideration is given by:

$$ß_m = (2\pi/\lambda)\, n_1\, \sin\theta_m \tag{17}$$

The index m indicates the order of each mode. The order decreases with increasing angle θ_m. Therefore the fundamental mode, $\theta_o = 90°$, is the fastest running wave propagating in a straight line through the wave guide. This becomes obvious by considering Fig. 45c. As a consequence of the dispersion of the various modes, the refractive index of planar waveguides becomes dependent on film thickness and is called the effective refractive indes $ß/k = n_{eff}$ [190, 191]. The propagation constant of light in vacuum is given by $k = 2\pi/\lambda_o$. The required minimum thickness t_{min} of a two-dimensional planar waveguide is given by:

$$t_{min} = \frac{m\pi + \tan^{-1} q\, [(n_2^2 - n_0^2)/(n_1^2 - n_2^2)]^{1/2}}{k\,(n_1^2 - n_2^2)^{1/2}} \tag{18}$$

$m = 0, 1, 2, \ldots$

$$q = \begin{cases} n_0^2/n_1^2 & \text{E vector of the light wave normal to the film plane: \quad TM mode} \\ 1 & \text{E vector parallel to the film plane: \qquad TE mode} \end{cases}$$

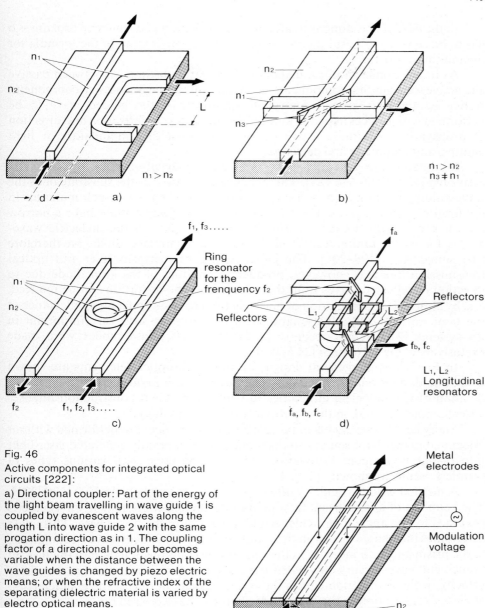

Fig. 46

Active components for integrated optical circuits [222]:

a) Directional coupler: Part of the energy of the light beam travelling in wave guide 1 is coupled by evanescent waves along the length L into wave guide 2 with the same progation direction as in 1. The coupling factor of a directional coupler becomes variable when the distance between the wave guides is changed by piezo electric means; or when the refractive index of the separating dielectric material is varied by electro optical means.

b) Coupler using partial reflection of the light beams from semitransparent mirror films with a refractive index n_3.

c) Frequency filter made by a ring resonator which separates frequency f_2 from a frequency mixture f_1, f_2, f_3.

d) Frequency filter made by longitudinal resonators.

e) Phase modulator: The modulation voltage changes the refractive index of the electro optical wave guide or substrate thus influencing the phase of the guided wave.

In the case of one-dimensional strip-shaped waveguides, the strip broadness b has to be considered, resulting in a considerable complication of the formula for the minimum thickness.

As in electronics in integrated optics, one has to distinguish between passive and active waveguiding components. A simple waveguide is a passive component. Active components enable controlled variations of the guided wave to be performed with respect to phase, amplitude, frequency, polarization and direction of propagation. All these active components are either modulators or light emitters and amplifiers and detectors.

Modulators are generally made of electro-, magneto- or acousto-optically active materials [192-196 187]. These materials are used as the substrate or as the wave-guiding medium. There are many problems when using such materials in the form of thin films, e.g. [197, 198]. The light emitters should have a narrow optical band width because of troublesome dispersion in the dielectric waveguides. Lasers and luminescent diodes or semiconductor laser diodes are therefore best suited, e.g. [199-208]. The close connection between laser and optical transmission line may, however, produce instabilities of the laser diode due to injection of reflected waves from the waveguide. The distortion can be reduced, e.g. by applying a combination of opto-electro feed back and a high reflectance of the laser mirror [209]. High-quality light detectors are also very important in i.o.-systems [210]. Until recently [211-215], such photo diodes were made exclusively of Ge [216] or Si [217-219].

In Fig. 46, some general designs of integrated optical elements are shown using one-dimensional strip guides. The designs were performed without consideration of technological difficulties in realization. Their construction imposes a tremendous demand on the micro-fabrication technology.

Many basic experiments in passive and active devices are performed without integrated planar light sources so that coupling of externally generated laser light into planar waveguides is required. This can be done with various types of couplers. Some are shown in Fig. 47.

A method used particularly often is the prism coupling technique [220, 221]. The coupling of a laser beam by a prism into the dielectric planar waveguide is governed by the angle θ of incidence of the light onto the prism base. This angle determines the phase velocity in the x direction of the incident wave with index n_3 in the prism and with index n_o in the gap. Strong coupling of light into the film by evanescent waves occurs only when θ is chosen in such a manner that the phase velocity in the x direction equals the phase velocity of one of the characteristic modes of propagation in the waveguide. This can be expressed by:

$$ß = k\, n_3\, \sin \theta_3 \tag{19}$$

with $k = 2\pi/\lambda = \omega/c$, ω = angular frequency and c = velocity of light in vacuum.

The coupling effect is reversible.

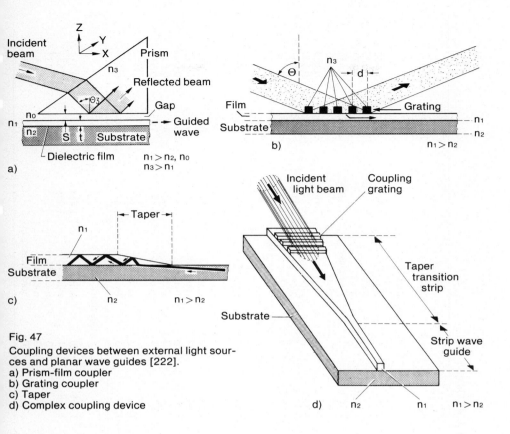

Fig. 47
Coupling devices between external light sources and planar wave guides [222].
a) Prism-film coupler
b) Grating coupler
c) Taper
d) Complex coupling device

9.10.2 PRESENT STATUS AND TREND

Glass fibres are available today in good and reliable quality [178]. The fabrication of glass fibres occurs often in combined processes of classical fibre manufacturing and modern vapour-deposition technologies or other wet chemical processes, e. g. [223-225]. Glass fibre transmission lines have become a reality and their application is increasing continuously for local area networks and also for inter-city lines.

In integrated optics, numerous methods have been used for the fabrication of planar wave guides of inorganic and organic materials. Most films were deposited on glass. The various methods used for the fabrication of waveguiding thin films are listed in Table 6 [183]. All these depositions on glass substrates are performed only for first fundamental studies of the general properties of materials and single two-dimensional micro-optical devices. The evolution of integrated optics depends strongly on the availability of materials and technologies suitable for the construction of film devices [178]. As well as low-loss glass-like waveguiding films, single crystal films are also needed for some active components such as

452

TABLE 6

METHODS OF FABRICATION OF WAVEGUIDING FILMS [183]

Method	Film	n_1	t (μm)	Substrate	n_2	α (dB/cm)	λ (μm)
				A. Inorganic films			
vacuum evaporation	ZnS	2.35	0.2	glass	1.51	>10	0.633
	CeO$_2$	~1.8	0.5	glass		>20	0.633
	Al$_2$O$_3$	1.63	2	glass		>30	0.633
sputtering	CAS 10 glass	1.465	1	FK6 glass	1.444	<5	0.633
	7059 glass	1.62	0.3	glass	1.51	1	0.633
	BaO+SiO$_2$	1.48–1.62	1	SiO$_2$	1.46	0.6	0.633
	ZnO	1.98	1.6	glass	1.51	>20	0.633
	glass:Nd	1.57	1.5	glass	1.53	0.45	1.06
CVD	SiON	1.48–1.99	0.5	SiO$_2$	1.46	<0.4	0.633
				B. Organic films			
thermal polymerization	polyurethane	1.56	0.8	glass	1.51	0.8	0.633
	epoxy	1.58	2.2	glass	1.51	0.3	0.633
	nitrocellulose	1.49	0.15	none			0.633
	polystyrene	1.586	0.5	glass	1.51	0.1	0.633
	SiO$_2$:PbO	1.64	0.9	glass	1.51	0.5	0.633
	PCHMA	1.505	10	Pyrex	1.48	0.2	0.633
photopolymerization			3.5	PMMA	1.49	4	0.633
	KPR	1.62	1	glass	1.51	7(2)	0.633
						1	1.06
rf polymerization	cinnamate resist	1.59	2.2	glass	1.51	1	0.633
	VTMS	1.49–1.53	3	Pyrex	1.47	0.04	0.633
				C. Surface films			
ion implantation	SiO$_2$:p	1.468	<30	silica	1.458		0.633
	SiO$_2$:Li	1.468	1	silica	1.458	<0.2	0.633
	GaP:p		3.1	n-GaP			0.633
ion migration	GaAs:p		3	n-GaAs		10	1.15
	glass:Tl		50	borosilicate glass		0.1	0.633

Method	Film	n_1	t (μm)	Substrate	n_2	α (dB/cm)	λ (μm)
C. Surface films							
oxidation	Ta_2O_5	2.21	0.8	glass	1.53	0.9	0.633
D. Single-crystal films							
molecular beam	GaAs	3.34	1	$Al_xGa_{1-x}As$	3.18	<4	1.15
	AgBr	2.14	3	NaCl	1.49	<5	10.6
CVD	ZnO	1.98	10	sapphire	1.77	4	0.633
						1	1.06
LPE	GaAs	3.27	>20	n-GaAs	2.97	<1	10.6
	GaAs	2.135	12	n-GaAs	1.96		1.15
	$Y_3Fe_{4.3}Sc_{0.7}O_{12}$	2.124	2.4	$Gd_3Ga_5O_{12}$	1.94	8	1.15
						<3	1.52
	YAG	1.818	>10	sapphire	1.77	~5	1.06
diffusion	KDP:ADP	1.5105	38	KDP	1.5095		0.633
	$ZnSe_{1-x}Cd_x$	> 2.6	3	ZnSe	2.6	3	0.633
	$(Li_2O)_x(Nb_2O_5)_{1-x}$	2.220	250	$LiNbO_3$	2.214	<1	0.633

modulators and lasers. The generation of such heteroepitaxial systems requires that the crystal lattice structures of the film can be closely matched with that of the substrate. Various systems of materials are studied for this purpose, such as:

AlGaAs

InGaAs

InGaAsP	Bell Labs. USA	1977/78
$Y_3 \begin{bmatrix} Fe \\ Ga \\ Al \end{bmatrix}_5 O_{12}$	Siemens FRG Philips NL	1972
Ti diffused LiNbO$_3$	Bell Labs. USA	1974/75.

The corresponding light wavelengths in that systems are between 0.6 and 1.5 μm. The crystal lattice of, for example, AlAs matches exactly that of GaAs at about 900°C, so that at this temperature a perfect waveguiding film of $Al_xGa_{1-x}As$ can be grown on a GaAs substrate. To prevent the undesired absorption of the light produced by a $Al_xGa_{1-x}As$–laser by GaAs waveguides, In and P are additionally introduced. A good approach may be to use the quarternary film compound InGaAsP on InP substrates. The problems to be solved in this case are numerous and the technology is quite difficult and expensive [226]. For the fabrication of such complicated structures, only molecular beam epitaxy, e.g. [227], is currently useful.

For some time, optical circuits will probably consist of devices made of different systems and in this case passive optical elements deposited onto glass substrates may be of considerable importance. However, the goal is to replace such hybride systems and to achieve a monolithic integrated optic. For immediate future, however, integration of optical and electronic elements to one circuit may be of great advantage.

9.11 SCIENTIFIC APPLICATIONS

Coatings on glass and glass-like substrates are also used often as tools for scientific research, both fundamental and applied. Structure and microstructure investigations of thin films, for example as a function of the applied deposition technology, the substrate temperature and the surrounding gas atmosphere are often performed by using of glass substrates. Thus, an attempt at correlation of optical, electrical, mechanical or chemical film properties with the observed film structure is made. Into this category also belong investigations on the optical anomalies, e.g. [228], or on electrical resistivity and electromigration, e.g. [229], of thin discontinuous films formed from isolated metal droplets or islands of different shape and size. Such investigations can be used to study the effectiveness of cleaning procedures and other surface pretreatment operations [230]. Studies of adherence, stress and abrasion resistance as well as of other mechanical properties such as density and hardness are also often made with films deposited on glass. An important scientific application of ultra-high vacuum-deposited films is in studies of the physical properties of uncontaminated surfaces using modern surface investigation techniques such as X-ray and electron diffraction, various electron and photon spectroscopic techniques, and others. Possible changes in the surface properties can be observed readily as various contaminating gases or vapours are introduced in a controlled way into such systems. Studies on adsorption behaviour as well as on catalytic activity of thin films for various chemical reactions are relevant here, e.g. [231]. It was found by Cremer et al. [232–235] that indium oxide films of about 1 μm thickness, prepared according to Auwärter and Rheinberger by metal deposition onto glass and subsequent oxidation of the deposit in atmosphere at elevated temperature, are suited for chromatographic analysis. In this new type of solid-liquid chromatography with coated surfaces, the thin film chromatography (TFC), a mobile phase formed by a liquid layer, moves along the highly active surface of the deposited metal oxide coating. Separations of various organic dyes could be performed ten times faster with this technique than by classical thin layer chromatography (TLC) using surfaces of about 50 μm thick sheets of silica gel made by wet chemical precipitation and tempering. The achieved detection sensitivity with TFC is in the nanogram region.

A further application is the use of initially pinhole-free films of soft metals such as lead and tin on glass plates to study possible micro-meteorite impingement in various space experiments. Optical inspection of such plates after exposure

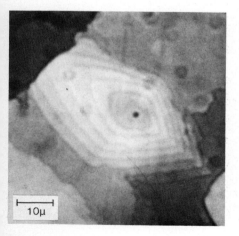

Fig. 48

Growth spiral of 2,3-benzofluoren obtained by crystal growth from a benzene/xylene solution in the interference filter [238]. The step height of the spiral was found to be 6 nm. (Panchromatic black-and-white copy of the original colour picture).

456

indicates the meteorite bombardment events as holes in the opaque film. Finally, let us mention some unconventional uses of interference filters for scientific investigations. They are all based on the high sensitivity in the relation between spacer layer optical thickness variation and corresponding peak transmission shift. An increase in optical thickness causes the peak to appear at longer wavelengths. A step height of about 0,3 nm can be detected by a wavelength shift of the peak of about 1 nm in a filter of lowest order. Auwärter, Haefer and Rheinberger [236] suggested the film thickness be measured by using a semi-manufactured Fabry Perot filter. After coating part of the area of the spacer layer with the film of unknown thickness and completing the filter by Ag-mirror deposition, the optical film thickness can be determined by spectral peak transmission measurement of both areas. Knowing the film refractive index, the corresponding geometrical thickness can be calculated.

Speidel [237] replaced the solid spacer layer by a liquid one and made small embedded thin phase objects visible. Using liquid solutions as the spacer layer, Pulker and Junger [238] studied crystallization phenomena, especially spiral growth of organic compounds in situ, by the application of the interference filter method together with a microscope and an optical spectrometer. Figure 48 shows a growth spiral of 2,3-benzofluorene.

The interference filter method was also used by Ruf and Winkler [239] for the detection of oil backstreaming in oil diffusion pumps with a pinhole-camera arrangement.

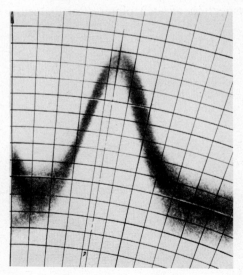

Fig. 49
Spectrogram of the velocity distribution of
SiO vapour obtained with the interference
filter method [240]. (Panchromatic black-
and-white copy of the original colour
picture).

An ingeneous application of such interference filters as a detector is found in gas-kinetic studies performed by Cremer, Kraus and Ehrler [240, 241]. In effusion experiments and particle velocity studies of silicon monoxide vapour, the shape of the vapour cloud and the intensity and density distribution of its molecules as well as their velocity distribution, shown in Fig. 49, can be fixed and reproduced within the interference filter.

Finally as the last example of scientific thin film applications sensor devices will be discussed. Thin film thermocouples and thin film bolometers are often applied in scientific experiments to measure and to control the surface temperature, e. g. [242]. These rather simple thin film devices are complemented today by various other, often more complex, sensor elements. These may use beside thermal also mechanical or other input signals such as e. g. adsorbed water vapour. Such sensores are often made from or contain thin films as an essential constituent. This is especially true when a small size of the device and a high response are required. The science of sensors is a very active area of research, development and manufacture and is in many ways an example of the use of thin film physics in technology [243]. Recently thin film miniature capacitance transducers of high sensitivitiy have been made by Jones et al. [244-246] which were incorporated into remote passive transponders applied for pressure measurements in biomedicine. For such applications, the displacement of primary sensing elements is small and it is important to have miniature devices. High capacitance

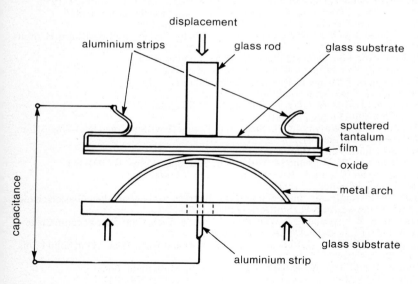

Fig. 50

Schematic representation of a metal-arch capacitance transducer according to Jones et al. [245]. This capacitance transducer is of the variable-area electrode type. A tantalum/tantalum oxide film structure is deposited onto a flat glass substrate to obtain a suitably flat surface for contact with a flexible electrode which may be a thin metal foil possibly deposited in the shape of a dome or an arch.

458

in a miniature device can be obtained if a thin film of dielectric materials is used to cover a capacitor plate and the area of the other plate in contact with this material is varied. The scheme of such a metal-arch capacitance transducer is shown in Fig. 50. The transducer is about 3 mm thick and has a diameter between 6 and 10 mm. The fixed electrode consists of a sputtered tantalum film on a glass substrate, the dielectric is a reactively sputtered tantalum pentoxide film, and the variable electrode consists of phosphor bronze and is either an arch shaped 15 μm thin foil obtained from a metal strip or a spherical dome shaped foil made by sputtering [245]. A maximum capacitance value of 340 pF could be obtained and short time and long time stabilities were better than 0.2 pF min^{-1} and 0.05 pF hour^{-1} respectively. The sensitivity of such transducers was between 2000 and 8000 pF mm^{-1}. Q factors have been obtained in the range between 25 and 200 at 1529 Hz. The temperature coefficient at 40°C was found to be much smaller than 1 pF °C^{-1} [245].

REFERENCES

[1] H. Mayer, Physik dünner Schichten I, Wissenschaftl. Verlagsges. m. b. H., Stuttgart, 1950.
[2] O. S. Heavens, Optical Properties of Thin Solid Films, Butterworths, London, 1955.
[3] A. Vašiček, Optics of Thin Films, North Holland, Amsterdam, 1960.
[4] H. Anders, Dünne Schichten für die Optik, Wissenschaftl. Verlagsgesellschaft. m. b. H.,Stuttgart, 1965.
[5] H. A. Macleod, Thin Film Optical Filters, A. Hilger, London, 1969.
[6] Z. Knittl, Optics of Thin Films, J. Wiley & Sons, London, 1976.
[7] a) J. D. J. Jackson, B. Rand and H. Rawson, Thin Solid Films, 77 (1981) 5.
    b) N. Jackson and J. Ford, Thin Solid Films, 77 (1981) 23.
    c) A. S. Sanyal and J. Mukerji, Glass Technol. (GB), 23 (6) (1982) 271.
[8] S. M. Budd, Thin Solid Films, 77 (1981) 13.
[9] a) R. D. Southwick, J. S. Wasyliyk, G. L. Smay, J. B. Kepple, E. C. Smith and B. O. Augustsson, Thin Solid Films, 77 (1981) 41.
    b) F. Geotti-Bianchini, Vetro Informazione, 1 (1981) 6.
    c) F. Geotti-Bianchini, B. Scalet and M. Verita, Riv. Staz. Sper. Vetro, 11 (1981) 1, 3.
    d) E. Guadagnino and B. Locardi, Riv. Staz. Sper. Vetro, 12 (1982) 2, 51.
[10] Y. Talmor, Thin Solid Films, 77 (1981) 21.
[11] H. Schröder and R. Kaufmann, in E. Schott (Ed.), Beiträge zur angewandten Glasforschung, Wissenschaftl. Verlagsges. m. b. H., Stuttgart, 1959, p. 355.
[12] D. L. Chambers, R. B. Love and C. T. Wan, Proc. 20th Ann. Tech. Conf. Soc. Vacuum Coaters, Atlanta, GA, 1977, p. 26, and F. Da Lauro, ibid. p. 48.
[13] a) J. P. Lloyd, M. C. Petty, G. G. Roberts, P. G. Lecomber and W. E. Spear, Thin Solid Films, 89 (1982) 395.
    b) Langmuir-Blodgett Films, W. A. Barlow (Ed.), Elsevier, Amsterdam, 1982.
[14] J. R. Hollahan, Th. Wydeven and C. C. Johnson, Appl. Opt., 13 (1974) 1844.
[15] J. C. Manifacier, J. P. Fillard and J. M. Bind, Thin Solid Films, 77 (1981) 67.
[16] a) F. Mezei, Commun. Physics, 1 (1976) 81.
    b) F. Mezei and P. A. Dagleish, Commun Physics, 2 (1977) 41;
    c) J. Schelten and K. Mika, Nucl. Instr. Meth., 160 (1979) 287;
    d) E. A. Spiller, Opt. Spectra, (March) (1981) 18; Ind. Res. Developm., March (1981) 68;
    e) M. R. Howells and E. A. Spiller, SPIE, Brookhaven, Conf. Synchrotron Rad., Nov. 16–20, 1981.

[17]   E. J. Daniels, Neues Opt. Journal, 8 (1974) 553, and
       L. Kaminski, a) Manufact. Opt. Int. (Sept.) (1978) 30;
                    b) Der Schweizer Optiker, 54 (1978) 219;
                    c) Der Augenoptiker, 34, (1979), Balzers DN 1899.
[18]   J. A. Thornton, Proc. 18th Ann. Tech. Conf. Soc. Vacuum Coaters, Key Discayne, Florida,
       USA, 1975, p. 8.
[19]   A. S. Forschirm, Proc. 23rd Ann. Tech. Conf. Soc. Vacuum Coaters, Chicago, IL, USA, 1980,
       p.15.
[20]   G. Dittmer, Proc. 21st Ann. Tech. Conf. Soc. Vacuum Coaters, Detroit, MI, USA, 1978, p. 64.
[21]   R. T. Sorg, Ref. [20], p. 66.
[22]   F. Adams, Ref. [19], p. 46.
[23]   D. L. Ross, RCA Rev., 39 (1978) 137.
[24]   G. Kaganowicz and J. W., Robinson, US Patent No. 4.072.985.
[25]   C. L. Hammermesh and L. W. Crane, J. Appl. Polym. Sci., 22 (1978) 2395.
[26]   T. Smith, D. H. Kaelble and C. L. Hammermesh, Surf. Sci., 78 (1978) 203.
[27]   D. D. Neiswender, Adv. Chem. Scr., 80 (1969) 338.
[28]   J. H. Hollahan and A. T. Bell (Eds.), Techniques and Application of Plasma Chemistry, Wiley,
       New York, 1974.
[29]   J. L. Vossen, Proc. 4th Int. Symp. Plasma Chemistry, Zürich, 1979.
[30]   P. K. Tien, G. Smolinsky and R. J. Martin, Appl. Opt., 11 (1972) 637.
[31]   H. A. Beale and R. F. Welonski, Proc. 24th Ann. Tech. Conf. Soc. Vacuum Coaters, Dearborn,
       MI, 1981, p. 56.
[32]   R. P. Howson, J. N. Avaritsiotis, M. I. Ridge and C. A. Bishop, Appl. Phys. Lett., 35 (2) (1979)
       161; R. P. Howson, C. A. Bishop and M. I. Ridge, Thin Solid Films, 90 (1982) 296.
[33]   a) M. I. Ridge, R. P. Howson and C. A. Bishop, in R. I. Seddon, (Ed.), Proc. SPIE, Vol. 325,
          Optical Thin Films, Los Angeles, CA, January 1982, p. 46.
       b) E. K. Hartwig, in R. I. Seddon, (Ed.), Proc. SPIE, Vol. 325, Optical Thin Films, 1982, p. 52.
[34]   W. C. Kittler jr. and I. T. Ritchie, in R. I. Seddon, (Ed.), Proc. SPIE, Vol. 325, Optical Thin
       Films, 1982, p. 61.
[35]   a) Neue Zürcher Zeitung, supplement: Forschung und Technik, wednesday, May 11 (1983) 69.
       b) R. A. Sprague (Ed.), Proc. SPIE Vol 329, Optical Disk Technology, Los Angeles, CA, USA,
          January 26–28, 1982.
       c) A. E. Bell (Ed.), Tutorial T 3, SPIE, Optical Data Storage Technology, Arlington, Virginia,
          USA, June 6, 1983.
       d) Technical Digest, Optical Data Storage, Winter '83, Incline Village, Nevada, USA, January
          17–20, 1983.
       e) Matsushita Comp., Electron. News, Vol. 29 No 1440 (1983) 27.
       f) E. Rothschild, Optical-Memory Media, Issue of BYTE-Magazine, March 1983, San
          Francisco, CA, USA.
       g) Rothschild Consultants, Optical Memory Newsletters, May–June 1983 and July–August
          1983, San Francisco, CA, USA.
[36]   Souheng Wu, Polymer Interface and Adhesion, M. Dekker, Basel, Switzerland, 1982.
[37]   A. V. Shatilov and L. P. Tyutikova, Example of the calculation of an interference filter using
       the method of successive synthesis, Opt. Spectr., 14 (1963) 227.
[38]   a) J. A. Dobrowolski, Completely automatic synthesis of optical thin film systems, Applied
          Opt., 4 (1965) 937;
       b) J. A. Dobrowolski, Versatile computer program for absorbing thin film systems, Applied
          Opt., 20 (1981) 74.
[39]   H. M. Liddell, Computer-Aided Techniques for the Design of Multilayer Filters, Adam Hilger,
       Bristol, 1981.
[40]   a) J. F. Tang and Q. Zheng, Automatic design of optical thin film systems merit function and
          numerical optimization method, J. Opt. Soc. Am., 72 (1982) 1522;
       b) I. J. Hodgkinson and R. G. Stuart, Thin film synthesis algorithm for realization of refractive
          indices and Airy summation, J. Opt. Soc. Am., 72 (1982) 396.
       c) Y. Zerem, C. Bear, E. Greenfield and E. Secemski, Automatic computer synsthesis of
          broadband antireflection coatings, Abstract in Thin Solid Films, 89 (1982) 286.
[41]   P. W. Baumeister, Design of multilayer filters by successive approximations, J. Opt. Soc. Am.,
       48 (1958) 955.

460

[42] a) A. M. Ermolaev, I. M. Minkov and A. G. Vlasov, Method for the calculation of a multilayer coating with a given reflectivity, Opt. Spectr., 13 (1962) 142;
b) N. Kimura, Automatic design of multilayer dielectric film (I), J. Appl. Phys. Jpn, 31 (1962) 739 (translated by the National Research Council of Canada, Tech. Translation No. 1085, 1963);
c) Z. N. Elsner, On the calculation of multilayer interference coatings with given spectral characteristics, Opt. and Spectroscopy, 17 (1964) 238;
d) N. Kimura, Automatic design of multilayer dielectric film (II), J. Appl. Phys., Jpn, 32 (1963) 199 (trans. National Research Council of Canada, Tech. Translation No. 1166, 1965).
[43] O. S. Heavens and Heather M. Liddell, Least squares method for the automatic design of multilayers, Opt. Acta, 15 (1968) 129. This appears to be potentially one of the most useful methods. Only the number and refractive indices of the layers need be specified as the initial design.
[44] a) H. Zycha, Refining algorithm for the design of multilayer filters, Appl. Opt., 12 (1973) 979.
b) H. S. Hou, Method for optimized design of dielectric multilayer filters, Appl. Opt., 13, (1974) 1863.
[45] a) J. P. Borgogno and E. Pelletier, Synthesis of dielectric stacks with prescribed optical properties by a Fourier transform method, Thin Solid Films, 34 (1976) 357;
b) J. H. Apfel, Optical coating design with reduced electric field intensity, Appl. Opt., 16 (1977) 1880;
c) P. Aufmuth, H. W. Brandt and J. R. Kiehl, Berechnung und Herstellung von Systemen dünner dielektrischer Schichten für optische Anwendungen, Optik, 50 (1978) 329.
[46] A. L. Bloom, Refining and optimization in multilayers, Appl. Opt., 20 (1981) 66.
[47] J. Strong, J. Opt. Soc. Am., 26 (1936) 73.
[48] A. Smakula, German Patent Nr. 685767, (1935).
[49] L. A. Catalan, J. Opt. Soc. Am.,52 (1962) 437.
[50] H. Vonarburg, Optik, 20 (1963) 43.
[51] J. T. Cox and G. Hass, in Physics of Thin Films, Vol. 2,G. Hass and R. E. Thun (Eds.), Academic Press, New York, 1964, p. 239.
[52] J. C. Lee, Appl. Opt., 17 (1978) 2645.
[53] W. Geffcken, German Patent Nr. 758764, (1944).
[54] P. Brauer, Kolloid Z., 110 (1948) 93.
[55] J. T. Cox, G. Hass and A. Thelen, J. Opt. Soc. Am., 52 (1962) 965.
[56] A. Thetford, Opt. Acta, 16 (1969) 37.
[57] T. Turbadar, Opt. Acta, 11 (1964) 195.
[58] M. J. Shadbolt, in Ref. [5], Chapter 3, p. 59.
[59] J. T. Cox and G. Hass, J. Opt. Soc. Am., 48 (1958) 677.
[60] J. T. Cox, G. Hass and G. F. Jacobus, J. Opt. Soc. Am., 51 (1961) 714.
[61] J. T. Cox, J. Opt. Soc. Am., 51 (1961 1406.
[62] A. Thelen, J. Vac. Sci. Technol., 20 (1982) 310.
[63] T. Turbadar, Opt. Acta, 11 (1964) 159.
[64] I. Lewin, US Patent No. 4 081 667, (1978).
[65] E. A. Small, US Patent No. 4 112 483 (1978).
[66] C. J. Snavely, US Patent No 4 173 778, (1979).
[67] E. A. Small, US Patent No. 4 225 908, (1980).
[68] K. Försterling, Phys. Z., 14 (1913) 265.
[69] W. Kofink and E. Menzer, Ann. Phys., 2 (39) (1941) 388.
[70] H. Schröder, Ann. Phys., 5 (39) (1941) 55.
[71] W. Geffcken, Ann. Phys., 5 (40) (1941) 385.
[72] R. Jacobsson and J. O. Martensson, Appl. Opt., 5 (1966) 29.
[73] G. Hass and N. W. Scott, J. Opt. Soc. Am., 39 (1949) 179.
[74] G. Hass and N. W. Scott, J. Phys. Radium, 11 (1950) 394.
[75] G. Hass, J. Opt. Soc. Am., 45 (1955) 945.
[76] G. Hass, W. R. Hunter and R. Tousey, J. Opt. Soc. Am., 47 (1957) 1070.
[77] G. Hass and R. Tousey, J. Opt. Soc. Am., 49 (1959) 593.
[78] P. H. Berning, G. Hass and R. P. Madden, J. Opt. Soc. Am., 50 (1960) 586.
[79] G. Hass and J. E. Waylonis, J. Opt. Soc. Am., 51 (1961) 719.
[80] D. W. Angel, W. R. Hunter, R. Tousey and G. Hass, J. Opt. Soc. Am., 51 (1961) 913.
[81] L. R. Canfield, G. Hass and J. E. Waylonis, Appl. Opt., 5 (1966) 45.
[82] J. T. Cox, G. Hass and J. E. Waylonis, Appl. Opt., 7 (1968) 1535.

[83] E. T. Hutcheson, G. Hass and J. T. Cox, Appl. Opt., 11 (1972) 2245.
[84] J. Strong, Proc. Experimental Physics, Prentice-Hall, New York, 1945, p. 171 ff.
[85] L. G. Schulz, J. Opt. Soc. Am., 37 (1947) 349.
[86] M. Auwärter, Z. techn. Phys., 18 (1937) 457; Die Umschau, 43 (1937); J. Appl. Phys., 10 (1939) 705.
[87] M. Auwärter, private communication, 1982.
[88] G. Hass and W. W. Erbe, J. Opt. Soc. Am., 44 (1954) 669.
[89] Leaflet: Replicated Optics, Balzers Optical Group, Waltham, MA, USA.
[90] H. E. Bennett, J. Opt. Soc. Am., 53 (1963) 1389.
[91] M. Auwärter, unpublished photographs.
[92] M. Banning, J. Opt. Soc. Am., 37 (1947) 686.
[93] D. L. Perry, Appl. Opt., 4 (1965) 987.
[94] H. K. Pulker, Thin Solid Films, 34 (1976) 343.
[95] K. H. Günther, H. L. Gruber and H. K. Pulker, Thin Solid Films, 34 (1976) 363.
[96] S. Penselin and A. Steudel, Z. Physik, 142 (1955) 21.
[97] P. W. Baumeister and J. M. Stone, J. Opt. Soc. Am., 46 (1956) 228.
[98] O. S. Heavens and H. M. Liddell, Appl. Opt., 5 (1966) 373.
[99] A. F. Turner and P. W. Baumeister, Appl. Opt., 5 (1966) 69.
[100] H. Schröder, Z. angew. Phys., 3 (1951) 53.
[101] A. F. Turner, J. Phys. Radium 11, (1950) 444.
[102] J. C. C. Fan and J. F. Bachner, Appl. Opt., 15 (1976) 1012.
[103] G. Haacke, Ann. Rev. Mater. Sci., 7 (1977) 73.
[104] P. Nath and R. F. Bunshah, Thin Solid Films, 69 (1980) 63.
[105] H. Köstlin, Philips Tech. Rundschau, 34 (1974/75) 240.
[106] Balzers Thin Film Product Information DN 9238.8008.
[107] A. Thelen, J. Opt. Soc. Am., 70 (1980) 118.
[108] Balzers, Thin Films Principle Curves, BAG, FL 9496-Balzers, 1980.
[109] Balzers, Thin Film Product Information DN 9240.8008.
[110] H. K. Pulker, Thin Solid Films, 13 (1972) 291.
[111] See Ref. [108].
[112] Laserschutzfilter und Laserschutzbrillen, DIN 58215, Sept. 1974; Filters and Eye Protectors against Laser Radiation ISO, Draft International Standard ISO/DIS 6161, submitted 1978.
[113] J. H. Apfel, Thin Solid Films, 73 (1980) 167.
[114] National Bureau of Standards Special Publication 620, ASTM, STP 759: Laser Induced Damage in Optical Materials 1980, Boulder Damage Symposium.
[115] R. C. Nairn, Fluorescent Protein Tracing, 4th edn., Churchill Livingstone, New York, 1976.
[116] M. Auwärter, Swiss Patent No. 410 498, (Oct. 31, 1966), and other countries. Priority claim: Austrian Patent A 9295/60, (Dec. 13, 1960), and Austrian patent No 265718, (Oct. 25. 1968). Priority claim: Swiss Patent No 16652/65, (Dec. 2, 1965).
[117] A. Ross, F.R.G. Patent No. 1.066.395, (March 17, 1960). Priority claim: Austrian patent A 3425/57 (May 23, 1957).
[118] A. Ross, Swiss Patent No. 346 666, (July 15, 1960).
[119] P. Rheinberger, F.R.G. Patent No. 1.216.165, (Nov. 17, 1966).
[120] W. Geffcken, German Patent 716 153 (1939), Schott u. Gen.
[121] L. N. Hadley and D. M. Dennison, J. Opt. Soc. Am., 37, (1947) 451.
[122] L. N. Hadley and D. M. Dennison, J. Opt. Soc. Am., 38, (1948) 483.
[123] M. L. Baker and V. L. Yen, Appl. Opt., 6 (1967) 1343.
[124] A. F. Turner, J. Phys. Radium, 11 (1950) 444.
[125] W. Geffcken, German Patent No. 913.005, (1944), Schott u. Gen.
[126] A. F. Turner, See Ref. [5], Appendix 2.
[127] S. D. Smith, J. Opt. Soc. Am., 48 (1958) 43.
[128] R. G. T. Nielson and J. Ring, J. de Physique, 28 (1967) C2-270, Suppl. to No. 3–4, March–April.
[129] P. Bousquet, A. Fornier, R. Kowalczyk, E. Pelletier and P. Roche, Thin Solid Films, 13 (1972) 285.
[130] E. Pelletier, R. Kowalczyk and A. Fornier, Opt. Acta, 20 (1973) 509.
[131] E. Pelletier, P. Roche and L. Bertrand, Opt. Acta, 21 (1974) 927.
[132] J. Meaburn, Appl. Opt., 5 (1966) 1757.
[133] H. K. Pulker, Optik, 32 (1971) 496.
[134] M. Peršin, H. Zorc and A. Peršin, Thin Solid Films, 57 (1979) 199.

462

[135] H. Dillings, J. Phys. Radium, 11 (1950) 407.
[136] H. Schröder, in Ergebnisse der Hochvakuumtechnik und Physik dünner Schichten I, M. Auwärter (Ed.), Wissenschaftl. Verlagsgesellschaft m. b. H.,Stuttgart, 1957, p. 225.
[137] X. Leurgans and A. F. Turner, J. Opt. Soc. Am., 37 (1947) 983 A.
[138] P. H. Berning and A. F. Turner, J. Opt. Soc. Am., 47 (1957) 230.
[139] P. W. Baumeister and F. A. Jenkins, J. Opt. Soc. Am., 47 (1957) 57.
[140] P. W. Baumeister, F. A. Jenkins and M. A. Jeppesen, J. Opt. Soc. Am., 49 (1959) 1188.
[141] P. Giacomo, P. W. Baumeister and F. A. Jenkins, Proc. Phys. Soc., 73 (1959) 480.
[142] L. I. Epstein, J. Opt. Soc. Am., 42 (1952) 806.
[143] a) E. Bosshard, Der Schweiz. Optiker, 33 (1957) 98 ;
       b) E. Bosshard, Der Augenoptiker, H. 4 (1960) 1;
       c) W. Köppen, Der Schweiz. Optiker, 55 (1979) 574.
[144] G. Littmann, Zeiss, Neues Opt. J. 25 No 2 (1983) 13.
[145] Verwendung von Sichtscheiben für Augenschutzgeräte (Schweisserschutzfilter) DIN 4647, Sept. 1972. ISO: TC 94/SC 6 EYE Protectors.
[146] G. J. Zinsmeister, Proc. SPIE, Vol. 333, Submicron Lithography, (1982) 40.
[147] R. Hoffmann, in Ergebnisse der Hochvakuumtechnik und der Physik dünner Schichten, Vol. II, M. Auwärter (Ed.), Wissenschaftl. Verlagsges. m. b. H., Stuttgart, 1971, p. 288.
[148] G. J. Zinsmeister, Hard Surface Mask Technology, IGC Conf. Amsterdam, 19–21 Sept. 1979.
[149] a) J. K. Hassan and H. G. Sarkary, Solid St. Tech., 25 (5) (1982) 49;
       b) M. J. Cowan, Solid St. Tech., 25 (5) (1982) 55.
[150] a) E. D. Liu, M. M. O'Toole and M. S. Chang, Solid St. Tech., 25 (5) (1982) 66;
       b) J. D. Buckley, Solid St. Tech., 25 (5) (1982) 77.
[151] Balzers, Dünnschicht Elektronik, BD 800 040 PE and
       Balzers, Specification Chrome Blanks, BD 800 004.
[152] Hoya Electronics, Japan, Tech. Report 810 318 (1981).
[153] Basic Microelectronics Inc., USA, Process Specifications, (1980).
[154] a) F. Zernike, Z. tech. Physik, 16 (1935) 554; Physica, 9 (1942) 686 and 974.
       b) H. Beyer, Theorie u. Praxis des Phasenkontrastverfahrens, Akadem., Verlagsges. Geest u. Portig, Leipzig, 1965.
[155] M. Auwärter, private communication, 1982.
[156] G. P. Montgomery jr., Opt. Eng. 21 (1982) 1039.
[157] E. Ritter, in Progress in Electro-optics, E. Camatini (Ed.), Plen. Press, New York, 1972, p. 181.
[158] a) G. H. Brown, J. Opt. Soc. Am., 63 (1973) 1505;
       b) J. Vogel, in Jahrbuch für Optik und Feinmechanik, D. Hacman (Ed.), Schiele u. Schön, Berlin, 1974, p. 72.
[159] N. Simpson, The Current Status of Liquid Crystal Displays, in Electronic Displays 82, ISBN: 0904999971, London, Oct. 82, p. 21.
[160] SID, Soc. for Information Display, Digest of Technical Papers, 1982, San Diego Meeting, Pub. Lewis Winner, Coral Gables, FL 33134, USA. See also the other Digest of the Annual Meetings of SID.
[161] a) H. R. Zeller, in Non-emissive Electro-optic Displays, A. R. Kmetz and F. K. von Willisen (Eds.), Plenum Press, New York, 1976, p. 149;
       b) D. Theis, H. Venghaus and G. Ebbinghaus, in Siemens Forschg. u. Entwickl.-Berichte, Vol. 11, (1982), No. 5, p. 265, Springer, Berlin, 1982;
       c) H. Venghaus, D. Theis, H. Oppolzer and S. Schild, J. Appl. Phys., 53 (1982) 4146.
[162] a) I. Hamberg, A. Hjortsberg and C. G. Granqvist, Appl. Phys. Lett., 40 (5) (1982) 362.
       b) R. P. Howson, M. I. Ridge and C. A. Bishop, in J. R. Jacobsson (Ed.), Proc. SPIE, Vol 401, Thin Film Technologies, 1983, paper 401–40.
[163] R. Groth, Glastechn. Ber., 50 (1977) 239.
[164] D. L. Antonson, U. S. Patent No. 3 290 203, (1966).
[165] H. Dislich et al. F.R.G. Patent No. DT 2263353, (1975).
[166] W. D. Münz and S. R. Reineck, in R. I. Seddon (Ed.), Proc. SPIE, Vol. 325; Optical Thin Films, (1982) 65.
[167] R. Groth and E. Kauer, Philips Tech. Rev., 26 (1965) 105.
[168] a) S. O. Hoffman, U.S. Patent No. 1 425 967 (1922);
       b) J. Brett, R. Fontana, P. Walsh, S. Spura, L. Parascandola, W. Thouret and L. Thorington, J. Illum. Eng. Soc., July (1981) 214.

463

[169] a) Catalog: Sunpak TM, Energy Products and Ventures, Group of Owens-Illinois Inc., P.O.Box 1035, Toledo, Ohio 43666, USA.
b) R. O. Doughty and D. W. Goodwin, US Patent No 4 142 511, (March 6, 1979);
c) C. M. Lampert and J. Washburn, Solar Energy Mat., 1 (1979) 81;
d) J. C. C. Fan, Thin Solid Films, 54 (1978) 139;
e) P. Call, Nat. Prog. Plan for Absorber Surfaces Research and Development, Report No. SERI/TR-31-103, (1979).
f) H. G. Craighead and R. A. Buhrman, J. Vac. Sci. Technol. 15 (1978) 269;
g) H. G. Craighead, R. Bartynski, R. Buhrman, L. Woijcik and A. J. Sievers, Solar Energy Mater., 1 (1979) 105.
[170] Balzers Thin Film Product Information DN 4435.
[171] A. Javan, W. R. Bennett jr. and D. R. Harriott, Phys. Rev. Lett., 6 (1961) 106.
[172] S. E. Miller, Bell System Techn. J., 48 (1969) 2059.
[173] H. G. Unger, Optische Nachrichtentechnik, Elitera, Berlin, 1976.
[174] Neue Zürcher Zeitung, supplement: Forschung und Technik, Feb. 23, (1977).
[175] W. Heitmann, Deutsche Bundespost, Forschungsinstitut FTZ, Techn. Bericht FI 465 TBr 17, Okt. 1979, DK 621.391.63.091.1.
[176] C. Baack, Frequenz, 32 (1978) 16.
[177] R. Lombaerde, Electronics, Dec. 18 (1980) 83.
[178] R. Th. Kersten, H. F. Schlaak and C. H. von Helmolt, in Proc. Summer School on Integrated Optics, Fiber Optics and Holography, Varna 28.9.–3.10.1981, Publishing House Bulgarian Academy of Sciences, 1982, p. 19–65.
[179] W. E. Martin and D. B. Hall, Appl. Phys. Lett., 21 (1972) 325.
[180] P. K. Tien, R. J. Martin and G. Smolinsky, Appl. Opt., 12 (1973) 1909.
[181] H. P. Weber, F. A. Dunn and W. N. Leibolt, Appl. Opt., 12 (1973) 755.
[182] R. K. Watts, M. de Wit and W. C. Holton, Appl. Opt., 13 (1974) 2329.
[183] R. Ulrich, J. Vac. Sci. Technol., 11 (1974) 156.
[184] R. Th. Kersten, H. F. Mahlein and W. Rauscher, Thin Solid Films, 28 (1975) 369.
[185] M. Kartzow, T. Le Hiep, and R. Th. Kersten, Opt. Commun., 29 (1979) 160.
[186] K. Sasaki, H. Takahashi, Y. Kudo and O. Hamano, Proc. 8th Internat. Vac. Congr., Cannes, France, 1980, Vol. I, Thin Films, p. 374; Suppl. à la Revue, Le Vide, les Couches Minces, Nr. 201, 1980.
[187] C. von Helmolt, J. Opt. Commun., 2 (1981) 142.
[188] N. S. Kapany, Fiber Optics, Academic Press, New York, 1967.
[189] R. E. Collin, Field Theory of Guided Waves, McGraw-Hill, New York, 1960.
[190] R. Ulrich and P. K. Tien, J. Opt. Soc. Am., 60 (1970) 1325.
[191] R. Ulrich and R. J. Martin, Appl. Opt., 10 (1971) 2077.
[192] R. V. Schmidt, IEEE Trans. Sonics and Ultrasonics., SU-23 (1976) 1.
[193] C. S. Tsai, IEEE Trans. Circ. and Syst., CAS-26 (1979) 1072.
[194] A. A. Oliner, Acoustic Surface Waves, Springer, Berlin, 1978.
[195] C. von Helmolt, H. F. Schlaak and R. Th. Kersten, Electron. Lett., 17 (1981) 446.
[196] C. von Helmolt, Electron. Lett., 17 (1981) 447.
[197] I. P. Kaminow and J. Carruthers, Appl. Phys. Lett., 22 (7) (1973) 326.
[198] P. L. Liu, J. Opt. Commun., 2 (1981) 2.
[199] J. G. Grabmaier et al., Phys. Lett., 43A (3) (1973) 219.
[200] R. Th. Kersten, Acta Physica Austriaca, 37 (1973) 335.
[201] H. G. Danielmeyer and H. P. Weber, IEEE QE, 8 (1972) 805.
[202] H. G. Danielmeyer et al., Appl. Phys., 2 (1973) 335.
[203] C. H. Gooch, Gallium Arsenide Lasers, Wiley Interscience, New York, 1969.
[204] P. G. McMullin et al., Appl. Phys. Lett., 24 (1974) 595.
[205] B. L. Sopori, W. S. C. Chang and C. M. Phillips, Appl. Phys. Lett., 29 (1976) 800.
[206] G. H. Olsen, SPIE Proc. 224 (1980) 113.
[207] G. H. Olsen, J. Opt. Commun., 2 (1981) 11.
[208] S. A. Gurevich, E. L. Portnoi and V. I. Skopina, J. Opt Commun., 3 (1982) 133.
[209] K. Kikushima, O. Hirota, M. Shindo, V. Stoykov and Y. Suematsu, J. Opt. Commun., 3 (1982) 129.
[210] H. Melchior, F. Arams and M. B. Fisher, Proc. IEEE, 58 (1970) 10.
[211] Y. Matsushima et al., Appl. Phys. Lett., 35 (1979) 446.
[212] T. P. Pearsall and M. Papuchon, Appl. Phys. Lett., 33 (1978) 201.
[213] L. R. Tomasetta et al., IEEE J. Quant. Electron., QE-14 (1978) 800.

464

[214]   K. Nishida, K. Taguchi and Y. Matsumoto, Appl. Phys. Lett., 35 (1979) 251.
[215]   Y. Takanashi and Y. Horikoshi, Jpn. J. Appl. Phys., 17 (1978) 2065.
[216]   H. Melchior and W. T. Lynch, IEEE Trans. Electron. Dev., ED-13 (1966) 829.
[217]   K. Nishida et al., Electron. Lett., 13 (1977) 280.
[218]   H. Kanbe et al., J. Appl. Phys., 47 (1977) 3749.
[219]   H. Melchior, Physics Today, Nov. (1977) 32.
[220]   P. K. Tien, R. Ulrich and R. J. Marti, Appl. Phys. Lett., 14 (1969) 291.
[221]   H. H. Harris, R. Shubert and J. N. Polky, J. Opt. Soc. Am., 60 (1970) 1007.
[222]   R. Th. Kersten, Elektron. Rundschau, Heft 12 (1973) 261.
[223]   D. Gloge, Appl. Opt., 13 (1974) 249.
[224]   H. Dislich and A. Jacobsen, F.R.G. Patent Nr. 14 94 872, (1965), Schott u. Gen.
[225]   Corning glass works, Telecommun. Products Dep., Corning, N. Y. 14830, USA.
[226]   a) T. Tamir (Ed.), Integrated Optics, 2nd. edn., Springer, Berlin, 1982.
        b) R. G. Hunsperger, Integrated Optics: Theory and Technology, Springer, Berlin, 1982.
        c) R. Th. Kersten, Einführung in die Optische Nachrichtentechnik, Springer, Berlin, 1983.
[227]   a) A. Y. Cho and J. R. Arthur jr., Prog. Solid St. Chem., 10 (1975) 157.
        b) A. C. Gossard, Thin Solid Films, 57 (1979) 3.
[228]   E. Hahn, in Proc. Second Coll. on Thin Films in Budapest, E. Hahn (Ed.), Vandenhoeck and
        Ruprecht, Göttingen, 1968, p. 283; and G. Devant and M. L. Theye, loc. cit., p. 306.
[229]   a) C. A. Neugebauer and R. H. Wilson, in Basic Problems in Thin Film Physics,
           R. Niedermayer and H. Mayer (Eds.), Vandenhoeck and Ruprecht, Göttingen, 1966, p. 579;
        b) P. Wissmann, Thin Solid Films, 13 (1972) 189;
        c) R. E. Hummel, Thin Solid Films, 13 (1972) 175.
[230]   H. K. Pulker, Glastechn. Berichte, 38 (1965) 61.
[231]   a) G. Wedler, Adsorption, Chemische Taschenbücher, Verlag Chemie, Weinheim/Bergstrasse,
           1970;
        b) G. H. Comsa, Thin Solid Films, 13 (1972) 185.
[232]   E. Cremer and H. Nau, Naturwiss. 55 (1968) 651.
[233]   E. Cremer and Th. Kraus, Swiss Patent No. 493 847, (1968).
[234]   E. Cremer, Th. Kraus and H. Nau, Z. Analyt. Chem., 245 (1969) 37.
[235]   E. Cremer and Th. Kraus, US Patent No 3.669,881, (June 13, 1972).
[236]   M. Auwärter, R. Haefer and P. Rheinberger, in Ergebnisse der Hochvakuumtechnik und der
        Physik dünner Schichten I, M. Auwärter (Ed.), Wissenschafliche Verlagsges. m. b. H., Stuttgart,
        1957, p. 22.
[237]   R. Speidel, Z. Phys., 160 (1960) 375.
[238]   H. K. Pulker and E. Junger, Optik, 24 (1966/67) 152.
[239]   J. Ruf and O. Winkler, in Ref. [236], p. 207.
[240]   E. Cremer, Th. Kraus and F. Ehrler, Z. Phys. Chem., Neue Folge, 49 (1966) 310.
[241]   F. Ehrler and Th. Kraus, Proc. 3rd Int. Vac. Congress, Vol. 2, Parts 1–3, Pergamon, London,
        1966, p. 131.
[242]   a) H. K. Pulker and H. Hilbrand, Proc. Second Colloquium on Thin Films in Budapest,
           Hungary, 1967, E. Hahn (Ed.), Akadémiai Kiadó, Budapest, 1968, p. 375.
        b) H. Ahrens, Thesis, Inst. of Physics, University of Technol., Hannover, Germany, 1974.
        c) H. K. Pulker, Thin Solid Films 34 (1976) 343.
[243]   R. F. Coe, Physics in Technol., 14 (1983) 3.
[244]   B. E. Jones and A. Solomons, Proc. IEE, 126 (1979) 717.
[245]   B. E. Jones and D. M. Chulatunga, Advan. Instrum., 35 (1980) 477.
[246]   B. E. Jones, Physics in Technol., 14 (1983) 4.

AUTHOR INDEX

The page numbers indicated are those on which the author's name is mentioned or a literature reference given. Pages on which multiple references occur are underlined.

470

472

478

Panovsis, N.T., 79
Panpuch, R., 38
Pantano, C.G. jr., 337
Panzera, C., 221
Papagno, L., 336
Papuchon, M., 450
Parekh, P.C., 129
Parikh, N.M., 39
Parisot, J.M., 288
Parascandola, L., 445
Parrat, L.G., 289
Parson, W.F., 169, 206
Pashley, D.W., 200, 322
Pask, J.A., 39
Pastor, R., 22
Patai, S., 127
Patano, C.G., 53
Pavlath, A.E., 272
Pawlewicz, W.T., 257, 268, 269, 318, 319, 371, 372, 373
Pawlyk, P., 132
Pearlstein, F., 95
Pearsall, T.P., 450
Pearson, J.P., 325, 328
Peek, H.L., 135, 137
Pela, C.A., 346
Pelletier, E., 288, 306, 307, 369, 398, 435
Pen'kov, I.A., 38
Penselin, S., 420, 421
Perny, G., 257
Perot, A., 4, 433, 435, 456
Perri, J.A., 223
Perry, A.J., 68, 84, 87, 88, 130
Perry, D.L., 420
Peršin, A., 436
Peršin, M., 436
Pestes, M., 68, 86
Peterson, D., 127, 129, 134
Petrov, D.A., 133
Petty, M.C., 392
Pfänder, H.G., 2, 44, 45
Pfefferkorn, G., 290
Philipps, B.F., 36
Phillips, C.M., 450

Phillips, J.C., 10
Pietenpol, W.B., 39
Pinsker, Z.G., 313
Pippard, A.B., 369
Pisarkiewicz, T., 257
Pitt, C.W., 116
Pivot, J., 320
Planck, M., 76
Platt, J.R., 188
Pliskin, W.A., 320, 321, 322
Plücker, J., 213
Poate, J.M., 335
Pócza, J.F., 322
Pohl, R., 4, 169
Poisson, S.D., 30
Pole, R.V., 59
Poley, M.M., 68
Polezhaev, Y.M., 38
Polky, J.N., 450
Polurotova, I.F., 122
Pommier, R., 122, 124
Pompei, J., 257
Ponjee, J.J., 384
Popel, S.I., 38
Popova, V., 257
Poppa, H., 48, 201, 322
Porai-Koshits, E.A., 10, 11
Porteus, J.O., 268
Portevin, A., 38
Portnoi, E.L., 450
Poulis, J.A., 288
Pound, G.M., 72, 73
Powell, C.F., 118, 130
Power, B.D., 156
Power, G.L., 336
Preisinger, A., 35, 206, 356, 367, 379
Prengle, H.W., 57
Prestridge, H.B., 130
Preuss, L.E., 290
Priest, J., 344
Priestland, C.R.C., 222
Pringsheim, P., 4, 169
Prokhavulov, A.I., 354
Prosser, J.L., 69, 72
Pukite, G.N., 288

480